Research Progress in Organic, Biological and Medicinal Chemistry

Research Progress in Organic, Biological and Medicinal Chemistry

volume 3

PART I

EDITORS:

Ulisse GALLO and Leonida SANTAMARIA

Società Editoriale Farmaceutica, Milan, Italy

1972

AMERICAN ELSEVIER PUBLISHING COMPANY INC., NEW YORK

Library of Congress Catalog Card Number 64-21126
ISBN North-Holland: 0 7204 4106 4
ISBN American Elsevier: 0 444 10384 8

Publishers:
NORTH-HOLLAND PUBLISHING COMPANY – AMSTERDAM
NORTH-HOLLAND PUBLISHING COMPANY, LTD – LONDON

Sole distributors for the U.S.A. and Canada:
AMERICAN ELSEVIER PUBLISHING COMPANY, INC.
52 VANDERBILT AVENUE, NEW YORK, N.Y. 10017

Printed in The Netherlands

EDITORS' PREFACE

According to our announcement Volume III is devoted to chemical, biological, pharmacological and physiopathological aspects of natural and synthesized pigments and dyes. In respect to the material presented in the previous volumes, this differs in that each article does not fulfil the ambition to review a particular field of research, but does contribute to this goal along with all the others published in the volume. This was possible not because of any particular leadership but primarily because of a fortunate gathering of experts and their decision to participate in a publication of their differing research experiences. This actually fulfilled the original scope of our serial publication, that is, to appeal to medicinal as well as to organic and biological chemists, and to provide a common ground of expression for the research workers in differing but interdependent fields. In addition, this time contributions were also received from experts in bio-medical sciences. Thus many, if not all, of the papers should appeal to pathologists and even to clinicians.

Most of the natural and synthesized pigments and dyes which are of common interest to research workers in chemistry, physics, biology, pharmacology and pathology are those usually known as photodynamic substances. They were found to be responsible for a biological phenomenon which was called photodynamic, as it was produced upon the light excitation of the above mentioned dyes present in definite biological systems. The latter vary from very simple to the most complex ones. This phenomenon was first discovered by physiologists and pharmacologists, and was later studied by pathologists and clinicians. As soon as the resolving of its mechanism of action became of fundamental importance the cooperation of chemists, physicists and physical chemists was necessary. This led to the development of new fields of research.

The articles published in Volume III are from specialists who developed the above new fields of research. However their results have been presented in such a way as to produce a publication of a highly interdisiplinary nature which could act as a working tool for young scientists who are eager to participate in such a stimulating area of investigation.

The contributions are preceded by a Table listing the photosensitizing dyes classified according to chemical criteria. The total number, more than 400, of these compounds may give some idea of the work so far carried out to identify these biologically active agents. One should be impressed by finding that common drugs such as antibiotics, phenothiazines, sulphonamides etc. are potent photosensitizers. At the same time this Table may be helpful to various researchers during their day to day work.

The volume is divided up into different sections. Section I deals with the fundamentals of photochemistry. Here the mechanisms of sensitized reactions are described. Also the basic information on light sources, dosimetry, methodology of irradiation, action spectra and chemi- and bio-luminescence is reported. Section II contains papers on the mechanisms of photodynamic action on molecules of biological importance. In addition it shows the energy transfer in dye-nucleic acid complexes, and explains how photodynamic action processes can be used as tools in the study of protein structure and molecular biology phenomena. Section III indicates cellular effects at the DNA level producing genetic damage by photodynamic action. Section IV follows directly from Section III since DNA damage is rediscussed in terms of cell repair mechanisms. Here most of the problems deal with UV-light effects but contribute to a better understanding of the material in the previous Section, and indicate new avenues of research. Section V describes damage by photosensitizing processes in neuro- and mitochondrial membranes. Section VI deals with photodynamic sensitivity observed in biological systems with no addition of photosensitizing dyes. This Section actually demonstrates the importance of pigments which are normal cellular components and thus tend to be of interest to biologists, physiologists and pathologists equally. Section VII reflects the concern of pahtologists and clinicians with the field of photodynamic action. Here it is interesting to realize how the discovery of the photodynamic phenomenon helps the physiopathologists to characterize diseases which were once joined together under general terms of skin disorders. The problem of phototoxicity and photoallergy brought on by drugs are discussed. In addition it might be surprising to learn of the importance of photodynamic action in relation to skin carcinogenic processes. Section VIII, the last one, is devoted to X- and UV-ray effects of possible interest to photodynamic action and the question posed to the basic philosophy of this field by the finding that cells of the inner ear contain melanin.

The contributors are all outstanding scientists; many of them also known for their extraordinary personalities. The dedication of the Volume to the memory of Douglas E. Smith was masterfully performed by Max Zelle. With much regret we must announce that Alfred Kleczkowsky is not any more with us.

We hope that the effort, as expressed by the publication of Volume III fulfils the task of attracting new and young workers to this field where chemistry, pharmacology and physiopathology are so closely interdependent.

For the most part credit for this success is due to the North Atlantic Treaty Organization which encouraged Holger Brodthagen, Ian A. Magnus, Leonida Santamaria, John D. Spikes and Adolf Wacker to initiate a Study Institute on Photochemistry and Photobiology of Photodynamic Action in 1969. Acknowledgement is made to Mrs. Jacqueline Locci for her invaluable help with editorial work.

Ulisse Gallo and Leonida Santamaria

Dr. Douglas E. Smith – Physiologist and Photobiologist

Dr. Douglas E. Smith, to whom this volume is dedicated, died following emergency surgery on August 8, 1966 at the age of 49.

Born in Nova Scotia, he obtained his undergraduate education at Trinity College and Boston University, earned a Masters in Biology at Boston University, and his Ph. D. in Physiology at the Ohio State University in 1945. Following two years of teaching physiology at St. Louis University, he came to the Biological and Medical Research Division of the Argonne National Laboratory in 1948 where he continued his work until his death.

Doug Smith was one of the fine scientists in radiation biology and photobiology. His early work reflected his training in physiology but throughout his career, he displayed an unusual versatility, an unusually wide interest in some of the more basic aspects of radiation biology, and an unusual knowledge and appreciation of modern chemical and physical techniques as applied to biology.

He was one of the codiscoverers of the protective action of cysteine against ionizing radiation. His interest in the physiological effects of radiation led to his work on the tissue mast cell in which he effectively employed electron microscopy. He was a leader in the use of hibernating animals in the study of radiation effects. He was long interested in energy transfer in biological systems, in photosensitized oxidations, and in free radicals in biology. He made many important contributions directly by his own work and many more indirectly by stimulating his colleaques.

Until I became Director of the Division at Argonne, I knew Doug Smith primarily through his published work. Soon after arriving at Argonne, I appointed him as Associate Director of the Division, a position which he filled with his customary excellence until his failing health forced him to relinquish it. It was during those two years of close association that I came to appreciate the many fine qualities of Doug Smith which extended far beyond his research. His help was invaluable; his loss irreparable.

It is fitting that this volume be dedicated to him and it is an honour to have been invited to make the dedicatory remarks.

M. R. ZELLE

LIST OF THE PHOTODYNAMIC SUBSTANCES*

Leonida SANTAMARIA and Giuseppe PRINO**

*Camillo Golgi Institute of General Pathology,
University of Pavia, Pavia, Italy*

FOREWORD

The discovery of the photodynamic effect was due to observations of a biological nature and, accordingly, the study of its mechanism has been carried out primarily by pathologists, pharmacologists and physiologists. Nevertheless, such an effect has characteristics in common with the photographic process, so that it might belong to the section of photochemistry dealing with photosensitized reactions. There are reasons, however, which allow one to consider photodynamic action as a typical biological process. Indeed, chemical or physico-chemical systems developed to study the mechanisms of the phenomenon are successful when based on biological parameters. Therefore, we suggest that the term photodynamic should be used instead of photosensitization in spite of the more exactness of the latter one. Thus, the biological nature of this phenomenon is characterized.

The history of the discovery and development of photodynamic action is referred to in various reviews [6, 97, 117]. Here, it should be recalled that the first report on a biological phenomenon produced by a chemical agent in the presence of light was made by Marcacci [65], even though it might be stated that he failed to reveal the importance of the phenomenon that he observed. The great interest started when Raab [89] interpreted the inconsistency of toxicity tests of different dyes on paramecia as the result of light effect. Tappeiner [123] called the phenomenon 'photodynamic' and promoted extensive studies on the subject. Thus, it was pointed out that such compounds are active and fluorescent when excited with the wavelengths corresponding to their absorption peaks; further, the fluorescence was considered a characteristic of the photodynamic substances although inactive per se in determining the effect. Then, the fact that oxygen is necessary to the reaction was observed by Ledoux-Lebard [58], but his hypothesis that the effect was connected with the toxicity of photo-oxidized products did not correspond to the experimental results of later investigations.

---

* Revision of the list published in Res. Progr. Org. Biol. Med. Chem. 1 (1964) 259–336.
** Permanent address: Crinos Biological Research Laboratories, Villaguardia, Como, Italy.

XI

The attempt to associate photodynamic effect with general organic reactions came from the observation that eosin does accelerate the oxidation of many chemicals in the presence of light. Upon this phenomenon, which occurs also with other photodynamic substances, the hypothesis was advanced that the photosensitized oxidation in living and non-living systems are identical [120]. Photodynamic action, however, can occur without involving oxygen. This is the case of the photodynamic lesions caused by furocoumarins [73].

In connection with the biological problem, the numerous investigations have ascertained that any living systems such as organisms, organs, cells, enzymes, viruses, proteins and amino acids are susceptible to photosensitize oxidation. The fact that substances of biological interest such as chlorophyll, porphyrins, cytochromes, vitamin K, flavins etc. are photodynamic suggests that photodynamic action might play an important role in physiological processes [108, 126]. Much attention at present is paid to photodynamic action in relation to the discovery that widely used drugs such as phenothiazines, furocoumarins, sulphonamides, alogenate salicylanilides and different antibiotics are photosensitizers and, accordingly, can bring about serious toxic syndromes following exposure. Recently quite a few investigations have been carried out in order to understand the meaning of the photodynamic action of many carcinogenic substances and the mechanism of the carcinogenic process initiated by photodynamic chemicals [97]. In this connection, energy transfer models have been applied to explain radical formations in photodynamic simple systems [114] and to propose schemes of general interest [98, 117].

Investigators involved in research problems connected with photodynamic action may be impressed by the fact that the photodynamic property is present in chemicals with entirely different structures. We thought, therefore, that it would be interesting to list all the photodynamic substances so far known, according to chemical criteria.

In our previous reports [73, 101] the number of the active compounds we listed was first 144 and later 380. Here we gather in table 1 a list of 417 compounds along with the first effects observed and the reference of the discoverer. As to the effects here reported we considered also the clinical observations, although their evaluations might not be so clear in relation to the mechanism of their action. The validity of these observations, however, is supported by the action of analogues tested in vitro, as in the case of phenothiazines and sulfanilamide groups.

Table 1
Photodynamic substances

No.	Compound	Observed effect	Reference
	Antibiotics		
1	Griseofulvin	Human skin photosensitization	Sams [95]
2	Demethylchlortetracycline	idem	Falk [23] Zöllner and Parrisins [133]
3	Chlorotetracycline	idem	Harris [35] Shaffer et al. [112]
4	Oxytetracycline	idem	Quoted by Kirshbaum and Beerman [50]
5	Daunomycin	Enhancement of antibiotic activity	Sanfilippo et al. [96]
	Vitamins		
6	Riboflavin	Hemolysis	Heiman [39]
7	Riboflavin-5-phosphate	Trypsin inactivation	Spikes [116]
8	Vitamin K (water-soluble)	Lethal effect on *B. coli*	Santamaria et al. [105]
9	4-Methylesculetin-6,7-disulfuric ester, disodium salt	Inhibition of the acetylcholine contraction of guinea pig ileum; lethal effect on *B. coli*; human blood serum photo-oxidation (electrophoretic test)	Santamaria et al. [102, 103, 104, 105]
	Sulfonamides		
10	Sulfanilic acid	Human skin photosensitization	Burckhardt [10]
11	Sulfanilamide	idem	Brunsting [9]
12	Sulfathiazole	idem	Burckhardt [10]
13	Sulfathiazyl formaldehyde deriv.	idem	Storck [119]
14	N′-dimethylbenzoylsulfanilamide	idem	Burckhardt [11]
15	Sulfacetamide	idem	Burckhardt [12]
16	Sulfaguanidine	idem	Peterkin [85]
17	Sulfapyridine	idem	Hallam [31]
18	Sulfapyridine calcium	idem	Burckhardt [10, 11]
19	N-4-benzylsulfanilamide	idem	idem
20	p-(2,4-Diaminophenylazo)-benzenesulfonamide	idem	Snodgrass and Anderson [115]
21	Sulfadiazine	idem	Peterkin [85] Burckhardt [12]
22	Sulfamethazine	Human skin photosensitization	Quoted by Kirshbaum and Beerman [50]
23	Sulfamerazine	idem	Burckhardt [12]

Table 1 (continued)

No.	Compound	Observed effect	Reference
24	Sulfamethoxypyridazine	idem	Perry and Winkelmann [84]
25	N-1-dimethylacroylsul-fanilamide	Human skin photosensi-tization	Mayr [66]
26	Sulfanilanilide-4'-sulfa-moyl derivatives (Uliron)	idem	Schölzke [109]
27	Carbutamide	idem	Ippen [47] Burckhardt et al. [14]
28	Tolbutamide	idem	Sams [95]
29	Chlorpropamide	idem	Sams [95] Hitselberger and Fosnaugh [42]
30	Blancophors	idem	Quoted by Pillsbury and Caro [86]
	Benzothiadiazines		
31	Chlorothiazide	Human skin photosensi-tization	Brodthagen [8] Harber et al. [33] Norins [77]
32	Hydrochlorothiazide	idem	Harber et al. [33]
	Polycyclic hydrocarbons		
33	Naphthalene	Polymethacrylic acid de-polymerization	Alexander and Fox [1]
34	Pyrene	Optical density variation of mitochondrial suspen-sion	Quoted by Guerrini [30] Santamaria and Fanelli [99]
35	1-Aminopyrene	Lethal effect on *paramecia*	Epstein et al. [22]
36	1-Methylpyrene	idem	idem
37	3-Methylpyrene	idem	Wulf et al. [131]
38	4-Methylpyrene	idem	Epstein et al. [22]
39	Chrysene	Polymethacrylic acid de-polymerization	Alexander and Fox [1]
40	6-Aminochrysene	Lethal effect on *paramecia*	Epstein et al. [22]
41	3,4-Benzphenanthrene	idem	idem
42	1'-Methyl-3,4-benz-phenanthrene	Lethal effect on *paramecia*	Epstein et al. [22]
43	6-Methyl-3,4-benz-phenanthrene	Human blood serum photo-oxidation (electro-phoretic test)	Santamaria et al. [107] Santamaria [97]
44	7-Methyl-3,4-benz-phenanthrene	idem	idem
45	8-Methyl-3,4-benz-phenanthrene	idem	idem
46	2-Methyl-3,4-benz-phenanthrene	Lethal effect on *paramecia*	Mottram and Doniach [69]
47	1-Methyl-3,4-benz-phenanthrene	Human blood serum photo-oxidation (electro-phoretic test)	Santamaria et al. [107] Santamaria [97]

Table 1 (continued)

No.	Compound	Observed effect	Reference
48	Anthracene	Lethal effect on *paramecia*	Jodlbauer and Tappeiner [46]
49	1,2-Benzanthracene	idem	Mottram and Doniach [69]
50	1′-Methyl-1,2-benz-anthracene	Human blood serum photo-oxidation (polaro-graphic test)	Santamaria [97]
51	2′-Methyl-1,2-benz-anthracene	Human blood serum photo-oxidation (electro-phoretic test)	Santamaria et al. [107] Santamaria [97]
52	3′-Methyl-1,2-benz-anthracene	Human blood serum photo-oxidation (polaro-graphic test)	Santamaria et al. [107]
53	4′-Methyl-1,2-benz-anthracene	Human blood serum photo-oxidation (eletro-phoretic test)	idem
54	3-Methyl-1,2-benz-anthracene	Polymethacrylic acid de-polymerization	Alexander and Fox [1]
55	3-Cyano-1,2-benz-anthracene	Lethal effect on *paramecia*	Epstein et al. [22]
56	4-Methyl-1,2-benz-anthracene	Human blood serum photo-oxidation (eletro-phoretic test)	Santamaria et al. [107] Santamaria [97]
57	10-Methyl-1,2-benz-anthracene	Polymethacrylic acid de-polymerization	Alexander and Fox [1]
58	10-Amino-1,2-benz-anthracene	Lethal effect on *paramecia*	Epstein et al. [22]
59	10-Octadecanoyl-1,2-benzanthracene	idem	idem
60	10-(2″,4″-Dimethyl phenyl)-1,2-benzan-thracene	idem	idem
61	10-(4″Carbomethoxy phenyl)-1,2-benzan-thracene	idem	idem
62	10-(4″-Hydroxy phenyl)-1,2-benzanthracene	idem	idem
63	5-Methyl-1,2-benzan-thracene	Polymethacrylic acid de-polymerization	Alexander and Fox [1]
64	6-Methyl-1,2-benz-antracene	Human blood serum photo-oxidation (eletro-phoretic test)	Santamaria et al. [107] Santamaria [97]
65	7-Methyl-1,2-benz-anthracene	idem	idem
66	8-Methyl-1,2-benz-antracene	Polymethacrylic acid de-polymarization	Alexander and Fox [1]
67	9-Methyl-1,2-benz-anthracene	Human blood serum photo-oxidation (electro-phoretic test)	Santamaria et al. [107] Santamaria [97]

Table 1 (continued)

No.	Compound	Observed effect	Reference
68	9-(2″-Pyridyl)-1,2-benzanthracene	Lethal effect on *paramecia*	Epstein et al. [22]
69	9-(3″-Pyridyl)-1,2-benzanthracene	idem	idem
70	9-(2″,3″-Dimethyl-phenyl)-1,2-benzan-thracene	idem	idem
71	9-(3″-Methyl-phenyl)-1,2-benzanthracene	idem	idem
72	9-(4″-Methyl-phenyl)-1,2-benzanthracene	idem	idem
73	4,5-Dimethyl-1,2-benzan-threne	idem	idem
74	9,10-Dimethyl-1,2-benz-anthracene	Polymethacrylic acid de-polymerization	Alexander and Fox [1]
75	9,10-Dione-1,2-benzan-thracene	Lethal effect on *paramecia*	Epstein et al. [22]
76	4′-Chloro-10-bromo-1,2-benzanthracene	idem	idem
77	1′-Fluoro-10-methyl-1,2-benzanthracene	idem	idem
78	2′-Fluoro-10-methyl-1,2-benzanthracene	idem	idem
79	3′-Fluoro-10-methyl-1,2-benzanthracene	Human blood serum photo-oxidation (electro-phoretic test); optical density variation of mito-chondrial suspension	Fanelli and Santamaria [24]
80	4′-Fluoro-10-methyl-1,2-benzanthracene	Lethal effect on *paramecia*	Epstein et al. [22]
81	3-Fluoro-10-methyl-1,2-benzanthracene	Human blood serum photo-oxidation (electro-phoretic test); optical density; variation of mito-chondrial suspension	Fanelli and Santamaria [24]
82	4-Fluoro-10-methyl-1,2-benzanthracene	idem	idem
83	6-Fluoro-10-methyl-1,2-benzanthracene	idem	idem
84	7-Fluoro-10-methyl-1,2-benzanthracene	idem	idem
85	8-Fluoro-10-methyl-1,2-benzanthracene	Lethal effect on *paramecia*	Epstein et al. [22]
86	Dichloroanthracene		Quoted by Guerrini [30]
87	1,2,5,6-Dibenzanthracene	Abnormal mitosis in chick-embryo cells in vitro	Lewis [60]
88	3,4-Benzpyrene	idem	idem
89	6-Methyl-benz(a)pyrene	Lethal effect on *paramecia*	Epstein et al. [22]
90	6-Amino-benz(a)pyrene	idem	idem

Table 1 (continued)

No.	Compound	Observed effect	Reference
91	Benzo(e)pyrene	idem	idem
92	Cholanthrene	idem	Mottram and Doniach [69]
93	20-Methylcholan-threne	Abnormal mitosis in chick-embryo cells in vitro	Lewis [60
94	Fluoranthrene	Lethal effect on *paramecia*	Epstein et al. [22]
95	4-Aminofluoranthrene	Lethal effect on *paramecia*	Epstein et al. [22]
96	Benzofluoranthrene	Human blood serum photo-oxidation (electro-phoretic test)	Santamaria et al. [107] Santamaria [97]
97	Benz(k)fluoranthrene	Lethal effect on *paramecia*	Epstein et al. [22]
98	Benz(c)fluoranthrene	idem	Wulf et al. [131]
99	Benz(g,h,i,)fluoranthrene	idem	Epstein et al. [22]
100	3,4,9,10-Dibenzpyrene	idem	idem
101	Benzo(g,h,i)perylene	idem	idem
102	Anthanthrene	idem	idem
103	Coronene	idem	idem
104	Pyranthrene	idem	idem
105	Naphtacene	idem	idem
106	1,2,8,9-Dibenzpen-tacene	idem	idem
107	Benz(e)acephenan-thrylene-9,12-quinone	idem	idem
108	Anthanthrone	idem	idem
109	7,16-Quinone-dibenzo-(a,o)-perylene	idem	idem
110	Benzanthrone	idem	idem
111	1,2,5,6-Dibenzfluorene	Human blood serum photo-oxidation (electro-phoretic test)	Santamaria et al. [107] Santamaria [97]
112	2-Acetylaminofluorene	Human blood serum photo-oxidation (polaro-graphic test)	Santamaria [97]
	Sulfonic acids of naphthalene and anthracene derivatives		
113	Anthracen-α-monosul-fonate (sodium salt)	Effect on yeast growth	Jodlbauer and Tappeiner [46]
114	Anthracen-disulfonate (sodium salt)	idem	idem
115	Dichloro-anthracendisul-fonate (sodium salt)	idem	idem
116	Naphthionic acid	Lethal effect on *paramecia*	idem
117	α-Naphthylaminesol-fonic acid	idem	idem
118	β-Naphthylaminesol-fonic acid	idem	idem
119	α-Naphthol-trisulfonic acid	idem	idem

Table 1 (continued)

No.	Compound	Observed effect	Reference
120	β-Naphthol-trisulfonic acid	idem	idem
121	2,5-Amino-naphthol-7-mono-sulfonic acid	idem	idem
122	Naphthosulfan-2,4-di-sulfonic acid	idem	Quoted by Tappeiner and Jodlbauer [124]
	Stilbene derivatives		
123	2'-Methyl-4-dimethyl-aminostilbene	Polymethacrylic acid de-polymerization	Alexander and Fox [1]
124	4'-Methyl-4-dimethyl-aminostilbene	idem	idem
125	2'-Methyl-4-amino-stilbene	idem	idem
126	4'-Methyl-4-amino-stilbene	idem	idem
127	Tetramino-stilbene chloride	Lethal effect on *paramecia*	Tappeiner and Jodlbauer [124]
	Anthraquinone derivatives		
128	Anthraquinone	Lethal effect on *paramecia*	Tappeiner and Jodlbauer [124]
129	Anthraquinon-α-mono-sulfonate (potassium salt)	idem	idem
130	2,7-Anthraquinon-disul-fonate (sodium salt)	idem	idem
131	Chrysophanic acid	idem	idem
132	Anthraquinone-2-sul-fonic acid	Lethal effect on *Proteus vulgaris*	Berg and Jacob [4]
	Phenothiazines		
133	Phenothiazin	Photosensitization in sheep	Swales [122]
134	Diethazine	Human skin and mice photosensitization; lethal effect on *paramecia*	Schulz et al. [111]
135	Fenethazine	Lethal effect on *paramecia*; human blood serum photo-oxidation (electro-phoretic test); mucine de-polymerization	Prino [87] Prino and Santamaria [88]
136	Promazine	idem	Santamaria and Prino [100]
137	Promethazine	Human skin photosensi-tization	Sidi et al. [113] Epstein and Rove [21]
138	Profenamine	Human skin and mice photosensitization; lethal effect on *paramecia*	Schulz et al. [111]
139	Thiazinamium	Lethal effect on *paramecia*; human blood serum photo-oxidation (electro-phoretic test); mucine de-polymerization	Prino [87] Prino and Santamaria [88]

Table 1 (continued)

No.	Compound	Observed effect	Reference
140	10-(3-Dimethylamino-2,2-dimethyl-propyl)-phenothiazine	Lethal effect on *paramecia*; human blood serum photo-oxidation (electrophoretic test); mucine depolymerization	Prino [87] Prino and Santamaria [88]
141	Trimeprazine	idem	idem
142	Aminopromazine	idem	idem
143	10-(3-pyrrolidyl-propyl) phenothiazine	idem	idem
144	Parathiazine	Human skin photosensitization	Quoted by Kirshbaum and Beerman [50]
145	Mepazine	Human skin and mice photosensitization; lethal effect on *paramecia*	Schulz et al. [111]
146	Perazine		Enss et al. [20]
147	Chloropromazine	Human skin and mice photosensitization; lethal effect on *paramecia*	Labhardt [55] Schulz et al. [111]
148	10-(3-pyrrolidyl-propyl)-2-chlorophenothiazine	Lethal effect on *paramecia*; human blood serum photo-oxidation (electrophoretic test); mucine depolymerization	Prino [87] Prino and Santamaria [88]
149	Prochlorperazine	idem	idem
150	Perphenazine	idem	idem
151	Tiopropazate	Human skin photosensitization	Pellerat and Rives [83]
152	Methopromazine	Lethal effect on *paramecia*; human blood serum photo-oxidation (electrophoretic test); mucine depolymerization	Prino [87] Prino and Santamaria [88]
153	Levomepromazine	idem	idem
154	Trifluoperazine	Human skin photosensitization	Quoted by Pathak [80]
155	Trifluopromazine *Oxazins*	Human skin photosensitization	Cahn and Levy [15]
156	Alizarin blue	Lethal effect on *paramecia*	Tappeiner and Jodlbauer [124]
157	Nile blue	idem	idem
158	Resofurine	idem	idem
159	Resorcin blue *Thiazins*	idem	Heinrich [40]
160	Thionin	Bacteriophage inactivation	Yamamoto [132]
161	Toluidine blue	idem	Suarez-Morales [121]
162	Azure A	Trypsin inactivation	Spikes [116]
163	Azure B	p-toluenediamine photo-oxidation	Oster et al. [78]

Table 1 (continued)

No.	Compound	Observed effect	Reference
164	Azure C	idem	idem
165	Methylene blue	Lethal effect on *paramecia*	Jodlbauer and Tappeiner [46]
166	New methylene blue N.	Nicotine photo-oxidation	Weil and Maher [128]
167	Azure I	Bacteriophage inactivation	Suarez-Morales [121]
168	Azure II	idem	idem
169	Phtalocyanine magnesium	ascorbic and oleic acid photo-oxidation	Krasnovskii and Brin [52]
170	Dehydro-thio-toluidine-sulfonate (sodium salt)	Lethal effect on *paramecia*	Tappeiner and Jodlbauer [124]
171	Isothipendyl	Human skin photosensi-tization.	Quoted by Kirshbaum and Beerman [50]
	Phenazines		
172	Phenazine	Lethal effect on *paramecia*	Tappeiner and Jodlbauer [124]
173	Phenazine metasulfate	Trypsin inactivation	Spikes [116]
174	Diaminophenazine	Lethal effect on *paramecia*	Tappeiner and Jodlbauer [124]
175	Dimethyl-diamino-tolu-phenazine	idem	idem
176	Safranine	Lethal effect on bacteria	T'ung [125]
177	Phenosafranin	Lethal effect on *paramecia*	Jodlbauer and Tappeiner [46]
178	Tolusafranin	idem	Tappeiner and Jodlbauer [124]
179	Rosindulin	idem	Jodlbauer and Tappeiner [46]
180	Azocarmin	idem	idem
181	Aposafranin	idem	Tappeiner and Jodlbauer [124]
182	Magdala Red	idem	idem
183	Fluorindindisulfonate (sodium salt)	enzymes inactivation	Jodlbauer and Tappeiner [46]
184	Pyocyanine		Quoted by Guerrini [30]
	Thiazoles		
185	Thioflavin TG	p-toluenediamine photo-oxidation	Oster et al. [78]
186	Thioflavin S	idem	idem
187	Primuline yellow	idem	idem
	Di- and triphenylmethane derivatives		
188	Methylene green	Trypsin inactivation	Spikes [116]
189	Methylene violet	idem	idem
190	Neutral red	Lethal effect on newt	Frankston [27]
191	Auramin		Quoted by Tappeiner and Jodlbauer [124]
192	Crystal violet		Quoted by Jodlbauer and Tappeiner [46]

Table 1 (continued)

No.	Compound	Observed effect	Reference
193	Fuxin		Quoted by Jodlbauer and Tappeiner [46]
194	Rosolic acid		Quoted by Jodlbauer and Tappeiner [46]
195	Malachite green		Quoted by Tappeiner and Jodlbauer [124]
196	Aniline blue	Bacteriophage inactivation	Suarez-Morales [121]
197	Water blue	idem	idem
	Quinoline derivatives		
198	Quinoline red	Lethal effect on *paramecia*	Tappeiner and Jodlbauer [124]
199	Quinoline blue	idem	idem
200	Quinoline yellow	Trypsin inactivation	Spikes [116]
201	Chloroquine		Quoted by Ippen [48]
202	Phenylquinaldina	Lethal effect on *paramecia*	Raab [89]
203	4-Nitroquinoline	Lethal effect on *paramecia*	Nagata et al. [76]
204	4-Nitroquinoline-1-oxide	idem	idem
205	7-Methyl-4-nitroquinoline-1-oxide	idem	idem
206	2-Methyl-4-nitroquinoline-1-oxide	idem	idem
207	8-Methyl-4-nitroquinoline-1-oxide	idem	idem
208	6-Methyl-4-nitroquinoline-1-oxide	idem	idem
209	7-Chloro-4-nitroquinoline-1-oxide	idem	idem
	Acridins		
210	2-Methylacridine	Skin diseases	Wulf et al. [131]
211	Acridine	Lethal effect on *paramecia*	Raab [89]
212	Benzoflavin	idem	Tappeiner and Jodlbauer [124]
213	Chrysanilin	idem	idem
214	Trypaflavine	Human skin photosensitization	Jausion and Marceron [43]
215	Proflavine	p-toluenediamine photooxidation	Oster et al. [78]
216	Quinacrine	idem	idem
217	2,7-Dimethoxy-3,6-diaminoacridine	idem	idem
218	4,5-Dimethyl-3,6-diaminoacridine	idem	idem
219	Acridine orange	idem	idem
220	Acridine yellow	p-toluenediamine photooxidation	Oster et al. [78]
221	2,5-Diamino-7-ethoxyacridine lactate	Thickening of pigment in pigmented epitelium of frog retina	Kyo [54]

Table 1 (continued)

No.	Compound	Observed effect	Reference
222	Acridine red	Lethal effect on mosquito larvae	Schuldmacher [110]
223	Diaminomethyl-acridine	Lethal effect on spirochetes	Hawking [38]
224	12-Methyl-benz(a)-acridine	Lethal effect on *paramecia*	Epstein et al. [22]
225	6,6-Dimethyl-11-keto-6,11-dihydro-benz(b)-acridine	idem	idem
226	Benz(c)acridine	idem	Wulf et al. [131]
227	1,4-Dimethyl-benz(c)-acridine	idem	Epstein et al. [22]
228	7,9-Dimethyl-benz(c)-acridine	idem	idem
229	7,10-Dimethyl-benz(c)-acridine	idem	idem
230	5,6-Dimethyl-benz(c)-acridine	idem	idem
231	9-Chloro-5,6-dimethyl-benz(c)acridine	idem	idem
232	10-Chloro-5,6-dimethyl-benz(c)acridine	idem	idem
233	11-Chloro-5,6-dimethyl-benz(c)acridine	idem	idem
234	7-Chloro-5,6-dimethyl-benz(c)acridine	idem	idem
235	7-Phenoxy-5,6-dimethyl-benz(c)acridine	idem	idem
236	7-Morpholino-5,6-di-methyl-benz(c)acridine	idem	idem
237	Dibenz(a,c)acridine	idem	idem
238	1,2,3,4-Tetrahydro-di-benz(a,c)acridine	idem	idem
239	1,2,5,6-Dibenz-acridine	Human blood serum photo-oxidation (polarographic test)	Santamaria [97]
240	Aurophosphine	Lethal effect on bacteria and *paramecia*	Graevskii [28] Graevskii et al. [29]
241	Coriphosphine	idem	idem
242	7,14-Dihydrodibenz(a,j)-acridine	Lethal effect on *paramecia*	Epstein et al. [22]
	Xantene derivatives		
243	Fluorescein	Lethal effect on *paramecia*	Jodlbauer and Tappeiner [46]
244	Tetrachlorofluor-escein	idem	idem
245	Eosine	idem	Raab [89]
246	Eosine Y	Trypsin inactivation	Spikes [116]

Table 1 (continued)

No.	Compound	Observed effect	Reference
247	Eosin BA extra	Trypsin inactivation	Spikes [116]
248	Methyleosin	idem	idem
249	Erythrosine	Lethal effect on *paramecia*	Jodlbauer and Tappeiner [46]
250	Dichloro-tetrabromo-fluorescein	idem	idem
251	Dichloro-tetrajodo-fluorescein	idem	idem
252	Dichlorofluorescein	idem	Tappeiner and Jodlbauer [124]
253	Dibromofluorescein	idem	idem
254	Dijodofluorescein	idem	idem
255	Tetrabromofluorescein ethyl ester	idem	idem
256	Dichloro-tetrabromo-fluoresceine ethyl ester	idem	idem
257	Dibromohydroxy-mercurifluorescein di-sodium salt	Trypsin inactivation	Spikes [116]
258	Dibromo-dimethyl-fluorescein	idem	idem
259	Rhodamine B	Lethal effect on *paramecia*	Heinrich [40]
260	Rhodamine IIIB	idem	idem
261	Phloxine	p-toluenediamine photo-oxidation	Oster et al. [78]
262	Pyronine Y	idem	idem
263	Thiopyronine	Growth inhibition of *Saccharomyces*	Lochmann and Stein [62]
264	Phenolphtalein		Quoted by Guerrini [30]
265	Rose bengal (tetra-bromotetraiodo-fluorescein)		Quoted by Tappeiner and Jodlbauer [124]
266	Erythrosine B	Inactivation of *E. coli* 15T	Bellin et al. [2]
267	Rhodamine 6 G	Inactivation of *E. coli* 15T	Bellin et al. [2]

Pyrone and xanthone derivatives

No.	Compound	Observed effect	Reference
268	Khellin	Lethal effect on bacteria	Fowlks et al. [26]
269	Visnagin	idem	idem
270	Quercetin	Mice skin photosensitization	Mélas-Joannides [67]
271	Hematoxylin		Quoted by Tappeiner and Jodlbauer [124]
272	Esculine	Lethal effect on *paramecia*	Jodlbauer and Tappeiner [46]
273	1-Hydroxyxanthone	idem	Quoted by Tappeiner and Jodlbauer [124]

Table 1 (continued)

No.	Compound	Observed effect	Reference
274	2-Hydroxyxanthone	idem	idem
275	1,3-Dihydroxyxanthone	idem	idem
276	1,6-Dihydroxyxanthone	Lethal effect on *paramecia*	Tappeiner and Jodlbauer [124]
277	1,7-Dihydroxyxanthone	idem	idem
278	2,3-Dihydroxyxanthone	idem	idem
	Azo dyes		
279	Methyl orange	Ovalbumin photo-oxidation	Clark [17]
280	Azofuchsin	idem	Jodlbauer and Tappeiner [46]
281	Victoria violet		Quoted by Tappeiner and Jodlbauer [124]
282	Azobordeaux		Quoted by Jodlbauer and Tappeiner [46]
283	Benzopurpurin		idem
284	Congo red	Ovalbumin photo-oxidation	Heinrich [40]
285	Trypan blue	Bacteriophage inactivation	Suarez-Morales [121]
286	Azophloxine	Trypsin inactivation	Spikes [116]
	Nitroso and nitro dyes		
287	Picric acid		Quoted by Tappeiner and Jodlbauer [124]
288	Naphtol Green		idem
	Furocoumarins		
289	Psoralen	Human skin photosensitization	Musajo et al. [71, 72]
290	4-Methylpsoralen	Albino guinea pig and human skin photosensitization	Pathak and Fitzpatrick [81]
291	5'-Methylpsoralen	Albino guinea pig skin photosensitization	Pathak et al. [82]
292	4'-Methylpsoralen	Human skin photosensitization	Musajo et al. [73]
293	4,4'-Dimethylpsoralen	Human skin photosensitization	Musajo et al. [74]
294	4,5',8-Trimethylpsoralen	Albino guinea pig and human skin photosensitization	Pathak and Fitzpatrick [81]
295	3,4,5',8-Tetramethylpsoralen	Albino guinea pig skin photosensitization	Pathak et al. [82]
296	5',8-Dimethylpsoralen	Albino guinea pig and human skin photosensitization	Pathak and Fitzpatrick [81]
297	5-Isopropyloxypsoralen	Human skin photosensitization	Rodighiero and Caporale [93]
298	5-n-Propyloxypsoralen	idem	Musajo and Rodighiero [70]

Table 1 (continued)

No.	Compound	Observed effect	Reference
299	5-n-Butyloxypso-ralen	idem	idem
300	5-Isoamyloxypso-ralen	idem	idem
301	8-n-Propyl-4,5'-dimethyl-psoralen	Guinea pig skin photo-sensitization	Pathak et al. [82]
302	3-n-Butyl-4,5',8-tri-methylpsoralen	idem	idem
303	8-Bromo-4,5'-dimethyl-psoralen	idem	idem
304	8-Acetamido-4,5'-di-methylpsoralen	idem	idem
305	8-Acetyl-4,5'-dimethyl-psoralen semicarbazone	idem	idem
306	3,4-Benzo-5',8-di-methylpsoralen	idem	idem
307	3,4-Cyclohexeno-5,8-dimethylpsoralen	idem	idem
308	8-Benzyloxypsoralen	Human skin photosensi-tization	Musajo et al. [73]
309	5-Ethoxypsoralen	idem	Rodighiero and Caporale [93]
310	Psoralen glucoside	Albino guinea pig and human skin photosensi-tization	Pathak and Fitzpatrick [81]
311	Xanthotoxin (8-Methoxy-psoralen)	Human skin photosensi-tization	Musajo et al. [71,72]
312	3-Methylxanthotoxin	idem	Musajo et al. [73]
313	4-Methylxanthotoxin	idem	idem
314	4'-Methylxanthotoxin	idem	Musajo et al. [74]
315	4',3-Dimethylxantho-toxin	idem	idem
316	4',4-Dimethylxantho-toxin	idem	Musajo et al. [73]
317	5',4-Dimethylxantho-toxin	idem	Musajo et al. [74]
318	5-Chloroxanthotoxin	idem	Musajo et al. [73]
319	Bergapten	idem	Kuske [53]
320	4-Methylbergapten	idem	Musajo et al. [74]
321	Bergapten-8-carboxylic acid, methyl ester	idem	Musajo and Rodighiero [70]
322	4-Methylallobergapten	idem	idem
323	Allobergapten	idem	idem
324	Angelicin	idem	Musajo et al. [71,72]
325	4-Methylangelicin	idem	Musajo et al. [73]
326	5'-Methylangelicin	Guinea pig skin photo-sensitization	Pathak et al. [82]
327	Oxypeucedanin	Human skin photosensi-tization	Kuske [53]

Table 1 (continued)

No.	Compound	Observed effect	Reference
328	Marmesin	Guinea pig skin photo-sensitization	Pathak et al. [82]
329	Anhydromarmesin	idem	idem
	Thiocoumarins		
330	Thiocoumarin	Human blood serum photo-oxidation (po-larographic test); lethal effect on *paramecia*	Ricci et al. [91]
331	7-Jodothiocoumarin	idem	idem
332	7-Aminothiocoumarin	idem	idem
	Alkaloids		
333	Armaline	Invertasi inactivation	Jodlbauer and Tappeiner [46]
334	Quinine	Fermentation inhibi-tion; plant growth in-hibition; inhibition of anourous egg segmenta-tion	Marcacci [65]
335	Cinchonamine	idem	idem
336	Morphine	idem	idem
337	Strychnine	idem	idem
338	Veratrine (mixture)	idem	idem
339	Atropine	idem	idem
340	Hydrastinine	Lethal effect on *paramecia*	Tappeiner and Jodlbauer [124]
341	Hydrastine		Quoted by Guerrini [30]
342	Phenylquinidine		idem
343	Dihydroquinidine	Lethal effect on *paramecia*; toxic effect on frog heart; human blood serum photo-oxidation (electrophoretic test)	Pace et al. [79]
344	Perloline	Lethal effect on *paramecia*; rabbit mice and sheep skin photosensitization	Cummingham and Clare [18]
	Tetrapyrrole derivatives		
345	Bilirubin	Hemolysis	Hausmann [37]
346	Chlorophyll	Lethal effect on *paramecia*	Hausmann [36]
347	Sodium magnesium chlorophyllin	Trypsin inactivation	Spikes [116]
348	Phylloerythrine	Photosensitization on sheep	Rimington and Quin [92]
349	Protoporphyrin	Hemolysis	Lepeschkin and Davis [59]
350	Hematoporphyrin	Hemolysis; lethal effect on *paramecia*	Hausmann [37]
351	Coproporphyrin	Photosensitization on mice (lethal effect)	Dankemeier [19]

Table 1 (continued)

No.	Compound	Observed effect	Reference
352	Deuteroporphyrin	idem	idem
353	Uroporphyrin	idem	idem
	Other heterocyclic compounds		
354	Benzo(a)carbazolo	Lethal effect on *paramecia*	Epstein et al. [22]
355	Quinolino (3,4-b)-1H-benz(g)indole	idem	idem
356	11H-Indeno(1,2-b)quinoxaline	idem	idem
357	11-Hydroxy-11-methyl-11H-indeno(1,2-b)quinoxaline	idem	idem
358	11-One-11H-indeno(1,2-b)-quinoxaline	idem	idem
359	5-Oxide-11-one-11H-indeno(1,2-b)quinoxaline	idem	idem
360	γ-Phenylquinaldine	Lethal effect on *paramecia*	Raab [89]
361	Prothipendyl	Human skin photosensitization	Jessel [44]
362	Pyridine		Quoted by Guerrini [30]
363	Brilliant cresyl blue	Bacteriophage inactivation	Yamamoto [132]
364	6,11-Dimethyl-benzo(b)-naphto(2,3-d)thiophene	Lethal effect on *paramecia*	Epstein et al. [22]
365	2,3,5,6-Dibenzothiophanthrene	idem	idem
366	4,9-Dimethyl-2,3,5,6-di-benzothiophanthrene	idem	idem
367	2,3,7,8-Dibenzothiophanthrene	idem	idem
368	4,9-Dimethyl 2,3,7,8-di-benzothiophanthrene	idem	idem
369	Benzoxathioles		Kelling and Eissner [49]
370	Purine		Quoted by Guerrini [30]
	Inorganic and Metallorganic compounds		
371	Arsenic trioxide	Hemolysis	Löhner [63]
372	Ferric sulfate	Plant growth acceleration	Niethammer [75]
373	Ferrous sulfate	idem	idem
374	Uranyl sulfate	idem	idem
375	Selenium colloidal	Lethal effect on *paramecia*	Henri and Henri [41]
376	Cadmium sulfide	skin diseases	Bjornberg [5]
	Miscellaneous organic compounds		
377	p-aminosalicylic acid	skin diseases	Quoted by Ippen [48]
378	p-aminobenzoic acid	idem	idem
379	Procaine	idem	idem

Table 1 (continued)

No.	Compound	Observed effect	Reference
380	Isonicotinylhydriazide	idem	idem
381	Azulene derivatives	skin diseases	Ippen [48]
382	Diphenhydramine	idem	Kleine-Natrop [51]
383	N-butyl-4-chlorosalicyl-amide	skin diseases	Ippen [48]
384	Bromosalicyl-isopropyl-amide	idem	Langhof [56]
385	3,5-Dibromosalicylic acid	idem	Harber et al. [34]
386	3,5-Dibromosalicyl-anilide	idem	idem
387	4',5-Dibromosalicyl-anilide	idem	Wilkinson [130]
388	3,4',5-Tribromosalicyl-anilide	idem	idem
389	3',4',5-Trichlorosalicyl-anilide	idem	idem
390	3,3',4',5-Tetrachloro-salicylanilide	idem	Wilkinson [129]
391	bis (2 hydroxy-3,5-di-chlorophenyl)sulfide (Bithionol)	idem	Jillson and Baughman [45]
392	Hexachlorofene	idem	Wilkinson [130]
393	Digalloil trioleato	idem	Quoted by Kirshbaum and Beerman [50]
394	Ethidium chloride	Virus inactivation	Sprecher-Goldberger [118]
395	Bampine	Skin diseases	Quoted by Burckhardt and Schmid [13]
396	Chlortalidone	Skin diseases	idem
397	Chlordiazepoxide	Skin diseases	idem
398	Nalidixic acid	Human skin photosensitization	idem
399	Trimethadione	Human skin photosensitization	Quoted by Kirshbaum and Beerman [50]
400	Phenytoin	Human skin photosensitization	idem
401	Phenylbutazone	Human skin photosensitization	idem
402	Oestrone	Lethal effect on *paramecia*	Calcutt [16]
403	Diethylstilbesterol	idem	idem
404	Dicyanine A	Lethal effect on bacteria	Liechti et al. [61]
405	Dichloro-dioxi-dyphenyl-sulfide	idem	Langhof and Gassen-Bathke [57]
406	Tetramethylthiuram di-sulfide	idem	Quoted by Ippen [48]
407	Phycoerythrine	Lethal effect on *paramecia*	Metzner [68]
408	Ipericin	Photosensitization in sheep and rabbit	Ray [90]

Table 1 (continued)

No.	Compound	Observed effect	Reference
409	Fagopyrin		Brochkmann et al. [7]
410	Cytochromes		Quoted by Bergamasco [3]
411	Bacterio-fluoresceins		idem
412	Bacterio-porphyrins		idem
413	Carotenes		idem
414	Anthocyanins		idem
415	Buckwheat	Skin photosensitization	Wedding [127]
416	Lime oil	Human skin photosensitization	Sams [94]
417	Lantanine	Sheep skin photosensitization	Low [64]

References

[1] P. Alexander and M. Fox, Photodynamic degradation of macromolecules sensitized by dyes and carcinogenic hydrocarbons. Proc. First Int. Photobiol. Congress, Amsterdam (1954) 336–339.

[2] J.S. Bellin, L. Lutwick and B. Jonas, Effect of photodynamic action on *E. coli*. Arch. Biochem. Biophys. *132* (1969) 157–164.

[3] A. Bergamasco, Le fotodermatosi. Atti Soc. Ital. Dermatol. Sifilog. *5* (1942) 390–458.

[4] H. Berg and H.-E. Jacob, Biologische Wirkungen photochemischer Wasserstoffperoxyd-bildung. 1. Mitt.: Cytotoxische Effekte von Anthrachinonsulfonsäuren in bestrahlten Suspensionen von *Proteus vulgaris*. Z. Naturforsch. *17b* (1962) 306–309.

[5] A. Bjornberg, Reactions to light in yellow tattoos from cadmium sulfide. Arch. Dermatol. *88* (1963) 267–271.

[6] H.F. Blum, Photodynamic Action and Diseases Caused by Light (Reinhold, New York, 1941).

[7] H. Brokmann, E. Weber and G. Pampus, Protofagopyrin und Fagopyrin, die photodynamisch Wirksamen Farbstoffe des Buchweizens (*Fagopyrum esculentum*). Ann. Chem. *575* (1951) 53–83.

[8] H. Brodthagen, Ugeskr. Laeg *120*, 1958, 1126 (quoted by Ippen, 1962).

[9] L.A. Brunsting, Sulfanilamide dermatitis: question of relation to photosensitivity. Proc. Staff Meet. Mayo Clin. *12* (1937) 614–616.

[10] W. Burckhardt, Untersuchungen über die Photoaktivität einiger Sulfanilamide. Dermatologica *83* (1941) 63–69.

[11] W. Burckhardt, Dermatologica *91* (1945) 248–252 (quoted by Ippen [48]).

[12] W. Burckhardt, Aus der städtischen Poliklinik für Haut- und Geschechtsckraukhleiten Zürick. Dermatologica *114* (1957) 291–295.

[13] W. Burckhardt and P. Schmid, Das Licht als pathogene Noxe. Schweiz. Med. Wschr. *99* (1969) 439–444.

[14] W. Burckhardt, K. Burckhardt and M. Schwarz–Speck, Photoallergic Eczemas caused by nadisan. Schweiz. Med. Wschr. *87* (1957) 954–956.

[15] N.M. Cahn and E.J. Levy, Ultraviolet light factor in chlorpromazine dermatitis. A.M.A. Arch. Dermat. *75* (1957) 38–40.

[16] G. Calcutt, The photosensitizing action of chemical carcinogens. Brit. J. Cancer *8* (1954) 177–180.

[17] J.H. Clark, The action of ultraviolet light on egg albumin in relation to the isoelectric point. Amer. J. Physiol. *61* (1922) 72–79.

[18] I.J. Cunningham and E.M. Clare, A fluorescent alkaloid in rye grass (*Lolium perenne*) V. Toxicity photodynamic action and metabolism of perloline. New Zealand J. Sci. Tech. *24B* (1943) 167–178; C.A. *38* (1944) 1074.

[19] W. Dankemeyer, Ueber die photosensibilizierende Wirkung von Porphyrinen. Inaug. Diss., Hamburg (quoted by Blum [6]).

[20] H. Enss, K. Hartmann, H. Hippius and H.E. Richter, (1960) Quoted by Ippen [48].

[21] S. Epstein and R.J. Rowe, Photoallergy and photocross-sensitivity to phenergan. I. Invest. Dermatol. *29* (1957) 319–326.

[22] S.S. Epstein, M. Small, H.L. Falk and N. Mantel, On the association between photodynamic and carcinogenic activities in polycyclic compounds. Cancer Res. *24* (1964) 855–862.

[23] M.S. Falk, Light sensitivity due to demethylchlortetracycline. J. Amer. Med. Ass. *172* (1960) 1156–1158.

[24] O. Fanelli and L. Santamaria, Attività fotodinamica di mono-fluoro-derivati del 10-metil-1,2-benzanthracene. Atti Soc. Ital. Patol. *7* parte II (1961) 853–865.

[25] W.L. Fowlks, The mechanism of the photodynamic effect. J. Invest. Dermatol. *32* (1959) 233–247.

[26] W.L. Fowlks, D.G. Griffith and E.L. Oginsky, Photosensitization of bacteria by furocoumarins and related compounds. Nature *181* (1958) 571–572.

[27] J.E. Frankston, The photodynamic action of Neutral Red on *Triturus viridescens viridescens* (Rafinesque). J. Exp. Zool. *83* (1940) 161–165.

[28] E.Y. Graevskii, Concerning the photodynamic effect with ultraviolet light. Doklady Akad. Nauk SSSR *83* (1952) 393 (quoted by Fowlks [25]).

[29] L.Y. Graevskii, G.K. Ochinskaya and M.V. Shaak, Mechanism of photodynamic processes. Zhur Obshchei Biol. *13* (1952) 211 (quoted by Fowlks [25]).

[30] G. Guerrini, La luce e la vita (Soc. Ed. Libraria, Milano, 1951).

[31] R. Hallam, Severe skin and general reaction following the administration of M and B 693 and exposure to ultraviolet light. Brit. Med. J., I (1939) 559–560.

[32] L.C. Harber and R.L. Baer, Classification and characteristics of photoallergy, In: The Biologic Effects of Ultraviolet Radiation, ed. F. Urbach (Pergamon Press, London, 1969) pp. 519–525.

[33] L.C. Harber, A.M. Lashinsky and R.L. Baer, Photosensitivity due to chlorothiazide and hydrochlorothiazide. New Engl. J. Med. *261* (1959) 1378–1381.

[34] L.C. Harber, H. Harris and R.L. Baer, Structural features of photoallergy to salicylanilides and related compounds. J. Invest. Dermatol. *46* (1966) 303–305.

[35] H.J. Harris, Aureomycin and chloramphenicol in brucellosis with special reference to side effects. J. Amer. Med. Ass. *142* (1950) 161–165

[36] W. Hausmann, Über die photodynamische Wirkung chlorophilhaltiger Pflanzen extracta. Biochem. Z. *12* (1908) 331.

[37] W. Hausmann, Über die sensibilisierende Wirkung tierischer Farbstoffe und ihre physiologische Bedeutung. Biochem. Z. *14* (1908) 275–278.

[38] F. Hawking, The chemotherapeutic reactions of relapsing fever spirochetes in vitro. Ann. Trop. Med. *33* (1939) 1 (quoted by Fowlks [25]).

[39] M. Heiman, Ueber die Möglichkeit einer photodynamischen Wirkung des Vitamins B_2. Wien Klin. Wschr. *49* (1936) 398–399.

[40] K. Heinrich, Ueber die Wirkung einiger noch nicht untersuchter fluoreszierender Farbstoffe auf Paramecien. Diss. München (quoted by Musajo et al. [73]).

[41] Mme V. Henri and V. Henri, Action photodynamique du sélénium colloïdale. C. R. Soc. Biol. *72* (1912) 326–328.

[42] J.F. Hitselberger and R.P. Fosnaug, Photosensitivity due to chlorpropamide. J. Amer. Med. Ass. *180* (1962) 62–63.

[43] M. Jausion and M. Marceron, Le 'coupe de lumière' acridinique. Son traitement preventif par la resorcine. Bull. Soc. Franc. Dermatol. Syphil. *32* (1925) 358–362.

[44] H.J. Jessel, Experiences with prothipendyl in psychiatry. Deut. Med. Wschr. *85* (1960) 192–196 (quoted by Ippen [48].

[45] O.F. Jillson and R.D. Baughman, Contact photodermatitis from bithionol. Arch. Dermatol (Chicago) *88* (1963) 409–418.

[46] A. Jodlbauer and H. Tappeiner, Die Beteiligung des Sauerstoffs bei der Wirkung fluoreszierender Stoffe. Deut. Arch. Klin. Med. *82* (1905) 520–546.

[47] H. Ippen, (1957), Intern. Photobiol. Kongr., Torino 1957, 289 (quoted by Ippen [48].

[48] H. Ippen, Lichtbeeinfluste Arzneimittel. Nebenwirkungen an der Haut. Dtsch. Med. Wschr. *87* (1962) 480–488 and 544–548.

[49] H.W. Kelling and H. Eissner, Möglichkeiten und Aufgaben bei einer Lokalbehandlung der Psoriaris Vulgaris mit einem Benzoxathiol-Derivat. Die Medizinische Welt (1959) 1298–1299.

[50] B.A. Kirshbaum and H. Beerman, Photosensitization due to drugs. Amer. J. Med. Sci. *248* (1964) 445–468.

[51] H.E. Kleine-Natrop, Allergie Asthma *6* (1960) 3 (quoted by Ippen [48]).

[52] A.A. Krasnovskii and G.P. Brin, Photosensitizing action of magnesium phtalocyanine and chlorophyll in solution. Doklady Akad. Nauk SSSR *58* (1947) 1087; C.A. *46* (1952) 4613f.

[53] H. Kuske, Experimentelle Untersuchungen zur Photosensibilisierung der Haut durch pflanzliche Wirkstoffe. Archiv. Dermatol. Syphilis *178* (1938) 112–123.

[54] S. Kyo, Photodynamic effect of photocatalyst on the retinal motor changes due to light of various wavelengths in the frog. Japan. J. Med. Sci. IV Pharmacol. *12* (1939) 9–21; C.A. *34* (1940) 7306.

[55] F. Labhardt, Schweiz. Arch. Neurol. Psychiatr. *73* (1954) 338 (quoted by Ippen [48]).

[56] H. Langhof, Dtsch. Gesundh. Wes. *9* (1954) 308 (quoted by Ippen [48]).

[57] H. Langhof and G. Gassen-Batke, Hauterkrankungen durch Lichtsensibilisation. Dermat. Wschr. *142* (1960) 972–976.

[58] Ledoux-Lebard, Action de la lumière sur la toxicité de l'éosine et de quelques autres substances pour les paramècies. Ann. Inst. Pasteur *16* (1902) 587–594.

[59] W.W. Lepeschkin and G.E. Davis, Hemolysis and solar spectrum. Protoplasma *20* (1933) 189–194.

[60] M.R. Lewis, The photosensitivity of chick-embryo cells growing in media containing certain carcinogenic substances. Amer. J. Cancer *25* (1935) 305–309.

[61] A. Liechti, E. Feistmann and L. Guggenheim, Über die biologische Wirkung von Sensibilizatoren im langwelligen sichtbaren Licht. Strahlentherapie *64* (1939) 353–367.

[62] E.R. Lochman and W. Stein, Zur Inaktivierung durch Tiopyronin mit und ohne Licht. Naturwissenschaften *51* (1964) 59.

[63] L. Löhner, Ueber die Beeinflussung der Strahlen und Wasserhämolyse durch Arsen. Biochem. Z. *186* (1927) 194–202.

[64] P.G.J. Low, Lantanin, the active principle of *Latana camara L.* I. Isolation and preliminary results of the determination of its constitution. Onderstepoort I. Vet. Sci. Animal. Ind. *18* (1943) 197–202; C.A. *38* (1944) 6297.

[65] A. Marcacci, Sur l'action des alcaloids dans le règne végétal et animal. Arch. Ital. Biol. *9* (1888) 2–4.

[66] J. Mayr, (1950) Quoted by Ippen, 1962.

[67] Z. Mélas-Joannides, Le pigment photoxique de *l'Hipericum crispum.* C.R. Soc. Biol. *3* (1930) 349–351.

[68] P. Metzner, Zur Kenntnis der photodynamischen Erscheinung. III: Mitteilung: Ueber die Bindung der wirksamen Farbstoffe in der Zelle. Biochem. Z. *148* (1924) 498–523.

[69] J.C. Mottram and I. Doniach, The photodynamic action of carcinogenic agents. Lancet *I* (1938) 1156–1158.

[70] I. Musajo and G. Rodighiero, The skin-photosensitizing furocoumarins. Experientia *18* (1962) 153–161.

[71] L. Musajo, G. Caporale and G. Rodighiero, Isolamento del bergaptene dal sedano e dal prezzemolo. Gazz. Chim. Ital. *84* (1954) 870–873.

[72] L. Musajo, G. Rodighiero and G. Caporale, L'activité photodynamique des coumarines naturelles. Bull. Soc. Chim. Biol. *36* (1954) 1213.

[73] L. Musajo, G. Rodighiero and L. Santamaria, Le sostanze fotodinamiche con particolare riguardo alle furocumarine. Atti. Soc. Ital. Patol. *5*, parte I (1957) 1–70.

[74] L. Musajo, G. Rodighiero, G. Caporale and C. Antonello, Ulteriori ricerche sui rapporti fra costituzione e proprietà fotodinamiche nel campo delle furocumarine. Il Farmaco, Ed. Sci. *13* (1958) 355–362.

[75] A. Niethammer, Ueber die Wirkung von Photokatalyzatoren auf das frühtreiben ruhender Knospen und auf die Samenkeimung. Biochem. Z. *158* (1925) 278–305.

[76] C. Nagata, K. Fujii and S.S. Epstein, Photodynamic activity of 4-nitroquinoline-1-oxide and related compounds. Nature *215* (1967) 972–973.

[77] A.L. Norins, Chlorothiazide drug eruptions involving photosensitization. AMA Arch. Dermatol. *79* (1959) 592–598.

[78] G. Oster, J.S. Bellin, R.W. Kimball and M.E. Schrader, Dye sensitized photoöxidation. J. Amer. Chem. Soc. *81* (1959) 5095–5099.

[79] G. Pace, G. Prino and L. Santamaria, Azione fotodinamica della diidrochinidina. a) azione letale sui parameci. b) modicazioni del tracciato elettroforetico su siero di sangue umano. c) effetto su cuore isolato. Atti Soc. Ital Patol. *6*, parte II (1959) 797–799.

[80] M.A. Pathak, Basic aspects of cutaneous photosensitization. In: Biologic Effects of Ultraviolet Radiation, ed. F. Urbach (Pergamon Press, New York, 1969) pp. 489–511.

[81] M.A. Pathak and T.B. Fitzpatrick, Relationship of molecular configuration to the activity of furocoumarins which increase the cutaneous responses following long wave ultraviolet radiation. J. Invest. Dermatol. *32* (1959) 255–262.

[82] M.A. Pathak, J.H. Fellman and K.D. Kaufman, The effect of structural alterations of the erythemal activity of furocoumarins: psoralens. J. Invest. Dermatol. *35* (1960) 165–183.

[83] J. Pellerat and H. Rives, Frequence des accidents cutanés induits par les neuroleptiques chez les malades mentaux. Bull. Soc. Franç. Dermatol. Syphil. *67* (1960) 838–839.

[84] H.O. Perry and R.K. Winkelman, Adverse reactions to sulfamethoxypyridazine (Kynex). J. Amer. Med. Assoc. *169* (1959) 127–132.

[85] G.A. Peterkin, (1945) Quoted by Ippen [48].

[86] D.M. Pillsbury and W.A. Caro, The increasing problem of photosensitivity. Med. Clin. N. Amer. *50* (1966) 1295–1311.

[87] G. Prino, Le fenotiazine come sostanze fotosensibilizzanti. Atti. Soc. Ital. Patol. *6*, parte II (1959) 821–825.

[88] G. Prino and L. Santamaria, Proprietà fotodinamiche di derivati fenotiazinici. Boll. Chim. Farm. *99* (1960) 355–360.

[89] V.O. Raab, Ueber die Wirkung fluoreszierender Stoffe auf Infusorien. Z. Biol. *39* (1900) 524–546.

[90] G. Ray, Note sur les effects toxiques du millepertuis à feuilles crispée (*Hipericum crispum*). Bull. Soc. Cent. Med. Vet. *68* (1914) 39–42.

[91] G. Ricci, G. Prino and L. Santamaria, Proprietà fotodinamica di composti tiocumarinici Effetto letale sui parameci e misura polarografica della fotossidazione su siero di sangue. Boll. Chim. Farm. *97* (1958) 655–659.

[92] C. Rimington and J.I. Quin, Studies on the photosensitization of animals in South Africa. VII. The nature of the photosensitizing agent in Geeldikkop. Onderstepoort J. Vet. Sci. An. Ind. *3* (1934) 137–157.

[93] G. Rodighiero and G. Caporale, Eteri del bergaptolo. Il Farmaco, Ed. Sci. *10* (1955) 760–765.

[94] W.M. Sams, Photodynamic action of lime oil (*Citrus aurantifolia*). Arch. Dermatol. Syph. *44* (1941) 571–587.

[95] W.M. Sams, Photosensitizing therapeutic agents. J. Amer. Med. Ass. *174* (1960) 2043–2048.

[96] A. Sanfilippo, G. Sioppacassi, F. Morvillo, M. Ghione, Photodynamic action of daunomycin. I. Effect on bacteriophage T2 and bacteria. Microbiol. *16* (1968) 49–54.

[97] L. Santamaria, Photodynamic action and carcinogenicity. Recent contributions to cancer research in Italy, eds. P. Bucalossi and V. Veronesi (CEA, Milano, 1960) pp. 167–288.

[98] L. Santamaria, Problems of energy transfer in photodynamic reactions. Bull. Soc. Chim. Belg. *71* (1962) 889–905.

[99] L. Santamaria and O. Fanelli, Photodynamic action of polycyclic hydrocarbons on isolated mitochondria in relation to their carcinogenity. Progress in Photobiology Proceed. of the 3rd Intern. Congress on Photobiology (Elsevier Publ. Comp., Amsterdam, 1961) pp. 448–453.

[100] L. Santamaria and G. Prino, Proprietà fotodinamiche della chloropromazina e promazina. Azione fotossidativa sul *Paramecium* e su siero di sangue umano. Boll. Chim. Farm. *97* (1958) 67–71.

[101] L. Santamaria and G. Prino, The photodynamic substances and their mechanism of action. In: Research Progress in Organic, Biological and Medicinal Chemistry, Vol. I, eds. U. Gallo and L. Santamaria (S.E.F., Milano, 1964) pp. 260–336.

[102] L. Santamaria, G. Prino and R. Bianco, Proprietà fotodinamica dell'estere disolforico del 4-metil-esculetolo. Studio sull'ileo isolato di cavia. Atti Soc. Ital. Patol. *5*, parte II (1957) 427–433.

[103] L. Santamaria, G. Prino and R. Bianco, Proprietà fotodinamiche della vitamina K e dell'estere disolforico del 4-metilesculetolo. Effetto letale sul *Bacterium coli.* Atti Soc. Ital. Patol. *5*, parte II (1957) 443–448.

[104] L. Santamaria, G. Prino and E. Calendi, Proprietà fotodinamica dell'estere disolforico del 4-metilesculetolo. Inattivazione della jaluronidasi. Atti Soc. Ital. Patol. *5*, parte II (1957) 483–486.

[105] L. Santamaria, G. Prino and L. Monaco, Proprietà fotodinamiche dell'estere disolforico del 4-metilesculetolo. Studio elettroforetico. Atti Soc. Ital Patol. *5*, parte II (1957) 477–481.

[106] L. Santamaria, E. Calendi and L. Monaco, Azione di idrocarburi policiclici cancerogeni e non cancerogeni su mitocondri in presenza di luce. Atti Soc. Ital. Patol. *6*, parte II (1959) 829–831.

[107] L. Santamaria, E. Calendi, L. Monaco, M. Neri and G. Prino, Studio dell'attività fotodinamica dei derivati monometilici dell'1,2-benzanthracene e dei derivati monometilici del 3,4-benzofenantrene. a) ricerche elettroforetiche su siero di sangue; b) indagine polarografica; c) cinetica delle reazioni di fotossidazione. Atti Soc. Ital. Patol. *6*, parte II (1959) 833–835.

[108] L. Santamaria, G. Prino and G. Pace, Aumento dell'ampiezza di contrazione di cuore isolato per effetto fotodinamico da ematoporfirina. Atti Soc. Ital Patol. *6*, parte II (1959) 843–845.

[109] K.H. Scholzke, Über das Verhalten der Haut gegen Lichtstrahlen nach Verabreichung Von Uliron. Dermatol. Wschr. *107* (1938) 1460–1462.

[110] H. Schuldmacher, Über Photosensibilisierung von Stechmückenlarvalen durch fluoreszierende Farbstoffe. Biol. Z. *69* (1950) 468–477.

[111] K.H. Schulz, A. Wiskemann and K. Wulf, Klinische und experimentelle Untersuchungen über die photodynamische Wirksamkeit von phenothiazinderivativen, insbesondere von Megaphen. Arch. Klin. Exp. Dermatol. *202* (1956) 285–298.

[112] J.M. Shaffer, L.W. Bluemle, V.M. Sborov and J.R. Neefe, Studies on the use of aureomycin in hepatic disease. IV. Aureomycin therapy in chronic liver disease (with a note on dermal sensitivity). Amer. J. Med. Sci. *220* (1950) 173–182.

[113] E. Sidi, R. Melki and R. Longueville, Dermites aux pommades histaminiques. Acta Allerg. (kbh) *5* (1952) 292–303.

[114] D.E. Smith, L. Santamaria and B. Smaller, Free radicals in photodynamic systems. In: Free Radicals in Biological Systems, ed. M.S. Blois (Academic Press, New York, 1961) pp. 305–310.

[115] W.R. Snodgrass and Y.T. Anderson, Prontosil in the treatment of erysipelas; a controlled series of 312 cases. Brit. Med. J. *II* (1937) 101–104.

[116] J.D. Spikes, (1963) private communication.

[117] J.D. Spikes, Photodynamic action. In: Photobiology, Vol. 3 (Acedemic Press, New York, 1968) pp. 33–64.

[118] S. Sprecher-Goldberger, Photosensitization of RNA viruses by ethidinium chloride. Acta Virol. *9* (1965) 385–391.

[119] H. Storck, Dermatologica *107* (1953) 265 (quoted by Ippen [48]).

[120] W. Straub, The action of eosin solution on oxidizable substances. Arch. Exp. Pathol. Pharmacol. *51* (1904) 383–390.

[121] O. Suarez-Morales, Photodynamic action of blue dyes on bacteriophage. Rev. Asoc. Bioquim. Argentina *8* (1942) 44. C.A. *37* (1943) 6699[3].

[122] W.E. Swales, Will phenothiazine cause photosensitization in sheep. Can. J. Comp. Med. Vet. Sci. *4* (1940) 164–165.

[123] H. Tappeiner, Ueber die Wirkung fluoreszierender Stoffe auf Infusorien nach Versuchen von O. Raab. Münch. Med. Wschr. *47* (1900) 5–7.

[124] H. Tappeiner and A. Jodlbauer, Die sensibilizierende Wirkung fluoreszierender Substanzen (F.C. Vogel, Leipzig, 1907).

[125] T. T'ung, The photodynamic action of safranine on Gram-negative bacilli. Proc. Soc. Exp. Biol. Med. *39* (1938) 415–417.

[126] G. Viale, Le Azioni Biologiche delle Radiazioni (Treves Editori, Milano, 1934).

[127] M. Wedding, Verandlung der Berliner Gesellschaft für Antrhopologie, Ethnologie und Urgeschichte. Z. Ethnol. *19* (1887) 67.

[128] L. Weil and J. Maher, Photodynamic action of methylene blue on nicotine and its derivatives. Arch. Biochem. Biophys. *29* (1950) 241–259.

[129] D.S. Wilkinson, Photodermatitis due to tetrachlorosalicylanilide. Brit. J. Dermatol. *73* (1961) 213–219.

[130] D.S. Wilkinson, Patch test reactions to certain halogenated salicylanilides. Brit. J. Dermatol. *74* (1962) 302–306.

[131] K. Wulf, P.J. Unna and M. Willers, Experimentelle Untersuchungen über die photodynamische Wirksamkeit von Steinkohlenteerbestandteilen. Der Hautarzt *14* (1963) 292–297.

[132] N. Yamamoto, Photodynamic action on bacteriophage. I. The relation between chemical structure and the photodynamic activity of various dyes. Virus *6* (1956) 510–521.

[133] N. Zöllner and G. Parrisins, Photodynamische Rekationen nach Verabreichung von Demethylchlortetracyclin. Med. Welt (1960) 1682–1687.

CONTENTS OF PART I

Section II. Mechanisms of photodynamic action

Section III. Genetic damage by photodynamic action

Section IV. Cell repair mechanisms of light damage

ABBREVIATED CONTENTS OF PART II

**Section VIII. X-, UC-, and visible ray effects of
interest to photodynamic action**

SECTION I

FUNDAMENTALS OF PHOTOCHEMISTRY

CHAPTER 1

LIGHT SOURCES AND DETECTORS

K.H.RUDDOCK

Applied Optics Section, Physics Department,
Imperial College, South Kensington, London S.W. 7, U.K.

The material in this chapter is restricted to light sources and detectors which operate within the range 200 nm to 800 nm. The question of radiometric and photometric definitions and units is discussed in the first section and some basic photometric relations have been stated. There is a separate section devoted to lasers, in addition to that dealing with classical light sources. Emphasis has been placed on the practical aspects of lasers, and little has been written regarding the physical processes involved in their action. Those interested in the physics of laser systems should refer to one of the standard texts [17, 18, 19]. Similarly, detectors have been dealt with in terms of their practical performance.

1. Radiometry and photometry

1.1. *The relationship between photometric and radiometric quantities*

Light is a form of radiant energy and thus may be measured directly in terms of energy units. Such a system of light measurement is termed radiometry, whereas photometry is concerned with the efficiency of light in stimulating the human visual system. The two systems of units are related in terms of flux, that is, the rate of flow of energy.

Radiometrically, flux is defined as the rate of flow of radiant energy and this applies equally to all wavelengths. Photometric radiant energy is defined as the rate of flow of radiant energy weighted to take account of its efficiency in stimulating human vision. The sensitivity of the visual system is dependent upon the wavelength of the stimulating radiation, and its sensitivity to an equal energy spectrum V_λ, is plotted in fig. 1. V_λ is standardized by international agreement [7] and the equal response criterion is determined by flicker photometry. The photometric equivalent of radiometric flux with spectral energy distribution, E_λ, is given by the expression:

$$F = K \int V_\lambda E_\lambda \, d\lambda. \tag{1}$$

Eq. (1) assumes that the contribution of each spectral component of E_λ can be added to give the total photometric flux F.

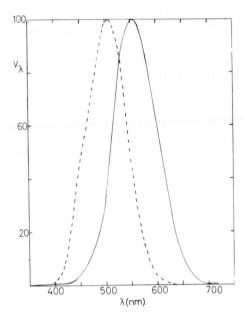

Fig. 1. Relative spectral luminous efficiency function, V_λ, plotted as a function of wavelength, λ. Continuous curve, photopic function; broken curve, scotopic function.

The limits of integration of (1) are effectively 380 nm to 700 nm, although V_λ is actually defined over the range 360 nm to 830 nm. K is a constant which normalizes the photometric and radiometric units. The relative sensitivity function of human vision is not unique, since it depends markedly upon the level of illumination and the location of the light stimulus in the visual field (see e.g. [3]). In general, photometric flux is calculated on the basis of the "daylight" (high luminance level) response of the eye. A different sensitivity function, V'_λ, is used if low illumination level performance is under consideration. V'_λ has also been standardized by the C.I.E. [9]. V_λ is referred to as the photopic relative sensitivity function of human vision, and V'_λ as the scotopic function. It should be noted that for the scotopic case a new value of K is required in eq. (1). Eq. (1) expresses the relationship between photometric and radiometric units, and this equation shows that it is impossible to make any estimate of the function E_λ, if only the value of F is known. In an extreme case, E_λ can take any set of values outside the effective limits of integration of (1) and a zero value of F always results.

1.2. Radiometric and photometric definitions

The following definitions based on those adopted by the C.I.E. [10], and definitions are given for both radiometric and photometric quantities.

Radiant flux, P (or F_e), is defined as the radiant energy emitted, transferred or received per unit time through a surface. Photometric flux, F, is defined in terms of the relative spectral luminous efficiency function, V_λ, by eq. (1). Flux may be integrated in time to give the total quantity of radiant energy or light energy.

Radiant intensity, I_e, is defined in a given direction for a point source or infinitis-simal element of a finite source, as:

$$I_e = \frac{dP}{dw} \quad \text{(i.e. flux/unit solid angle)},$$

where dP is the flux emitted by the source within a cone of solid angle dw, the cone containing the given direction.

A similar definition holds for luminous intensity. The radiant intensity per unit area, L_e, at a point on a surface and in a given direction, is known as the *radiance*. It is defined as the ratio of the radiant intensity of a surface element containing the point to the area of the orthogonal projection of the surface area on to a plane perpendicular to the direction of viewing.

i.e. $$L_e = \left(\frac{dP}{dw}\right) \frac{1}{A \cos\theta},$$

where dP/dw is the radiant intensity of an element of the surface of area of A, and θ is the angle between the surface element and the plane perpendicular to the direction of measurement (fig. 2).

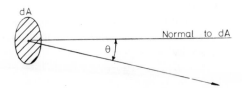

Fig. 2. Definition of radiance of an element of source dA, in direction θ (see text).

The equivalent photometric quantity is luminance, L or B, which again is defined at a point on a surface and in a given direction. The *irradiance*, E_e, at a point on a surface, is defined as the radiant power incident on an infinitessimal element of surface, dA, containing the point, divided by the area of that element

$$E_e = \frac{dP}{dA} \quad \text{(i.e. flux per unit area)}.$$

The equivalent photometric quantity is the *illumination* at a point on a surface.

The *radiant emittance*, M_e, from a point on a surface, is the radiant power emitted by an infinitessimal element of a surface containing the point, divided by the area, dA, of the element.

$$M_e = \frac{dP}{dA} \quad \text{(i.e. flux per unit area)}.$$

The equivalent photometric quantity is *luminous* emittance.

It is the *luminance* of a surface which correlates most closely with its "brightness". A discussion of these definitions may be found in Walsh (1958), although the terminology has since been revised.

1.3. *Radiometric and photometric units*

Radiometric flux is measured in *watt*, and luminous flux in *lumen*. The value of the constant, K, of eq. (1) is 684 lumen/watt (photopic V_λ function) or 1745 lumen/watt (scotopic V'_λ function). The radiometric unit of energy is the *joule* and the photopic unit is the *lumen per second*.

The unit of radiant intensity is the *watt per steradian* and that of photometric intensity is the *candela*.

The units of the other defined radiometric and photometric quantities all derive from the above units. Radiance is expressed in watt/m²/steradian and both irradiance and radiant emittance are expressed in terms of watt/m².

The unit of luminance is the candela/m² and that of illumination and luminous emittance the lumen/m². Many special units have been defined for luminance (e.g. the stilb = candela/cm², nit = candela/m²; apostilb = $1/\pi$ candela/m²) and for illumination (e.g. lux = lumen/m², phot = lumen/cm²). However, it is preferable that these quantities should be defined in terms of the fundamental units, the candela and the lumen.

1.4. *Some photometric relationships*

The *inverse square law* and *cosine law* of illumination state that the illumination at a point on a plane, at a distance r from a *point source* having intensity I in the direction of the point, is given by

$$E = \frac{I \cos \theta}{r^2},$$

where θ is the angle between the plane containing the point and the plane normal to the direction in which I is measured.

The *cosine law* of emission states that the intensity I at an angle θ to the normal to the emitting surface is given by:

$$I(\theta) = I(O) \cos \theta.$$

Such an emitter is known as a Lambertian radiator. A similar law holds for reflection from a diffuse scattering surface.

For such a surface, the luminance B is given by:

$$B = \frac{I(O) \cos \theta}{dA \cos \theta} = \frac{I(O)}{dA},$$

that is, it is not a function of θ, and the Lambertian emitter appears equally "bright" from all directions. The luminous emittance of a Lambertian radiator is $\pi I(O)$. The luminance of an optical image, B', is related to the luminance of the object, B, by the relationship:

$$B' = T\left(\frac{n'}{n}\right)^2 B,$$

where T is the transmission of the optical system and n and n' are the refractive indices in the object and image spaces. Usually $n' = n = 1$ (in air), and thus, apart from transmission losses, the luminance of an optical image is constant and equal to that of the object. Note, this applies only to paraxial systems or corrected optical systems.

Although the relationships derived in section 1.4, have been expressed in photometric terms, identical expressions are valid for the equivalent radiometric variables.

2. Light sources

2.1. *General emission properties*

The quality of the light emitted by a source is described in terms of its *spectral distribution*, its *coherence length* and its *state of polarization*. The spectral distribution is of primary interest, and light sources may be classified into two groups on the basis of their spectral properties, viz. those which emit a spectral continuum and those which emit line spectra. Some sources, such as heavy current discharge lamps, yield a spectral distribution consisting of a line spectrum superimposed on a continuum and this is known as a mixed spectral distribution.

The temporal coherence of a light beam is a measure of the degree of monochromacy of the radiation. For light of nominal wavelength λ and spectral bandwidth $\Delta\lambda$, the coherence length is defined as $\lambda^2/\Delta\lambda$. The frequency bandwidth $\Delta\nu$ corresponding to $\Delta\lambda$ defines the coherence time T, which is equal to $1/\Delta\nu$. The spatial coherence of a light source depends upon the phase relations which exist in the output beam of a source. If spatial coherence is required it is generally necessary to focus the source on to a narrow slit (one dimensional coherence) or on to a pinhole (two dimensional coherence). The coherence properties of a source are particularly important if it is used in image forming systems.

The polarization of the emitted light may be of importance in any system in which asymmetry exists e.g. absorption in orientated molecular systems. Sources rarely emit polarized light (see, however, the section on lasers), but polarization of the light is frequently introduced in its passage through optical systems, and this must be taken into consideration.

Measurement of light quantity was discussed in section 1, and it is the luminance or irradiance by which a source is usually specified. Sometimes the total flux output is also specified. A white light spectral distribution, E_λ, may be specified in terms of the spectral irradiance function, or alternatively by one of the methods discussed in section 2.3. The directional distribution of the emitted radiation may also be specified, and this can be completely uniform around the source axis (e.g.

gas discharge lamps) or it may approximate to a Lambertian cosine distribution (e.g. a strip filament tungsten lamp).

2.2. *Black-body radiation*

The term black-body radiation is applied to the radiation which exists under conditions of thermal equilibrium within a cavity formed by a material which absorbs all radiation incident upon it. The spectral radiant emittance of a black-body radiator is given by the expression:

$$M(\nu, T) = \frac{2\pi h \nu^3}{c^2} \cdot \frac{1}{(e^{h\nu/kT} - 1)}, \tag{2}$$

where ν is the frequency of the radiation, h Planck's constant, k Boltzmann's constant, c the velocity of E.M. radiation in the cavity and T the equilibrium temperature within the cavity.

The corresponding expression in terms of wavelength, λ, is:

$$M(\lambda, T) = \frac{2\pi h c^2}{\lambda^5 (e^{hc/\lambda kT} - 1)}. \tag{3}$$

Eqs. (2) and (3) were first derived by Planck. Black-body radiators are Lambertian radiators, that is, the radiance $L(\lambda, T)$ varies as cosine of the angle between the direction of emission and the normal to the radiator surface.

The total radiant emittance of a Planckian radiator at temperature T is given by:

$$M(T) = \int_\lambda M(\lambda, T)\, d\lambda = \sigma T^4, \tag{4}$$

where σ is a constant. This is known as the Stefan–Boltzmann radiation law.

The wavelength λ_{max} for which $M(\lambda, T)$ is maximised is determined by the relationship:

$$\lambda_{max} T = \text{constant} \tag{5}$$

and correspondingly

$$M(\lambda_{max}, T) \propto T^5.$$

Eq. (5) is known as Wien's displacement law, and shows that as the black-body temperature increases, so the wavelength at which maximum output occurs decreases (the transition from red heat to white heat).

The black-body function $M(\lambda, T)$ is plotted for several values of temperature, T, in fig. 3, and this figure also illustrates relations (4) and (5).

Published tables of black-body functions are often expressed in terms of a function normalized at peak value. The tabulated function normalized at peak value is thus:

$$\frac{M(\lambda, T)}{M(\lambda_{max}, T)} = \text{fn}(\lambda T),$$

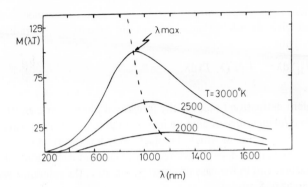

Fig. 3. Black-body radiant emittance function $M(\lambda, T)$, plotted as a function of wavelength, λ.
The broken curve illustrates Wien's law (see text).

that is, the normalized function may be expressed as a function of the single variable (λT).

Sources with properties which are closely described by eqs. (2) to (5) may be obtained experimentally, if care is taken to maintain a state of thermodynamic equilibrium. Such sources are usually only used as absolute standards, and operate at the
temperature of solidification of a metal (e.g. gold 1336°K, platinum 2042°K and
iridium 2716°K).

2.3. Specification of practical white light sources

White light sources may usually be described in terms of the black-body radiation laws (eqs. 2–5). However, as the source can never be in thermal equilibrium
with its surroundings, and most materials demonstrate spectrally selective absorption and emission characteristics, the practical white light sources cannot be specified merely in terms of the source temperature T. The emissivity, e_λ, is defined by
the relationship,

$$e_\lambda = \frac{M'(\lambda, T)}{M(\lambda, T)},$$

where $M'(\lambda, T)$ and $M(\lambda, T)$ are the radiant emittances of the source and the black-
body respectively, T being the real temperature of the source.

In general, e_λ is a function of wavelength, λ. If e_λ is not a function of wavelength,
the source is called a grey body, and the relative spectral distribution of light
emitted by a grey body is specified exactly by the relative distribution of the black-
body function $M(\lambda, T)$. Any source could be specified by tabulating e_λ wavelength
by wavelength, but this requires extensive radiometric investigation of the source.
Such tables are available for tungsten [12] and carbon [13]. In practice, however,
what is required is some shorthand means of relating the source output to the black-
body functions, and there are several methods by which this can be done.

The *total radiation temperature* of a white light source is defined as the temperature, T_R, at which a black-body radiator gives the same integrated radient emittance as the source. Thus:

$$\int_\lambda M(\lambda, T_R)\, d\lambda = \int_\lambda M'(\lambda, T)\, d\lambda.$$

An equivalent defination holds for the integrated photometric emittance.

The *radiance temperature* of a white light source for a specified wavelength is defined as the temperature at which the black-body has the same spectral radiance at the given wavelength as does the source. An equivalent definition holds for luminance temperature, and in the case of visual pyrometry, the reference wavelength is usually taken as 655 nm.

The *colour temperature*, T_o, is defined as the temperature at which a black-body radiator has the same colorimetric specification as the source. Colour can be specified in a two-dimensional (chromaticity) chart, such as that shown in fig. 4. The chart of fig. 4 is that standardized internationally [8], and it shows the loci of the spectral colours and of the black-body sources. If the (x, y) specification of a source lies on the black-body locus, the corresponding B-B temperature defines the colour temperature of the source. If the (x, y) specification does not lie on the B-B locus, the correlated colour temperature is defined by the point on the B-B locus which lies closest to (x, y) (see Wyszecki and Stiles [5]).

The colour temperature gives an approximate measure of the relative spectral distribution of a source *within the visible spectrum*, this distribution being equivalent to $M(\lambda, T_c)$. T_c may be greater or less than the source temperature T, de-

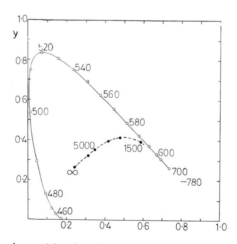

Fig. 4. C.I.E. 1931 xy chromaticity chart. Outer line: spectrum locus, the numbers representing the spectral wavelengths. Broken line: locus of black-body sources, the numbers representing the source temperature.

pending on whether the source emits greater energy in the short wavelength end
than the long wavelength end of the spectrum relative to $M(\lambda, T)$, or vice versa.

2.4. Practical white light sources

Tungsten lamps are made with a number of different filament and envelope configurations. The ribbon filament tungsten lamp is widely used for the illumination
of the entrance slits of optical systems, although only the central portion of the
filament should be so used, as the ends run at a lower temperature.

Tungsten lamps have nominal power ratings from 50 watt to 1000 watt, with a
corresponding range of colour temperature from 2500°K to 3400°K and luminance
from 200 cd/cm^2 to 2000 cd/cm^2. Their life time may be considerably increased by
running them at below the rated voltage, but the total light output and colour
temperature are reduced in consequence.

Tungsten lamps are frequently provided as calibrated secondary standards of
luminance and colour temperature.

Tungsten-quartz-iodine lamps utilize the dissociation and recombination of WI$_2$
to prevent evaporated tungsten depositing on the envelope. The rated colour temperature is between 3200°K and 3400°K, and luminance up to 5×10^3 cd/cm^2. As the
quartz envelope can withstand high temperatures, it can be much smaller than the
glass envelope of an ordinary tungsten lamp. Thus, a small aperture condenser lens
is capable of collecting its light output over a relatively large solid angle.

Arc discharge lamps which work at medium or high pressure are an important
class of white light source.

Xenon discharge lamps may be long arc, medium pressure (with arcs centimetres
in length and working pressures of about one atmosphere) or short arc, high pressure (arc lengths of millimetres and working pressures of some ten atmospheres).
The high pressure lamps yield large radiation fluxes with mixed spectral characteristics (see fig. 5). The visible spectral distribution approximates closely to that of
daylight ($T_c \sim 6500$°K) and thus xenon lamps are frequently used to simulate day-

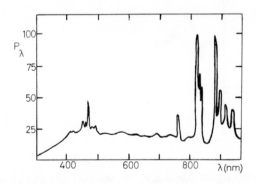

Fig. 5. Emission spectrum, P_λ, of a high pressure xenon arc lamp.

light. However, xenon lamps give a relatively higher output in the near UV and near IR spectral ranges than that of daylight.

The power rating of xenon arcs is typically in the range 400 watt to 7 kwatt with a corresponding luminance range of about 10^4 to 10^5 cd/cm^2 for an arc of about a square millimetre in area. The total power output for a lamp of luminance 10^5 cd/cm^2 is about 10 watt per 10 nm bandwidth throughout the visible spectrum. The arc may be A.C. or D.C. excited, A.C. excitation giving a more uniform intensity distribution across the arc, but D.C. excitation giving far greater arc stability. Xenon arcs sometimes require considerable auxiliary electrical supply equipment to provide the large arc currents and water cooling may also be necessary.

Mercury discharge lamps give a line spectrum, in the range 200 nm to 600 nm, this being superimposed on a weak continuum. The spectral character of the emitted radiation changes markedly with the lamp pressure. The high pressure lamps consist of a small, very high luminance arc (e.g. a $\frac{1}{4}$ mm square arc of luminance, 2×10^5 cd/cm^2). Strong emission lines occur at wavelengths 254, 313, 366 and 577–579 nm, and total power of tens of watts per emission line can be achieved. High pressure mercury lamps are available as radiometric standards of UV radiation, the output of one or more of the UV lines being calibrated.

As mercury spectra are rich in blue-green, blue and near UV radiation, the lamps are well matched to the sensitivity of photo-electric detector devices.

Hydrogen and deuterium discharge lamps yield continuous spectra within the wavelength range 190 nm to 500 nm (see fig. 6). Deuterium gives rather higher output at shorter wavelengths than does hydrogen. UV monochromating systems are frequently supplied with hydrogen or deuterium lamps, and a typical output after

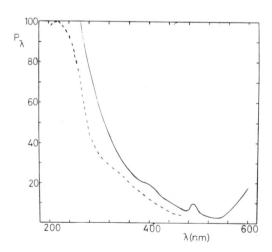

Fig. 6. Relative spectral energy distribution, P_λ, of radiation emitted by a hydrogen arc (full curve) and a deuterium arc (broken curve), plotted as a function of wavelength, λ.

passage of the light through a monochromator would be 0.1 milliwatt per 10 nm bandwidth.

A review of available UV sources is given by Hirt et al. [14].

Flash sources are of two types. One form consists of a long coiled, metallic wire filament contained in an oxygen filled bulb. Excitation by an electric current causes the filament wire to burn, resulting in a brief light flash. The metallic wire is usually of zirconium or aluminium, giving a colour temperature of about 3800°K and luminous output of 5000 to 100,000 lumen/sec in a flash duration of tens of milliseconds.

Electronic flashes consist of discharge tubes, excited by discharge of a capacitor through the tube. The flash duration depends upon the time constant of the discharge circuit and may be anywhere between a μsec and a msec. The time constant may be increased by the addition of a resistance in series with the discharge. The flash tube usually contains an inert gas. Xenon tubes give an output flash of colour temperature between 6000°K and 7000°K with output energy of up to 10^5 lumen/sec or 2000 J. Xenon tubes are also frequently used as stroboscope sources (see e.g. Rutkowski [15] for a discussion of stroboscopes).

2.5. *Spectral sources*

These usually consist of low pressure discharge tubes which contain inert gases and/or metallic vapour. Excitation by electron discharge results in the emission of strong lines, the frequency of the emitted radiation being determined by the energy levels involved in the electron transition and thus by the emitting atom. Low pressure discharge tubes are of low radiance, although "hollow-cathode" discharge sources have been developed, and these give strong but broadened resonance emission lines.

The coherence length of a source provides a measure of the bandwidth of an emission line (see section 2.1), the principal factor in line broadening of low pressure discharge lines being the Doppler effect. This arises from the one-dimensional Gaussian distribution of velocities in a gas, and results in a Gaussian spectral distribution of line intensity. The bandwidth of a spectral line is usually quoted as the wavelength (or frequency) separation between the two half-peak intensity points (see fig. 7).

Near spectral stimuli may be obtained by using a monochromating system in conjunction with a white light source and the monochromator may simply consist of a series of narrow band interference filters. Such filters are available with peak transmission at any point in the range 200 nm to 800 nm (and beyond) with typical bandwidth values of from 2 nm to 20 nm. Interference filters are also produced which isolate specific emission lines of "mixed" spectra. It should be noted that interference filters which are nominally of the same bandwidth and peak transmission wavelength may, none-the-less, differ considerably in the manner in which their transmission cut-off occurs. The properties of interference filters depend upon the direction of incidence of the light, and in order to achieve the specified per-

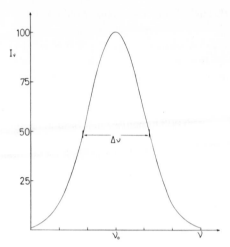

Fig. 7. Spectral distribution of intensity, I_ν, for a Doppler broadened line. $\Delta\nu$: frequency half width.

formance a filter should always be placed with its face normal to a collimated light beam. When monochromatic light stimuli free from all stray light are required, a double monochromating system should be employed (see e.g. Wright [16]).

For experimental purposes, low pressure discharge lamps normally serve only as standards for wavelength calibration, or as sources of long coherence length in optical interference experiments. Examples of this type of lamp are sodium (coherence length L, $\simeq$ 85 mm and emission wavelengths 589 and 589.6 nm), cadmium (L equal to 250 mm for the 643.8 nm line) and mercury (L equal to 400 mm for the 546.1 nm line).

Hollow cathode lamps are used in atomic absorption spectrophotometry, and a large number of sources giving resonance lines with wavelengths in the range 190 nm to 700 nm (and beyond) are available.

3. Lasers

3.1. *Introduction*

Lasers depend upon the electron transition mechanism in which an incident photon of suitable frequency causes an excited atom to emit a photon. A photon emitted in this way has the same frequency and phase as the incident photon. This process, known as stimulated emission of radiation, is the converse of optical absorption by electron transition and the two processes occur simultaneously. If the gain due to stimulated emission exceeds the loss by absorption, then an incident light beam is effectively amplified on its transmission through the laser material. It can be shown that this situation only arises if there is a non-equilibrium distribution of the atoms of the laser material between different excitation levels. In order to produce the required non-equilibrium state, it is necessary to pump energy into the laser material.

A practical laser system consists of the laser material contained within an optical cavity. The cavity, which may be simply a ruby rod, or the discharge tube of a gas laser, is usually terminated by highly reflecting surfaces. The cavity may have a number of geometric configurations, and its structure contributes significantly to the properties of the emitted radiation (see section 3.2.5). It should be noted that at optical frequencies, the stimulated emission process cannot in general be used simply to amplify external light incident upon the laser cavity. Spontaneous light emission occurs in the laser material and is then amplified by the laser.

In addition to the laser material and the cavity, a power supply is essential for the production of the required non-equilibrium state. As most laser "pumping" methods are very inefficient (typically $< 0.1\%$ conversion into light energy), the more powerful laser systems incorporate an exciter and some, such as the gas ion lasers, also require water cooling systems.

3.2. Properties of laser emission

3.2.1. *The wavelength* of the emitted radiation is determined by the pair of atomic levels between which the electron transition occurs. Although the wavelength bandwidth, $\Delta\lambda$, of the transition is not reduced for stimulated emission relative to spontaneous emission, the bandwidth of laser lines are many times smaller than those of the corresponding atomic emission lines. This is because the laser cavity acts as a highly tuned resonator, and amplifies one ore more wavelengths which lie within the total bandwidth $\Delta\lambda$.

At present there is only a limited number of lasers which emit radiation in the near UV and visible spectrum. Neutral atom gas lasers generally emit in the infra-red (He−Ne is a notable exception), as also do ion crystal lasers, in this case ruby being an exception. Again, semi-conductor injection lasers are exclusively infra-red emitters (e.g. Ga−As laser, giving an output wavelength of 8500−9000 Å). Gas ion lasers are important sources of near UV and visible lines, and recently dye laser have been developed, which provide visible radiation capable of being tuned over a wavelength range of up to 500 nm. The output of flashlamp excited dye lasers consists of a brief pulse of duration between 5 and 100 nsec, with peak power levels of some 10^5 watt. A recent review of organic dye lasers is given by Snavely [20]. One possible method of producing additional laser lines is to use frequency doubling techniques. The experimental arrangement is illustrated in fig. 8, the crystal C being anisotropic (e.g. KDP). The anisotropy of the crystal results in the generation of waves with frequencies which are harmonics of the frequency of the incident radiation. Using carefully prepared crystals, the conversion efficiency for production of the first harmonic may be as high as 15% and thus the method is capable of producing significant power output. For example, the 694.3 nm emission of a ruby laser may be used to produce radiation of wavelength 347.2 m. However, as the generation of harmonics is a non-linear process a very large input power level is re-

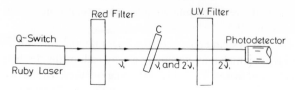

Fig. 8. Diagrammatic representation of an experimental set-up for frequency doubling. C: frequency doubling crystal.

quired to give a significant harmonic conversion efficiency, and a Q-switch pulse is usually used as the input. Frequency doubling (ADP) crystals have been employed to generate UV radiation of wavelength 257.3 nm from an input line of a C–W Argon laser. The power obtained is of the order of $4p^2$ milliwatt, P being the power of the incident 514.5 nm beam in watt.

3.2.2. The temporal characteristics of the output beam may be of two kinds, depending on whether the laser is pulse-type or continuous working (C–W). In the latter case, a steady light output is provided by the laser, and this may typically have long term fluctuations of ± 5% r.m.s. with a noise amplitude of ± 1% r.m.s.

An output trace of a pulse from a ruby laser, plotted as a function of time, shows a series of irregular spikes of μsec duration, separated by 2–3 μsec (fig. 9). These spikes are usually maintained for a period of $\frac{1}{2}$ msec to 3 msec. The laser emission is preceeded by fluorescence, typically of about 1 msec duration, and on termination of the laser action, further fluorescence occurs.

A more regular output pulse may be obtained by Q-switching a pulse laser (see e.g. Lengyel [18], chapter 6, for discussion of the methods of Q-switching). Essentially, Q-switching stores the laser energy until a chosen time, when all the output is released in a single pulse. The laser action is "spoiled" by inserting e.g. a chemical dye within the laser cavity, for which purpose one end of the ruby rod is made non-reflecting, and an external mirror placed beyond the dye-cell. The duration of a Q-switch pulse is extremely brief, typically of about a nano second, but none-the-less the pulse may be quite energetic.

3.2.3. The coherence length L of the laser beam is related to the spectral bandwidth by the relationship $L = \lambda^2/\Delta\lambda$ (see section 2.1). With special care, coherence lengths of the order of a kilometre may be obtained, but generally, very much shorter coherence lengths (e.g. of about one metre) are obtained with commercial laser sys-

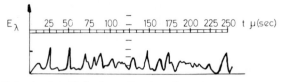

Fig. 9. Output pulse of ruby laser, E_λ, plotted as a function of time, t.

tems. The relatively short values of the coherence length result from lack of thermal and mechanical stability or, additionally, from failure to isolate individual modes of oscillation (see section 3.2.5). This latter is particularly important in ruby laser systems.

The output of a laser is also spatially coherent across the beam e.g. the output of a He—Ne laser has a phase front which coincides with the mirror reflector surfaces. Laser beams may thus be used to produce optical interference effects.

3.2.4. *Polarization.* Although the transition processes which give rise to laser action are not polarization selective, the output beam of a laser system is in many cases plane polarized. Thus, gas lasers usually give plane polarized output radiation because the laser gas is contained in a glass or quartz tube, terminated by Brewster windows. It is the polarization dependent reflection properties of these windows which render the output beam plane polarized. Crystal lasers may give a plane polarized output beam, depending upon the crystal properties. Thus, a ruby rod cleaved with its axis at $60°$ or $90°$ to the optic axis gives a plane polarized output, whereas if the two axes are parallel, the output is unpolarized. It is the anisotropy of the crystal gives rise to the polarization.

3.2.5. *Mode structure.* As has already been indicated, the laser cavity acts as a sharply tuned resonator and as such, it is responsible for the narrow bandwidth of the laser output. However, the cavity may support simultaneously a number of different frequencies of oscillation, and these constitute modes of oscillation of the cavity. Further, there may be modes which have the same frequency, but different spatial configuration – these are known as degenerate modes. The mode structure present in a particular laser system depends upon the form of the laser cavity e.g. whether it is terminated by plane or sperical reflecting surface. A general review of the analysis of laser mode characteristics is given by Toraldo di Francia [21].

Experimentally, different mode structures are apparent as different radiation patterns on the end mirrors of the laser. Some mode configurations are shown in fig. 10. It should be noted that modes of different frequencies may have the same spatial configuration in the cavity, and may thus be represented by a single spot of light on the laser mirrors.

In practice, gas lasers are usually constructed to oscillate in a single axial mode, or at least in a purely axial mode configuration. Ruby lasers are more difficult to control, however, and unless special precautions are taken, a multi-mode pattern is excited.

3.2.6. *The output energy* of most laser systems is large. Thus, C-W gas ion lasers give typical output flux values of the order of watt, and even the smallest He—Ne gas laser yields about half a milliwatt output flux. In the case of pulsed ruby lasers, the total output may be $1-200$ J, contained in a pulse of duration about 1 msec.

It should be noted that the output flux from a laser is not uniformly distributed

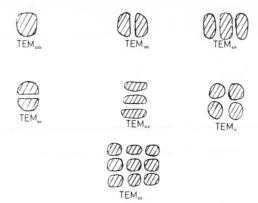

Fig. 10. Some transverse electro-magnetic mode configurations as seen on the terminating mirror of a helium−neon laser cavity.

across the output beam, e.g. a laser cavity with spherical mirrors gives a Gaussian
distribution of illumination across the beam. The diameter of a laser beam is usually
quoted as the value at which the illumination drops to $1/e^2$ of its central, peak value.

An important feature of laser beams is that they are highly collimated, with typical beam divergence values of a few milliradians. The laser is equivalent to a diffraction limited pinhole source, and in comparing it with conventional light sources,
this must be borne in mind. Thus, the conventional thermal source gives much
lower flux output when compared with the laser, if we consider the illumination
produced at a diffraction limited stop. On the other hand, the conventional source
is not diffraction limited, and very much larger areas of source are available.

For general illumination, therefore, thermal sources are preferable to lasers, but
if a powerful beam of collimated light is required, then a laser should be used.

3.3. Some experimental lasers

A list of some commonly used lasers and their characteristics are given in table 1.

4. Detectors

4.1. Introduction

Light measurement requires that the electro-magnetic energy be converted into
some more readily measurable form of energy. This conversion process may be performed in a number of different ways, e.g. into heat (thermopiles, bolometers), or
into an electric current (photo-electric cells, photomultipliers). Thus, different light
detectors may involve quite different physical principles. It is not the object of this
chapter to examine the different processes of light detection, but to discuss charac-
terisitcs of photo-detectors which are of general importance, and to list some commonly used detectors and their characteristics.

Table 1

Laser material	Output wavelength (nm)	Temporal characteristics	Output power	Mode structure	Polarization	Power
Helium–neon	632.8	Continuous output (CW)	0.1 to 50 mW	Axial	Usually linear	Electrical A.C. or D.C.
Helium–cadmium	441.6 and 325.0	CW	50 mW (441.6 line) 10 mW (325.0 line)	Axial	Linear	Electrical
Argon ion	Principally 514.5 and 488; also 351.1; 363.8; 457.9; 465.8; 472.7; 476.5; 501.7 (λ can be selected)	Usually CW	mW to 1 W uni-mode; up to 10 W multi-mode.	Axial	Linear	Electrical + water cooling
Krypton ion	Principally 647.1; 568.2; 530.8; 520.8; 476.2; also 676.4; 482.5; 468.0; 461.9; 350.7; 356.4 (λ can be selected)	Usually CW	mW to $\frac{1}{2}$ W uni-mode; up to 1 W multi-mode	Axial	Linear	Electrical + water cooling
CO$_2$ (+ nitrogen)	10.6 μm	CW or pulse (in Q-switch mode)	50 W per metre, up to ~ kW CW	Axial (unimodal)	Unpolarized	Electrical + water cooling
Ruby	694.3 (at room temperature)	Pulse 0.5 to 3 msec duration	0.1 to ~ 250 J per pulse	Multimode	Linear or unpolarized	Flash lamps
Q-switch ruby	694.3	Giant pulse of duration ~ msec	Peak power, 10^9 W Total energy, tens joule.	Multimode	Linear or unpolarized	Laser spoiled by dye switch Kerr cell, etc.
Neodymium–glass	1370; 1061 or 918	pulse (of ~ msec duration	~ 100 J	Multimode	Linear	Flash lamps
Gallium–arsenide diode	in range 800 to 900	pulse (~ μsec) or continuous	pulse: 100 W peak: CW: 1 W	Multimode	Linear	Electron current
Organic dyes	in range 430 to 650, continuously tuneable over 30–50 nm for any one dye	Pulse (~ $\frac{1}{2}$ μsec)	Total energy per pulse, 150 mJ (blue) to 300 mJ (red)	Multimode	Linear	Flush lamp
Organic dyes	in range 500 to 700, continuously tuneable	CW	> 5% conversion of the Argon laser output power for a 0.1 nm output	Axial (unimodal)	Linear	Argon laser as pump source

Photo-detectors may be broadly classified into two groups, namely, those which detect the total flux incident upon them and those which detect the distribution of illumination incident upon their photo-sensitive surfaces. The two groups are not distinct; for instance, a photomultiplier is essentially a total flux detector, but if a narrow slit is placed before its photo-sensitive cathode, it may be employed to scan a one-dimensional distribution of light energy.

In addition to the detector itself, light measurement also requires a meter to determine the detector output and, frequently, a power supply must be provided for the detector.

4.2. Characteristics of light detectors

4.2.1. *The absolute sensitivity* is defined as the ratio of the detector's output signal to the flux of the incident radiation, and may be quoted either in terms of radiant flux or photometric flux sensitivity. In the case of photo-electric devices, it is often expressed in terms of the quantum efficiency, which is defined as the number of photo-electrons emitted per incident light photon. In general, the sensitivity of photo-detector devices is a function of wavelength, and must be defined wavelength by wavelength. In such cases it is usual for a detector to be specified by its peak absolute sensitivity and its relative spectral sensitivity function.

Those detectors which are uniformly sensitive to all wavelengths are described as non-selective, and invariably depend in their action upon the heating effect of the incident radiation. Although such detectors are essential for calibration purposes, they are not suitable if the measurement requires high sensitivity or fast response. Many detectors are wavelength-selective in response, and some common spectral sensitivity functions are shown in fig. 11.

4.2.2. *The incident light flux-output signal* characteristics are important in that ideally they should be linear. Although many classes of detector are intrinsically linear (see table 2), it must be remembered that the measurement is also dependent upon the meter used in conjunction with the detector. Thus, linearity should always be checked for the detecting system as a whole.

4.2.3. *The temporal frequency response* of a detector is defined as:

$$\frac{\text{the amplitude of modulation of the output signal}}{\text{the amplitude of modulation of the light signal}}$$

for a sinusoidally modulated light signal, and is expressed as a function of the modulation frequency v. This is directly related to the *impulse response*, which defines the time distribution of the detector output signal in response to a limitingly short input signal. The more restricted in time is the impulse response, the broader is the frequency response function, and the more able is the detector to follow time fluctuations in the light signal. The temporal frequency response characteristics of a detector may be readily determined, and assume great importance in the light of

Table 2
Total flux detectors

Detector	Wavelength response	Sensitivity	Detection area	Frequency response	Linearity	Power pack	Special features
Thermopile	Uniform	50 μV/μW	< mm^2	10^2 c/s	Yes	No	Calibrating device
Golay cell	Uniform	0.2 μV/μW	~ 20 mm^2		Yes	Chopper for light signal	Sensitive in wavelength range from UV to μ-wave
Photo-electric cell	Selective (peaks ~ 400 nm)	0.002 μA/μW (20 A/lumen)	~ cm^2	10^5 c/s	Yes	Yes: 100–300 V	Current saturation, noise limited
Photo-multiplier	(as above)	~ 1 A/μW output	~ cm^2	10^9 c/s	Yes	Yes: 1–2.5 kV (stabilized)	High sensitivity; wide frequency response; significant dark current
Photo-conductive cell (selenium)	Decreases gradually from 400 nm to 800 nm	0.01 μA/μW	~ cm^2	10^3 c/s	No	Yes	Wide wavelength range
Barrier-layer cell (selenium)	Peaks sharply at 560 nm	400 μA/lumen	~ cm^2	10^3 c/s	Approx.	No	Relative spectral sensitivity almost the same as the eye
Photo-transistor	Peaks at ~ 800 nm	0.3 A/lumen	~ 1 mm^2	10^3 c/s	Yes	Dry battery 20 V	Marked directional sensitivity
Integrating photo-diode	(as above)	5000 A/lumen (variable)	~ 0.01 mm^2	Integrates for sec pulses	Yes	Variable frequency trigger device	High sensitivity, small area

Table 2 (continued)
Image forming detectors

Detector	λ response	Resolution	Image formation	Linearity	Special features
Eye	Peaks at 560 nm (photopic); 510 nm (scotopic)	1' arc (in central field)	Non-linear (edge effects such as mach bands occur)	No	Adaption, colour discrimination pattern recognition; accommodation
Photographic film	Varies depending on film type	10^6 image points per cm^2	Diffusion effects (e.g. Eberhardt effect)	No (logarithmic over a restricted range)	Low quantum efficiency, reciprocity failure (for illumination $\times$ time)

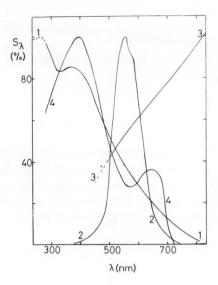

Fig. 11. The *relative* spectral sensitivity function, S_λ, for a number of classes of photo-detectors, plotted as a function of wavelength, λ. Curve 1: photo-electric cathode. Curve 2: selenium photo-voltaic cell. Curve 3: semi-conductor (phototransistor). Curve 4: photographic film (panchromatic).

the mathematical techniques which have been developed for the analysis of time-varying signals.

4.2.4. *Noise* is the term applied to any output signal from a photo-detector which does not result from the input signal. It becomes significant when weak light signals or small changes in a light signal are to be detected. There are three principal sources of noise which arise in photo-detection.

Shot noise results from the statistical nature of the emission and absorption of photons. It can be shown that the average flux rate of a photon beam, $\bar{n}$, is subject to fluctuations of standard deviation $n^{\frac{1}{2}}$. A corresponding fluctuation occurs, for example, in the output current of a photo-electric device, and the general expression for the amplitude of the shot noise fluctuations in a D.C. current, i_0, may be written as:

$$\langle \Delta i_0^2 \rangle = 2ei_0 \int_0^\infty G(\nu) G^*(\nu) \, d\nu ,$$

$$= 2ei_0 \Delta \nu .$$

(6)

Where $\langle \Delta i^2 \rangle$ is the root mean square deviation in the current i_0, $G(\nu)$ the complex frequency response of the detector, with complex conjugate $G^*(\nu)$, ν the frequency, $\Delta \nu$ the frequency bandwidth of the detector circuit and e the electronic change.

Thermal (or Johnson–Nyquist) noise arises from the Brownian motion of charge carriers in resistors. This is important in photo-detection, as photo-currents generated by a light signal are measured in terms of the corresponding voltage in a load resistance. The thermal fluctuations of voltage in a resistance, R, at temperature T, are given by:

$$\langle \Delta v^2 \rangle = 4kTR\Delta v , \tag{7}$$

where Δv is again the frequency bandwidth of the detection circuit and k is Boltzmann's constant.

The *dark current* is an important source of noise in photomultiplier detection, and is due to the emission of thermal electrons from the cathode, which are amplified by the photomultiplier in the same manner as are the photo-electrons. The dark current, i_d, is a D.C. current, and as such does not constitute noise, because it may be measured and subtracted from the signal. However, there is shot noise associated with the dark current.

The dark current i_d from a semi-conductor surface is governed by the relationship:

$$i_d \propto T^2 e^{-K/T} ,$$

where T is the absolute temperature, and K a constant which depends upon the semi-conductor material.

Thus, it is advantageous to cool photomultipliers if very weak light signals are to be detected. By careful choice of dynode material, dark current levels as low as 10^{-13} amp under working conditions may be achieved.

Shot noise and thermal noise are both "white" noise i.e. the noise amplitude is uniformly distributed through the frequency spectrum. A further type of noise which occurs in photo-detectors, *flicker noise*, is a low frequency noise. It is often important in components of electronic circuits, but it is usually neglected in discussions of noise in photo-detection, and may be ignored completely for detection of signals modulated at more than 500 c/s.

The *signal to noise ratio* (S/N ratio) is defined as the ratio of the output signal to the noise occuring in the measurement. Consider fig. 12, which shows a typical photo-detector and load resistance combination used to measure a light signal. The

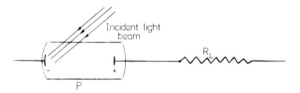

Fig. 12. A photo-electric light detector, P, acting as a source of shot noise current fluctuations and a load resistance, R_L, in which thermal fluctuations of voltage occur.

S/N ratio for the shot noise is given by eq. (6) as:

$$\frac{i_0}{(2ei_0\Delta v)^{\frac{1}{2}}} = \left(\frac{i_0}{2e\Delta v}\right)^{\frac{1}{2}},$$

where i_0 is the D.C. photo-current. The S/N ratio for the thermal fluctuations voltage in the resistance R is given (eq. (7)) by:

$$\frac{i_0 R_l}{(4kTR\Delta v)^{\frac{1}{2}}} = i_0 \left(\frac{R_l}{4kT\Delta v}\right)^{\frac{1}{2}}.$$

Thus, as the load resistance R_L is reduced, the thermal noise become less significant, and the great advantage of photomultipliers is that they give significant voltage signals for small load resistance values R_L. Thus the Johnson can be kept to a minimum.

A useful discussion of noise in electronic circuits is given by Robinson [22].

4.2.5. *Spatial resolution* is the fundamental property which defines the performance of illumination sensitive detectors, and it is customarily defined in terms of the number of line pairs per mm which can be resolved by the detector. A more general description of image formation is provided by the spatial frequency response characteristics, but is beyond the scope of this chapter (see e.g. Beckman [25]).

4.2.6. *Auxiliary equipment.* As was discussed in the section 4.1 light measurement requires auxiliary apparatus to be used with the photo-detector. In some cases, the performance of the equipment may significantly effect the accuracy of the measurements.

Photo-electric detectors require an applied voltage between the photo-sensitive cathode and the anode, in order that the emitted photo-electrons are collected at the anode. The variation of the *output current* with the value of the anode voltage is important, and is illustrated in fig. 13 for a vacuum photo-cell and for a photomultiplier. As can be seen, the former demonstrates saturation of output, and if the anode voltage is 150–200 volt, it may vary significantly without affecting the output current. On the other hand, the photomultiplier output is critically dependent upon the anode–cathode voltage, and in this case a stabilized power supply is necessary.

The *measuring device* used to determine the detector output may affect the performance in a number of ways. It may not be linear, even though the detector may itself be linear (section 4.2.2); it may have limited frequency response and thus be unable to follow time functions in the detector output; and it may introduce sources of noise in addition to those inherent on the detection process. In general, care must be taken to match the input impedance of the measuring device to the output impedance of the detector.

Certain procedures have been developed in recent years in order to achieve high photometric accuracy, and these are of some general interest (e.g. Budde [26]).

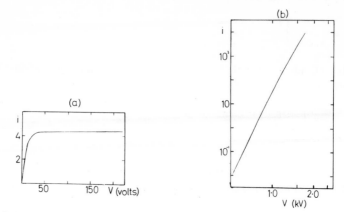

Fig. 13. Output photo-current, *i*, plotted as a function of applied voltage, *V*; (a) for a photo-electric cell and (b) for a photomultiplier.

Acknowledgements

I am grateful to Dr. L.Planskoy and Dr. W.T.Welford, both of this department, for reading sections of the original manuscript and for suggesting additions and alterations.

References

The list of references has been restricted, whenever possible, to books or articles of a review nature. Further, references have been classified as "specific" or "general" indicating whether or not they are concerned with a single aspect of the subject. Not all the references given are included in the text.

Sections 1 and 2: Radiometry, Photometry and Light Sources.

General:

[1] E.J. Gillham, Notes on Applied Science No. 23, Radiometric Standards and Measurements. London, H.M. Stationery Office (1961).

[2] O.C. Jones and J.S. Preston, Notes on Applied Science No. 24, Photometric Standards and the Unit of Light, London, H.M. Stationery Office (1969).

[3] Y. Le Grand, Light, Colour and Vision, translated by R.W.G. Hunt, J.W.T. Walsh and F.R.W. Hunt (Chapman and Hall, London, 1957).

[4] J.W.T. Walsh, Photometry, 3rd ed. (Constable, London, 1958).

[5] G. Wyszecki and W.S.Stiles, Color Science (Wiley, New York, 1967).

[6] International Recommendations of standard data and terminology are to be found in the Compte Rendu of the Commission Internationale de l'Eclairage

[7] C.I.E. Proc. 1924 (Cambridge University Press, 1926).

[8] C.I.E. Proc. 1931 (Cambridge University Press, 1932).

[9] C.I.E. Proc. 1951, I, Paris, Bureau Central, C.I.E.

[10] C.I.E. International Lighting Vocabulary, 2nd ed., I, Publication C.I.E. 1.1 (1957) II Publication C.I.E. W.I.I. (1959) Paris, Bureau Central, C.I.E., 57, rue Cuvier.

[11] C.I.E. Proc. 1963 (Vienna) Paris, Bureau Central, C.I.E., 57, rue Cuvier (1964).

Specific:

[12] J.C. De Vos, A new determination of the emissivity of tungsten ribbon, Physica *20* (1954) 690–720.

[13] J. Euler, Der Graphitbogen als spektralphotometrisches Strahldichtenormal in Gebiet von 0.25 bis 1.8 μ. Ann. Phys. *11* (1952–1953) 203–224.

[14] R.C. Hirt. R.G. Schmitt, N.D. Searle and A.P. Sullivan, Ultraviolet spectral energy distributions of natural sunlight and accelerated test light sources, J. Opt. Soc. Amer. *50* (1960) 706.

[15] J. Rutkowski, Stroboscopes for Industry and Research (Pergamon Press, London, 1966).

[16] W.D. Wright, A photoelectric spectrophotometer and tristimulus colorimeter designed for teaching and research.

Section 3: Lasers.

General:

[17] G. Birnbaum, Optical Masers, Advances in Electronics and Electron Physics, Supplement 2 (Academic Press, New York, 1964).

[18] B.A. Lengyel, Introduction to Laser Physics (Wiley, New York, 1966).

[19] D. Röss, Laser, Lichtverstärker and Oszillatoren (Akademische Verlaggesellschaft, Frankfurt-am-Main, 1966).

Specific:

[20] B.B. Snavely, Flashlamp-excited Organic Dye Lasers, Proc. I.E.E.E. *57* (1969) 137.

[21] Toraldo di Francia, Theory of Optical Resonators, Proc. International School of Physics, Enrico Fermi, *Course 21*, Quantum Electronics and Coherent Light (Academic Press, New York, 1964).

Section 4: Detectors.

General:

[22] F.N.H. Robinson, Noise in Electrical Circuits (O.U.P.) 1962.

[23] Ed. S. Larach, Photo-electric Materials and Devices (Von Nostrand, Princeton, 1965).

[24] V.K. Zworykin and E.G. Ramberg, Photo Electricity and its Applications. 3rd ed. (Wiley, New York, 1956).

Specific:

[25] J.E. Beckman, Application of information theory to the evaluation of two image intensifier tubes. Advan. Electron. Electron Phys. *22A* (1966) 364–379.

[26] W. Budde, A photomultiplier circuit for precision spectrophotometry. Appl. Opt. *3* (1964) 69–71.

CHAPTER 2

SOME PRINCIPLES OF DOSIMETRY IN PHOTODYNAMIC ACTION EXPERIMENTS

C.W. HIATT

Department of Bioengineering, University of Texas
Medical School, San Antonia, Texas 78229, USA

1. Introduction

The technology of photodynamic action began in what may have seemed to be a very simple way, when Raab unwittingly allowed the ambient light intensity in the laboratory to become a variable in his experiments with paramecia and acridine dyes. In retrospect it is evident that these initial experiments, far from being simple, were indescribably complex in terms of incident light intensity, wavelength, and degree of penetration. While photodynamic action has remained, to a large extent, a phenomenological playground, there has been a steady improvement in the technology, much of it in the direction of simplifying the control and measurement of light so that quantum energy and quantum dose may be dealt with as independent variables. A wide variety of irradiation apparatus is available to the modern experimenter, including continuous and line-spectrum lamps, absorption and interference filters, monochromators, flash irradiators, and lasers. Light intensity may be measured by a variety of sensitive and dependable photoelectric transducers and may be recorded precisely as a function of wavelength or time. In designing and interpreting photodynamic action experiments, however, the investigator must draw upon some basic physical principles which are equally applicable to elaborate apparatus and to the beaker on Raab's window-sill.

2. Distribution of light in irradiated fluids

If the intensity (irradiance) of light impinging upon the surface of a layer of liquid is determined by measurement to be I_0, and if there is no reflection from the supporting surface, then the intensity at any wavelength in a place x cm below the entering surface is given by the Beer-Lambert law as

$$I_x = I_0 e^{-\alpha x} \tag{1}$$

in which e is the base of natural logarithms and α is a constant known as the *absorption coefficient* (fig. 1).

Table 1

Average relative light intensity in a layer of liquid illuminated from one side only and supported by surfaces of various reflection coefficients

Transmittance ($e^{-\alpha L}$)	Reflection coefficient									
	0	0.1	0.2	0.3	0.4	0.5	0.6	0.7	0.8	0.9
0.1	0.391	0.395	0.399	0.403	0.406	0.410	0.414	0.418	0.422	0.426
0.2	0.497	0.507	0.517	0.527	0.537	0.547	0.557	0.567	0.577	0.587
0.3	0.581	0.599	0.616	0.634	0.651	0.669	0.686	0.704	0.721	0.738
0.4	0.655	0.681	0.707	0.733	0.760	0.786	0.812	0.838	0.864	0.891
0.5	0.721	0.757	0.793	0.830	0.866	0.902	0.938	0.974	1.010	1.046
0.6	0.783	0.830	0.877	0.924	0.971	1.018	1.065	1.112	1.159	1.206
0.7	0.841	0.900	0.959	1.018	1.077	1.135	1.194	1.253	1.312	1.371
0.8	0.896	0.968	1.040	1.111	1.183	1.255	1.326	1.398	1.470	1.542
0.9	0.949	1.035	1.120	1.205	1.291	1.376	1.462	1.547	1.632	1.718

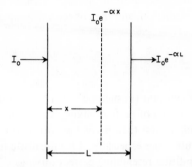

Fig. 1. Schematic representation of a layer of liquid illuminated from one side.

If the fraction of the light reflected by the supporting surface is the *reflection coefficient R,* and the total thickness of the irradiated liquid is L, then

$$I_x = I_0(e^{-\alpha x} + Re^{-\alpha L}e^{-\alpha(L-x)}) \tag{2}$$

$$= I_0(e^{-\alpha x} + Re^{-2\alpha L}e^{\alpha x}).$$

The mean intensity, $\overline{I}$, may be calculated as [3]

$$\overline{I} = \frac{I_0}{L}\int_0^L (e^{-\alpha x} + Re^{-2\alpha L}e^{\alpha x})\mathrm{d}x \tag{3}$$

$$= \frac{I_0}{\alpha L}\{1 - e^{-\alpha L} + Re^{-2\alpha L}(e^{\alpha L} - 1)\}.$$

As may be seen from an inspection of table 1, the relative mean intensity (I/I_0) depends upon $e^{-\alpha L}$ (the *transmittance*) and R (the *reflection coefficient*). By using reflective surfaces, such as polished aluminum ($R = 0.85$) one can improve the efficiency and uniformity of the irradiation. Possible interactions between metal surfaces and sensitive biological materials [6] should first be ruled out, however.

3. Calculation of average dose rate

As Morowitz [5] noted in 1950, the effective dose rate in a finite volume of liquid depends upon the degree of mixing, and can be defined readily for the two extreme cases: that in which perfect mixing occurs during irradiation, and that in which there is no mixing at all until after the irradiation is complete.

In the former case, the condition of 'perfect mixing' is fulfilled if every substrate molecule is equally likely to be in any part of the volume of fluid at any instant. By spending equal time in all parts of the volume, a molecule thus, in effect, receives a dose of radiant energy equal to the average intensity $(\bar{I})$ multiplied by the duration time (t) of the irradiation, and the concentration ratio (c/c_0) for a 1st-order reaction is

$$\frac{c}{c_0} = e^{-k\bar{I}t},\tag{4}$$

in which k is the velocity constant (in such units that the product $k\bar{I}t$ is unitless).

Unfortunately, the condition of perfect mixing is easier to imagine than to achieve, and degrees of mixing less than perfect are difficult to assess or compensate for. For short exposure times especially, it is unrealistic to depend upon stirring devices to approach a condition resembling perfect mixing.

The other extreme, therefore, in which no mixing at all occurs, becomes of interest. Following the reasoning of Morowitz [5], we may visualize the irradiated fluid as composed of a series of infinitely thin strata which are not disturbed during irradiation. The concentration ratio in each stratum is

$$\left(\frac{c}{c_0}\right)_x = \exp(-kI_0 te^{-\alpha x}),\tag{5}$$

in which x is the distance of the xth stratum from the plane of incidence. If the fluid is then mixed after irradiation (as it normally would be) the average concentration ratio becomes

$$\frac{c}{c_0} = \frac{1}{L}\int_0^L \exp(-kI_0 te^{-\alpha x})dx.\tag{6}$$

With the additional complication of single-pass reflection from a supporting surface of reflection coefficient R, eq. (6) becomes [3]:

$$\frac{c}{c_0} = \frac{1}{L}\int_0^L e^{-kI_0 t(e^{-\alpha x} + Re^{-2\alpha L}e^{\alpha x})}dx.\tag{7}$$

Eqs. (4) and (7) converge as αL decreases, i.e. for fluid volumes thin enough or transparent enough, mixing becomes unimportant insofar as it affects light penetration.

When the photodynamic substrate is a living microorganism, log c/c_0 is usually not determined with an uncertainty less than ± 0.1, so mixing ceases to be important when its effect falls below this level. Table 1, bases upon this criterion [3], lists the values of transmittance $(e^{-\alpha L})$ above which mixing does not appreciably affect the reaction rate.

It is important to note that mixing prior to irradiation may serve the useful purpose of maintaining a uniform dispersion of suspended substrate particles, and the mixing may as well be continued during the irradiation, but it is unlikely that a uniform dose of light will be achieved by mixing of thick or highly absorbent layers of liquid.

4. Other rate-controlling factors

4.1. Slow sensitization

The combination of sensitizing dye with substrate may be very rapid, even when living microbes are involved, or it may be a very slow reaction which dominates the kinetics of the photodynamic process [2]. The progressive photosensitization of T2r⁺ coliphage with toluidine blue is shown in fig. 2, and Arrhenius plots for the reaction with toluidine blue and proflavine are shown in fig. 3. T3 coliphage, in contrast, is sensitized so rapidly that kinetic studies by conventional means are impossible [1].

In experiments with novel substrates, particularly viruses or cells, it is advisable

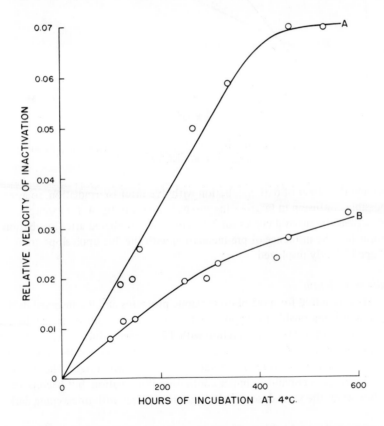

Fig. 2. Progressive photosensitization of T2r⁺ coliphage during incubation in the dark at 4°C with 6 mg/ml (A) and 3 mg/ml (B) of toluidine blue. (From Hiatt [7].)

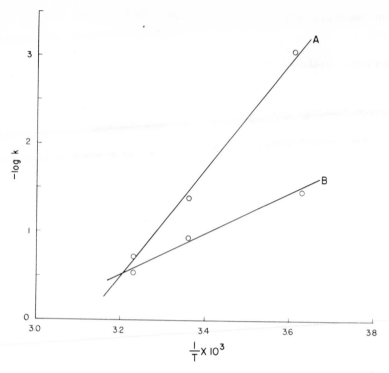

Fig. 3. Arrhenius plots for photosensitization of T2r⁺: (A) with toluidine blue; (B) with proflavine. (From Hiatt [7].)

to determine the effect of dark incubation with dye prior to irradiation, so as to
avoid possible confusion in interpreting survival curves. In fig. 4, for example,
which describes the survival curves for T2r⁺ coliphage irradiated after incubation
with proflavine, the influence of pre-incubation with the dye upon slope and
curve shape is clearly displayed.

4.2. *Oxygen depletion*

Since O_2 is required for most photodynamic processes and is consumed during
the reaction, it is reasonable to suppose that under some conditions it may be a
rate-controlling factor. Oxygen depletion with T2 coliphage during the photodynamic process has in fact been offered as an explanation for some of the 'tailing effect'
observed in survival curves at high dose rates [4]. As shown in fig. 5, the survival
curve for T2 phage is curvilinear under continuous illumination, but becomes a
straight line when the same dose of light is given as pulses with intervening dark
periods.

According to the explanation offered, O_2 is depleted within the phage particle
during continuous illumination, but has time to re-equilibrate if the light is pulsed

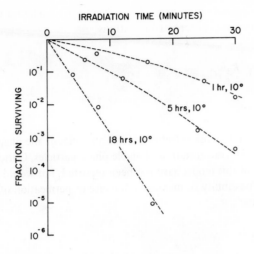

Fig. 4. Survival curves for T2r⁺ coliphage sensitized by preincubation with proflavine (4 mg/ml) for various periods of time in the dark at 4°C. (Unpublished data obtained by the author in collaboration with Eleanor Kaufman.)

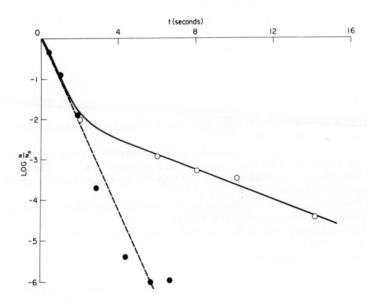

Fig. 5. Comparative effects of pulsed light (solid circles) and continuous light (open circles) on shape of the T2 survival curve. The pulsed light was on for 0.045 sec/sec, and the abscissa values were scaled accordingly. (From Hiatt, C.W. (1967) *loc. cit.*)

and the dark periods are long enough. The mathematical model for this system is given by two simultaneous differential equations:

$$\frac{dc_i}{dt} = k_2 c_o - k_2 c_i - k_3 \bar{I} c_i, \tag{8}$$

$$\frac{dN}{dt} = -k_1 c_i \bar{I} N, \tag{9}$$

in which c_i and c_o are O_2 concentrations inside and outside the phage particle, respectively and N is the concentration of viable phase particles. Critical experiments to test the validity of this model have not been reported, but would be of interest if they lead to the possibility of measuring the rate of permeation of O_2 through the phase 'membrane'.

5. Summary

   Kinetic studies of photodynamic action provide one source of information about the reaction mechanism and the properties of sensitizer and substrate, but they require careful attention to estimation of the effective dose rate. Attempts to assure a uniform dose of light by mixing during irradiation may be less effective than the expedient of using layers of liquid so thin or transparent that mixing becomes unimportant. Equilibration of the substrate with the sensitizer and O_2 in the system may occur at a slow rate and therefore dominate the kinetics of the overall reaction.

References

[1] J.J. Helprin and C.W. Hiatt, Photosensitization of T2 coliphage with toluidine blue. J. Bacteriol. 77 (1959) 502–505.
[2] C.W. Hiatt, E. Kaufman, J.J. Helprin and S. Baron, Inactivation of viruses by the photodynamic action of toluidine blue. J. Immunol. 84 (1960) 480–484.
[3] C.W. Hiatt, Theoretical considerations in the ultra-violet irradiation of micro-organisms. Nature 189 (1961) 678.
[4] C.W. Hiatt, Kinetics of virus inactivation by photodynamic action. Radiation Research (North-Holland, Amsterdam, 1967) pp. 857–868.
[5] H.J. Morowitz, Absorption effects in volume irradiation of microorganisms. Science 111 (1950) 229–230.
[6] N. Yamamoto, C.W. Hiatt and W. Haller, Mechanism of inactivation of bacteriophages by metals. Biochim. Biophys. Acta 91 (1964) 257–261.
[7] C.W. Hiatt, Photodynamic inactivation of viruses. Trans. N.Y. Acad. Sci. 22 (1960) 66–78.

CHAPTER 3

SINGLET OXYGEN AS OXIDIZING INTERMEDIATE AND PHOTODYNAMIC ACTION

Pierre DOUZOU

Institut de Biologie physico-chimique, Paris, France

1. Introduction

Primary mechanism of the sensitization act between a dye (sensitizer S), molecular oxygen ($^3O_2$) and an oxidizable substrate (AH) consists in the electronic excitation of S:

$$S_o + h\nu \to {}^1S^* \to {}^3S^* ,$$

where S_o is the sensitizer in its ground state, $^1S^*$ and $^3S^*$ are the lowest excited singlet and triplet states respectively fluorescent and phosphorescent.

At least five alternative mechanisms can be involved from $^3S^*$. These are:

$$(1)\ {}^3S^* + AH \to A^{\cdot} + R \xrightarrow{+{}^3O_2} Aox + S_o$$

$$(2)\ {}^3S^* + AH \to A^{\cdot} + R \xrightarrow{\quad AO_2^{\cdot} \quad} S_o + AO_2H \text{ (Gollnick–Schenck)}$$
$$\downarrow {}^3O_2$$
$$AO_2^{\cdot}$$
$$\downarrow AH$$
$$AO_2H + A^{\cdot}$$

$$(3)\ {}^3S^* + {}^3O_2 \to S_o + {}^1O_2^* \xrightarrow{AH} AO_2H \text{ (Kearns)}$$

$$(4)\ {}^3S^* + {}^3O_2 \to O_2^- + X \xrightarrow{AH} S_o + A^{\cdot} \xrightarrow{O_2} A_{ox} \text{ (Lindquist)}$$

$$(5)\ {}^3S^* + {}^3O_2 \to SO_2 \xrightarrow{AH} S_o + AO_2H \text{ (Schenck) .}$$

Mechanisms (1) and (2) involving a sensitizer–substrate interaction should mainly characterize the photosensitized alteration of macromolecules, that is, of biopolymers. Among such biopolymers, nucleic acids admitting a physical interaction with sensitizers (intercalation) could obey mechanisms (1) and (2), with the eventual formation of transient peroxides.

Mechanisms (3), (4) and (5) should occur in the presence of polyatomic substrates, dependent on media conditions (polarity of solvents, concentration of solutes, oxygen pressure, ...).

Mechanism (5) involving the formation of a dye—oxygen complex (moloxide-type SO_2) has received the strongest support from the work of Schenck, but also from Smith, Santamaria and Smaller who observed an intermediate of this type as evidenced by long-lived esr signals [1] after illumination of an aerobic solution of haematoporphyrin in phosphate-buffer at $-20°C$. Santamaria and Prino referred to this complex as an oxy-radical to distinguish it from the moloxide in which the paramagnetism of $^3O_2$ is lost [2].

In fact, all mechanisms listed above depend on the dye, on the substrate and as we said on media conditions. They can possibly be competitive in the photodynamic action, that is, in every dye-sensitized photo-autooxidation of biological substances.

On an other hand, most of the recent work dealing with photosensitized oxidation of organic molecules focus attention on the fact that molecular oxygen may act mainly as an energy-carrier from excited dye-molecules and oxidizable substrates according to the following process:

$$^3S^* + {}^3O_2 \rightarrow S_0 + {}^1O_2^*,$$

where S_0 is the dye in its ground state after acting as a catalyst and where $^1O_2^*$ is the excited singlet molecular oxygen [3].

We will now examine the main properties of such a species, its involvement in oxidation processes and its chemical as well as luminescent properties.

2. Excited singlet oxygen

Excited singlet molecular oxygen can exist in two different states corresponding to two different molecular orbitals: the highest singlest state $^1\Sigma_g^+$ in which the electrons in the Π-orbitals are antiparallel and the lowest one $^1\Delta_g$ with paired electrons in separate orbitals.

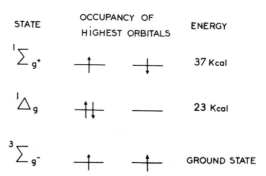

Fig. 1. Electronic states of molecular oxygen.

Corresponding energies for $^1\Sigma_g^+$ and $^1\Delta_g$ are 37 kcal/mole and 23 kcal/mole.

The $^1\Sigma_g^+$ state is very short-lived according to its rapid relaxation to the $^1\Delta_g$ state which is long-lived since it can survive at least 10^8 collisions with methanol in the vapor phase [4].

The $^1\Sigma_g^+$ state should act as an electrophilic reactant and the $^1\Delta_g$ as a nucleophilic one: for instance, singlet oxygen in the $^1\Sigma_g^+$ state could abstract one electron from polycyclic aromatic hydrocarbons (RH) giving R^+ (radical-cation) and O_2^-, and should in the $^1\Delta_g$ state give hydroperoxides and (or) endoperoxides [5].

In gas phase, excited singlet oxygen generated by microwave and electrical discharge leads to luminescent emissions due to the formation of singlet oxygen dimers, to their radiative deactivation and even eventually to an energy transfer to accurate emitters [6].

Singlet oxygen dimers (or pairs) can be the following in an order of decreasing probability:

$$^1\Delta_g + {}^1\Delta_g \rightarrow 2\ ^3O_2 + 633 \text{ nm}$$

$$^1\Delta_g + {}^1\Sigma_g^+ \rightarrow 2\ ^3O_2 + 480 \text{ nm}$$

$$^1\Sigma_g^+ + {}^1\Sigma_g^+ \rightarrow 2\ ^3O_2 + 360 \text{ nm}\ .$$

Singlet oxygen is thus characterized by emissive processes, eventually in condensed phase [7] where it can be detected by esr in its $^1\Delta_g$ state. Such a state generated in solution can be responsible for the formation of epoxides (with cholesterol for instance) and the $^1\Sigma_g^+$ state gives enone products in the same conditions [8].

Thus $^1O_2^*$ is among the species in which an oxygen molecule can react and which are listed in the following scheme

We can see that the dissociation to atomic oxygen as well as the oxidation (an electron removed) are the less probable events:

Even at 500°K there is little dissociation which only becomes appreciable towards 4000°K, and the high ionization potential (12.1 eV) of molecular oxygen makes the oxidation chemically very difficult.

Finally, of all the reactions of molecular oxygen perhaps the most interesting are those in solution, involving the reduction of oxygen to water. On an other hand, an increasing amount of work points out the possible role of $^1O_2^*$ in photochemistry and even in chemistry and maybe in biology.

Singlet oxygen can be chemically generated by the reaction between NaOCl and H_2O_2 [3] and also by microwave discharges, in condensed phases. It is possible to utilize $^1O_2^*$ (in the state $^1\Delta_g$) in a vessel other than that in which it is generated, by carrying $^1O_2^*$ in a stream of nitrogen gas, or by diffusion through layers of stearate. A two-phase reaction mixture can also separate $^1O_2^*$ from generating compounds.

···So, $^1O_2^*$ can be used as a tool for the study of reaction mechanisms involving the activation of molecular oxygen and we will see later a possible interesting application of such a procedure.

Since we are interested in photodynamic action mainly at the level of biopolymers, we have to consider binding of sensitizers and therefore sensitizer–substrate interactions even when molecular oxygen is in high concentration. Energy transfer, electron and hydrogen atom transfer could occur. As stated in the mechanisms (1) and (2), intermediate radicals resulting from sensitizer–substrate interactions can react immediately with $^3O_2$ to form peroxy-radicals giving in turn hydroperoxides.

Mechanisms (1) and (2) could be operative in the case of polycyclic aromatic hydrocarbons and dyes in physical interaction (intercalation) with polypeptides and nucleic acids (DNA). It is now known that some hydrocarbons in interaction with the DNA undergo a chemical reaction by physical and chemical oxidizing agents and contract a covalent bond with DNA [9]. Such a transformation of a physical interaction to a chemical one is attributed to an electron abstraction from the hydrocarbon giving a radical-cation undergoing in turn an arylation [10].

The question is to know if singlet oxygen is or is not involved in the formation of the radical cation (R^+). According to Cusachs and Steele, $^1O_2^*$ is able to perform the following reactions when acting upon polycyclic aromatic hydrocarbons:

(1) In its $^1\Delta_g$ state.

$$^1O_2^* + \begin{matrix} RH_2 \\ RH \end{matrix} \rightarrow \begin{matrix} R\text{-OOH} \overset{H}{} \\ R\text{-OOH} \end{matrix} \rightarrow \begin{matrix} R\text{-OH} \\ R=O \end{matrix} + H_2O + {}^1O_2^* \overset{\text{chain reactions}}{\underset{\text{luminescence}}{}}$$

The singlet oxygen is thus a chain sustaining species and determines the chemiluminescence.

(2) In the $^1\Sigma_g^+$ (much less probable since $^1\Sigma_g^+$ is very short lived),

$$^1O_2^* + RH \longrightarrow R^\cdot + O_2^-$$

$$R^\cdot + O_2 \longrightarrow ROO^\cdot$$

$$ROO^\cdot + RH \longrightarrow ROOH + R^\cdot$$

$$ROOH + HOOH \rightarrow ROH + H_2O + {}^1O_2^* \, (^1\Sigma_g^+) \rightarrow \text{chain...}$$

Such reactions could lead again to a chemiluminescence.

In fact, such schemes are yet hypothetical for corresponding reactions were performed in conditions where the states $^1\Sigma_g^+$ and $^1\Delta_g$ are mixed. It is impossible to establish the respective contribution of each state and we do not know yet if $^1\Delta_g$ is really able to give hydroperoxides with the polycyclic aromatic hydrocarbons.

3. Chemiluminescence

It is to be noted that the above reactions can lead to chemiluminescence.

Such a process can be effectively confirmed, and emission might be due to the radiative deactivation of $^1O_2^*$ dimers or to the deactivation of fluorescent molecules after energy transfer from dimers. In most of cases, chemiluminescence is very feeble and then difficult to analyze. However, the identification of its emitters is very important to check mechanisms leading to an emission.

In fact, chemiluminescence if mastered and correctly analyzed might be a useful tool for the study of reaction mechanisms involving $^1O_2^*$. Also, it is a proof of the intervention of molecular excited species in reactions, including as we know enzymic reactions (leading to bioluminescence) [11].

Chemiluminescence has known a sort of renaissance during the past 15 years and a symposium was organized on this topic in 1965 [12]. Many chemical and some biochemical redox systems display luminescence within the optical spectral region and as we saw in the above schemes, such an emission can result from intermediates of reactions leading to hydroxylations and ring cleavage. Unfortunately, quantum yields are in most of the cases quite low, and chemiluminescence has to be measured with coincident circuit scintillation counters, and the chemiluminescent spectra are recorded with difficulty.

Since chemiluminescence is a direct evidence of the involvement of electronically excited species and indirect proof of the existence of intermediates during the course of redox reactions chemically and possibly photochemically generated, it is of importance to enhance its quantum yields. Such a goal might be reached by accumulation of intermediates leading to chemiluminescent emission and this accumulation might be obtained in supercooled but still fluid solutions containing reactants participating in chemiluminescence.

4. Thermoluminescence

Most of chemiluminescent processes are thermoluminescent-like, i.e. they are temperature dependent. Lets consider for instance the reaction between the derivative flavine mononucleotide (FMN) and hydrogen peroxide in the presence of copper ions and EDTA: such a reaction leading to the oxidation of the FMN is luminescent but it can be seen in fig. 2 that the emission is feeble. In the same figure, it can be seen that when performed at $0°C$ (in the still fluid aqueous solution) the reaction is not spontaneously luminescent but that the emission is triggered by warming up.

It is possible to check that the oxidation of FMN is going on at $0°C$, for instance by recording the evolution of the fluorescence of FMN (progressive quenching) fig. 3.

So, the ultimate emission of luminescence by warming up seems to be due to the decomposition of a complex intermediate between FMN and the oxidizing species resulting from the decomposition of H_2O_2 by copper ions.

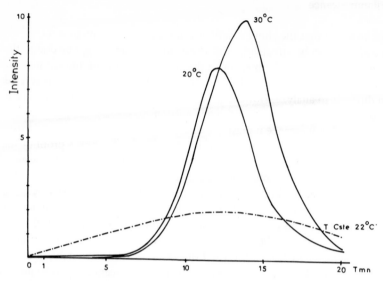

Fig. 2. Luminescence of FMN as a function of temperature. System FMN-H_2O_2: 2 ml,
FMN, 5×10^{-5} M; 2 ml Fe^{2+}; EDTA, 5×10^{-3} M; 2 ml H_2O_2, 0.7 M; pH 8.

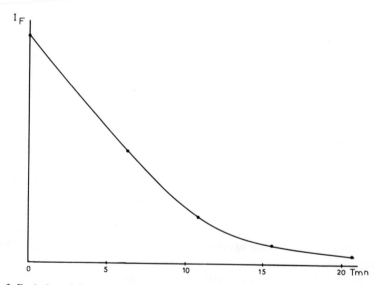

Fig. 3. Evolution of the fluorescence of the FMN during the "dark" reaction. System
FMN-H_2O_2, variation of the fluorescence at 0°C: 2 ml FMN, 5×10^{-5} M, 2 ml Fe^{2+}; EDTA,
5×10^{-3} M; 2 ml H_2O_2, 0.7 M; pH 8.

Many other reactions leading normally to luminescence can be studied as a function of temperature, for instance supercooled and still fluid organic and hydro-alcoholic solutions.

In our laboratory, we used such a procedure and we can see in figs. 4 and 5 some of the devices we built to get and control any temperature between + 80 and − 196°C.

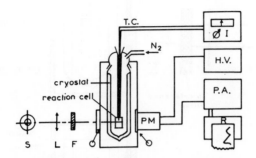

Fig. 4. Photochemiluminescence apparatus: S = mercury arc lamp; L = lens; F = filter; O = obturators; T.C. = thermocouple; PM = photomultiplier; I = temperature control and indicator (controls flow of hot or cold gaseous nitrogen to cryostat); HV = high voltage power supply; P.A. = picoamperemeter; R = recorder.

In the above scheme we remark on the possibility of exciting optically the cooled samples possibly through interferential filters, and also the possibility of analysing roughly the emission by such filters.

We checked in this way several hydrocarbons and dyes, used both as photosensitizers (S) of $^3O_2$ and as acceptors (A) of $^1O_2^*$. The corresponding thermoluminescent emissions are reported in table 1.

One can see that hydrocarbons and some dyes emit at about 480 nm, and that some other dyes emit their own fluorescence. The emissions centered upon 480 nm consists of a large spectrum and are quite different from the emission resulting from the deactivation of the dimers $(^1\Sigma_g^+ \cdot {}^1\Delta_g)$.

At presently it is impossible to determine the exact origin of such emission at 480 nm. Moreover, it can be demonstrated in the case of hydrocarbons that the thermoluminescent-like emission is due to the decomposition of hydroperoxides and endoperoxides [13]. In the case of dyes, the formation of a transient complex of the type $(S^+...O_2^-)$ postulated by Koisumi and Osui [14] could be responsible for the emission, in competition with the reaction $^3S^* + {}^3O_2 \rightarrow S_o + {}^1O_2^*$.

Finally, it can be established that the pathway by which $^1O_2^*$, both photochemically and chemically generated, gives an intermediate (peroxide, or complex $S^+...O_2^-$) responsible for the luminescence is insignificant compared to direct oxidation by $^1O_2^*$ as an "energy carrier": when a fluid oxidizable substrate (A) is added

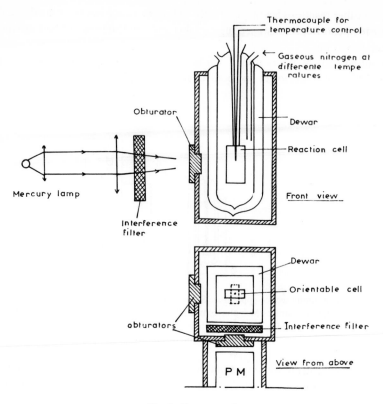

Fig. 5. Cryostat cell.

to the supercooled fluid solution containing an oxygenated intermediate and the latter is discomposed by warming, the yield of oxidation of A is negligible compared to the yield obtained under the selective excitation of S_o leading to the reaction [15]

$$S_o + h\nu \rightarrow {}^1S^* \rightarrow {}^3S^* \xrightarrow{+\,{}^3O_2} S_o + {}^1O_2^* .$$

In spite of such results, we think that studies of thermoluminescent reactions resulting from ${}^1O_2^*$ in fluid solutions at low temperature could serve as models for the study of mechanisms leading, for instance, to the bioluminescence which is temperature-dependent and might involve ${}^1O_2^*$ as an intermediate and (or) as an oxidizing species [11].

Table 1
Thermoluminescent emission of hydrocarbons and dye, experimental results

Sensitizer or acceptor molecule	Solvent	λ irrad.	λ_{max} abs. (nm)	λ_{max} fluor. (nm)	λ_{max} emission (nm ± 10 nm)
Pyrene	Ethanol	UV	335	385	480
1, 2 Benzo-pyrene	Chloroform	UV	330	385	480
Anthracene	Chloroform	UV	360, 380	400	480
Perylene	Acetone	Visible	400, 430	440, 470	480
Acridine orange	Acetone	Visible	490	520	*480*
Fluorescein	Ethanol	Visible	500	520	520
4'5' di-bromo-fluorescein	Ethanol	Visible	530	535	535
4, 5 di-iodo-fluorescein	Ethanol	Visible	530	540	540
Eosine	Acetone	Visible	535	555	555
Erythrosine	Acetone	Visible	540	560	560
Rhodamine B	Acetone	Visible	560	575	*480*
Bengal rose	Acetone	Visible	560	570	*480*
Methylene blue	Acetone	Visible	665	680	520

5. Singlet oxygen and photodynamic-like effects

We have seen above that singlet oxygen produced by the hypochlorite-H_2O_2 reaction reacts with hydrocarbons or dyes to give intermediates and products identical to those of the photosensitization (photo-oxygenation) where $^3S^* + {}^3O_2 \rightarrow S_o + {}^1O_2^*$.

Under these conditions, one might induce photodynamic-like action by chemically generated $^1O_2^*$. We tried to reach such a goal by using a special device in which $^1O_2^*$ is chemically generated in a separate compartment and next taken to the sample by a nitrogen stream, through a porous glass, fig. 6. Under these conditions, only $^1O_2^*$ in its $^1\Delta_g$ state can reach the sample and perform the oxidation of organic substrates. The absence of gaseous Cl_2 and the presence of $^1O_2^*$ in the gas stream can be ensured by appropriate tests. The above device was first used to perform the oxidation of oxyhaemoglobin (HbO_2) and to compare such an oxidation with the photochemical evolution under prolonged illumination in presence of O_2. A

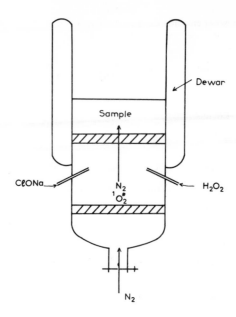

Fig. 6. Production of $^1O_2^*$ by chemical reaction.

solution of HbO_2 treated for a few minutes with $^1O_2^*$ showed qualitatively all, and the same, modifications as resulting from the action of O_2 under several hours of illumination [16].

Since then, P. Debey and myself have experimented under the same conditions, with the aromatic amino acids (including tryptophan and histidine), proteins such as human serumalbumin, cytochrome c, enzymes such as trypsin [17].

It is known that the aromatic amino acids are the principal loci of photodynamic damage in proteins, and that tryptophan and histidine are typically destroyed most rapidly but that, in fact, such a behavior depends upon their location at the surface or in the bulk of proteins.

The photoautooxidation of these residues is characterized by the disappearance of their absorption and (or) emission spectra. It is difficult to investigate the chemical nature of their probable intermediates and thus make progress in the organic chemistry of the dye-sensitizer photo-oxidation of small organic components of biological importance.

By using the method already described leading to the thermoluminescence (both inducible by photosensitization and by chemically generated $^1O_2^*$) it is possible to obtain an insight on such intermediates [17].

In aqueous solution at room temperature, it is easy to induce the inactivation of enzymes such as trypsine. Cytochrome c is also irreversibly oxidized. Human serumalbumin is denatured and loses its dissolving power towards insoluble substrates.

In summary, one can easily induce with "pure" $^1O_2^*$ $(^1\Delta_g)$ a variety of physico-chemical changes, and produce a number of changes in the biological properties of proteins, and such modifications appear to be qualitatively the same as those obtained in the dye-sensitized photoautoxidation.

Paradoxically, we could conclude this report devoted to some molecular mechanisms of the action of light by saying that we are now able to produce some photo-dynamiclike effects "without light", using directly the oxidizing species $^1O_2^*$ which seems to be the common denominator in many oxidative processes produced both in light and in the dark.

Acknowledgements

We thank very much our friends J. Canva, C. Balny and P. Debey for help, discussions and advice.

References

[1] D.E. Smith, L. Santamaria and B. Smaller, in: Free radicals in biological systems, ed. M.S. Blois (Acad. Press, New York, 1961) pp. 305–310.

[2] L. Santamaria and G. Prino, in: Research progress in organic, biological and medicinal chemistry, eds. U. Gallo and L Santamaria, Vol. 2 (Soc. Ed. Farm. Milan, 1964) pp. 259–336.

[3] C.S. Foote, Science *162* (1968) 963.

[4] S.J. Arnold, M. Kubo and E.A. Ogryzlo, Advan. Chem. Ser. *71* (1968) 133.

[5] L.C. Cusachs and R.H. Stelle, 2nd Symposium on carcinogenetic processes, Jerusalem, 1968, ed. Israel Acad. Sc. in press.

[6] R.J. Browne and E.A. Ogryzlo, Can. J. Chem. *43* (1965) 2915.

[7] A.U. Khan and M. Kasha, J. Amer. Chem. Soc. *33* (1966) 1574.

[8] C.S. Foote, Account. Chem. Res. *1* (1968) 104.

[9] R. Umans, A. Lesko and P.O.P. Tso, Nature *221* (1969) 763.

[10] M. Wilk and W. Girke, 2nd Symposium of Jerusalem, see ref. [5].

[11] J.M. Hastings, Ann. Rev. Biochem. *37* (1968) 597.

[12] Symposium on Chemiluminescence. Photochem. Photobiol. *4* (1965) 956.

[13] C. Balny, J. Canva, J. Bourdon and P. Douzou, in preparation.

[14] M. Koisumi and Y. Usui, Tetrahedron Lett. *57* (1968) 6011.

[15] C. Balny, J. Canva, J. Bourdon and P. Douzou, Photochem. Photobiol., in press.

[16] L. Possani, R. Banerjee, C. Balny and P. Douzou, Nature, in press.

[17] P. Debey and P. Douzou, in preparation.

CHAPTER 4

ACTION SPECTRA AND THEIR INTERPRETATION

A. KLECZKOWSKI

Rothamsted Experimental Station,
Harpenden, Hertfordshire, England

1. Definition of action spectra

An action spectrum of a given effect of irradiating a material with ultraviolet or visible radiation is a graphic presentation of the dependence of the efficiency of *incident* energy of a monochromatic radiation in causing the effect, on the wavelength of the radiation. Efficiency of the radiation is defined as the constant of the rate of its effect $\{k_{in}$ of eq. (1)$\}$; or a quantity proportional to the constant. The radiation will be assumed to fall as a parallel bundle of rays perpendicular to the surface of irradiated material.

The kinetics of the effect may be of any kind, but will be assumed that the rate of the effect can be expressed generally as

$$\frac{dy}{dx_{in}} = k_{in} f(y),\tag{1}$$

where y is the quantity in which we are concerned and which alters as a result of irradiation, x_{in} is the amount of incident radiation energy, k_{in} is the constant of the rate of its effect and $f(y)$ is a function of y (not involving the constant k_{in}) that depends on the kinetics of the process. The value of k_{in} is constant only at a given wavelength, and may vary from one wavelength to another. Plotting the values of k_{in}, or quantities proportional to them, against the wavelength results in an action spectrum. It is assumed that k_{in} is the only part of eq. (1) that may alter when the wavelength is altered. A practical way of obtaining an action spectrum, especially when the kinetics of an effect is uncertain or beyond the competence of the experimenter, is to plot some quantities proportional to the reciprocals of the amounts of incident radiation energy needed to cause a certain fixed amount of the effect, against the wavelength of the radiation.

Obviously, to obtain an action spectrum either a monochromator must be used, or a suitable combination of filters followed by a processing of results to allow for usually rather wide bandwidths of filters (a method of doing this was described by Hoover [14]). It is also essential that experimental conditions must be such that the amount of the effect of irradiating a material, and also the amount of incident radiation energy are measured with a reasonable precision.

It will be shown below that the shape of an action spectrum as defined above, depends on the optical density of the irradiated layer of the tested material. It may, therefore, seem desirable to limit the definition to effects of irradiating materials that are so diluted in non-absorbing media, or irradiated as layers that are so thin, that they are nearly transparent to the radiations. No such limitation will, however, be made in this article because it would exclude effects of irradiating not only solutions or suspensions of materials with which, for some technical reasons, these conditions cannot be fulfilled, but also surfaces of macro-organisms or parts of them, such as, for example, the skin of an animal or the leaf of a plant.

2. Introduction

The purposes for which action spectra are obtained, are various. An action spectrum may, for example, simply show how best to promote, or to avoid, a reaction that depends on exposure of a material to a radiation; or it may be used as a clue that may help to identify the component of a material that is directly involved in an effect of irradiating the material; or it may provide evidence of a photochemical interaction between two substances.

Getting a clue for identification or evidence for an interaction are probably the most frequent purposes for which action spectra are obtained. Similarities are looked for between the shapes of action spectra and absorption spectra of substances that are expected to be directly involved as absorbers of radiation energy that causes a given effect. There is some confusion at the moment as regards the similarities. They exist only when certain rather stringent conditions are fulfilled, whereas there is a tendency to expect them in any circumstances. This is so possibly because people of many different disciplines use action spectra, and some are not qualified to appreciate the importance of the conditions. A discussion of the conditions is the purpose of this article. The subject will be dealt with mainly theoretically, and only a few experimental examples will be given to illustrate the principles. We shall not be concerned here with whether a given effect of irradiation occurs immediately where energy is absorbed, or whether the effect is a final result of a chain of reactions started by absorption of radiation energy. In other words, the reasoning and the conclusions in this article apply equally to direct and indirect effects of radiations.

Interest in the dependence of effects of radiations on their wavelengths seems to have started at the end of the eighteenth century when Senebier [23] expressed the view that the blue-violet part of the light spectrum is responsible for photosynthesis. In the last century the problem of dependence of photosynthesis on the colour of light was tested experimentally by many workers, for example, Daubeny [6] and Draper [7] who came to the conclusion that the yellow part of the spectrum is the most active, Timiriazeff [27] who demonstrated the effectiveness of the red part part of the spectrum and Engelmann [9] who illuminated algal strands with light split into its spectrum and used his celebrated bacterial method to locate parts of the spectrum where oxygen was excreted at higher rates than in other parts. In this way he found another maximum of photosynthetic activity in the blue-violet part of the spectrum in addition to that in the red part.

Although the result obtained by these workers, especially those of Engelmann [9], can be considered as fore-runners of action spectra, they did not produce anything approaching action spectra as defined above because methods of precise measurement of light intensity and of photosynthesis were not available. The bolometer was invented by Langley only in the second half of the century, and was not easily available for some time. As regards photosynthesis, probably the most accurate method then available for measuring its rate, was by counting bubbles of oxygen produced by a plant immersed in water and illuminated.

In the early part of this century methods of measuring light intensities became easily available, and Warburg and Negelein [28] developed an accurate manometric method of measuring the rate of photosynthesis, which enabled them not only to compare quantitatively with great precision the efficiencies of light of different parts of the spectrum in promoting photosynthesis, but also to compute the numbers of quanta of light energy that must be absorbed by a suspension of *Chlorella* cells to assimilate one molecule of CO_2, thus estimating the quantum yield of a photochemical reaction. As, however only three rather broad regions of the spectrum (red, green and blue) were tested, only an outline of an action spectrum for photosynthesis in *Chlorella* can be obtained from the results of these tests. The outline is not suitable for any meaningful comparisons with the absorption spectra of any substances that may be expected to be absorbers of radiation energy that is used for photosynthesis, because it applies to irradiation of an optically dense material (a dense suspension of *Chlorella* was irradiated so that all the supplied radiation energy was absorbed).

In a later work Warburg and Negelein [29] deduced theoretically that such comparisons should be made when action spectra apply to irradiated materials that are nearly transparent to the radiations. When, in addition to this, the quantum yield of a given effect is known and does not vary with the wavelength, absorption coefficients of the compound that is directly involved in the effect of irradiation can be computed for all the tested wavelengths from the rates of the effect with respect to the amount of incident radiation energy. Thus, Warburg and Negelein [29] recognised the basic principles that are involved in comparison between action spectra for various effects of radiations and absorption spectra of compounds directly involved in the effects. They applied these principles to results of measurements of photochemical decomposition of the carbonyl form of the respiratory enzyme of yeast cells by visible light of different wavelengths, and were thus able to obtain indirectly the absorption coefficients of the enzyme. Kubowitz and Haas [19] expounded the principles still further and used them when they found that the action spectrum for inactivation of a sufficiently dilute solution of urease by ultraviolet radiation was exactly parallel to the absorption spectrum of a purified preparation of the enzyme. They concluded that the quantum yield for the inactivation was the same at all tested wavelengths and estimated its magnitude.

What is now called 'action spectra', was at first called by terms that specified particular effects of radiations. Thus, the authors described their results as dependence of such and such effects of radiations on the wavelengths of the radiations, then such terms as 'destruction spectra', 'inactivation spectra' and so on, were used. The term 'active absorption spectra' was used briefly by some workers, and the term 'action spectra' became finally established in the nineteen forties as a general term applicable to all effects of ultraviolet and visible radiations.

The number of action spectra for a great variety of effects of radiations obtained up to date by workers in a variety of disciplines is much too great to be reviewed comprehensively, and such a review would not serve any useful purpose even if it were possible. It must be stated, however, that the principles deduced by Warburg and Negelein [29] and by Kubowitz and Haas [19] never quite became common knowledge among workers concerned with action spectra, possibly because the principles are based on mathematical argument. Action spectra have been discussed by several authors, for example Blum [4], Setlow [24], McLaren and Shugar [21] and Allen [1].

3. Two conditions for parallelism between action and absorption spectra

The conditions will now be discussed that must be fulfilled if an action spectrum for an effect of irradiation is to be exactly parallel to the absorption spectrum of the component of irradiated material that is directly involved in the effect (i.e. which absorbs radiation energy which causes the effect). The component will be referred to below as the 'relevant' component. The action and absorption spectra will be considered 'exactly parallel', when their curves can be made to coincide by a suitable adjustment of scales.

It must be made clear at the start that there are two basically different kinds of circumstances. (1) An irradiated material is a solution or a suspension of molecules or of microscopic particles in a liquid medium which may be transparent to the radiation. The concentration, the optical density and the thickness of the layer of the irradiated material can, therefore, be known to the experimenter and controlled by him. It is also possible to stir the liquid during irradiation thus ensuring a uniform exposure of all the molecules or particles to the radiation. (2) The irradiated material is a macroscopic object, such as a macro-organism, whose surface is exposed to the radiation. In this kind of circumstances the optical density and the thickness of irradiated layer, and the uniformity of exposure or lack of it, are all beyond the control of the experimenter and may not even be known to him with a degree of accuracy required for a suitable treatment of results of measurements of effects of irradiation. We shall be mainly concerned with the first kind of circumstances.

Complete parallelism between an action spectrum for a given effect of radiation and the absorption spectrum of the relevant component can only be expected when it is the only relevant component of an irradiated material, although the material may contain several other components that also absorb radiation energy. This is quite obvious because when several different components absorb radiation energy for a given effect of radiation, each component will influence the shape of the action spectrum, and so it will not be able to follow the absorption spectrum of any one of the components, although it may then follow a combined absorption spectrum to which the different relevant components may contribute.

Two conditions must be fulfilled if an action spectrum is to parallel exactly the absorption spectrum of the relevant component when it is the only such component, or a combined absorption spectrum when there are several such components:

Condition no. 1: The action spectrum must apply to the effects of irradiating a layer of a material that is nearly transparent to the radiation of all the tested wavelengths.

Condition no. 2: The irradiated material must be such that the efficiency of radiation energy absorbed by anyone of the relevant components of the material in causing a given effect, is the same at all the tested wavelengths.

A discussion of the two conditions, which will follow, is the main purpose of this article. They will be referred to below as conditions nos. 1 and 2.

Let us assume that an irradiated material contains n components that absorb radiation energy. Let $\beta_1, \beta_2, \beta_3,...,\beta_n$ be the absorption coefficients (not including any contributions from scatter) and $c_1, c_2, c_3,...,c_n$ the concentrations of the components, τ the coefficient of turbidity due to any scatter of radiation energy by any molecules or particles of the irradiated material and l the thickness of the layer of the material. We shall assume that the component no. 1 whose absorption coefficient and the concentration are β_1 and c_1, is the only relevant component. We shall also assume that the absorption coefficient of any of the components did not alter as a result of exposure to the radiation.

If the amount of incident radiation energy was x_{in}, the amount of energy transmitted through the layer of the material was

$$x_{tr} = x_{in}\exp[-(\beta_1 c_1 + \beta_2 c_2... + \beta_n c_n + \tau)l].$$

Consequently, the amount of energy that was not transmitted, which means the amount of energy that was absorbed and scattered, was

$$xX_{(ab + sc)} = x_{in}\{1 - \exp[-(\beta_1 c_1 + \beta_2 c_2... + \beta_n c_n + \tau)l]\}.$$

This energy was divided between scatter and absorption by different components of the material in such a way that the amount of it absorbed by the relevant component was

$$x_{ab} = \frac{\{1 - \exp[-(\beta_1 c_1 + \beta_2 c_2... + \beta_n c_n + \tau)l]\}\, c_1 \beta_1 x_{in}}{\beta_1 c_1 + \beta_2 c_2... + \beta_n c_n + \tau}. \tag{2}$$

The condition no. 1 is fulfilled (i.e. the irradiated layer is nearly transparent to the radiation) when the value of $(\beta_1 c_1 + \beta_2 c_2...\beta_n c_n + \tau)l$ is very small, say smaller than about 0.05. Then eq. (2) becomes

$$x_{ab} = \beta_1 c_1 l\, x_{in},$$ (3)

so that

$$dx_{ab} = \beta_1 c_1 l dx_{in}.$$ (4)

To discuss any consequences of the fulfilment of the condition no. 2, a definition is needed of the efficiency of energy absorbed by the relevant component of the irradiated material in causing a given effect. The efficiency is defined as the value of the constant k_{ab} of the rate of the effect in the equation

$$\frac{dy}{dx_{ab}} = k_{ab} f(y),$$ (5)

where y is the quantity in which we are concerned and which alters as a result of irradiation, x_{ab} is the amount of energy absorbed by the relevant component, and $f(y)$ is a function of y (not involving the constant k_{ab}) that depends on the nature of the irradiated material. This equation differs from eq. (1) only in that it contains x_{ab} instead of x_{in} and the constant k_{ab} instead of k_{in}. It is assumed that when the irradiated material is the same, $f(y)$ of eq. (5) is the same as that of eq. (1).

Combining eqs. (4) and (5) we get

$$\frac{dy}{dx_{in}} = \frac{dy}{dx_{ab}} \beta_1 c_1 l = \beta_1 c_1 l k_{ab} f(y).$$

A comparison of this with eq. (1) shows that

$$k_{in} = \beta_1 c_1 l k_{ab}.$$ (6)

Now, the fulfilment of the condition no. 2 means that the value of k_{ab} is the same at all wavelengths. As the values of c_1 and l are kept constant by the experimenter, the values of k_{in} are proportional to the values of the absorption coefficient of the relevant component (β_1) and, therefore, both vary with the wavelength in the same way. Therefore, the action spectrum, which is a curve obtained by plotting some values proportional to those of k_{in} against the wavelength, will be exactly parallel to the absorption spectrum of the relevant component, which is a curve obtained by plotting the values of β_1.

It is obvious, therefore, that when condition no. 1 and/or no. 2 is not fulfilled, k_{in} is not directly proportional to β_1 and, therefore, the action spectrum will not be parallel to the absorption spectrum of the relevant component. When the two conditions are not far from being fulfilled, the shapes of the curves of the action and of the absorption spectra, though not exactly parallel, may still resemble each other more or less, but there may be no similarity at all between the two curves when either of the conditions is far from being fulfilled.

Let us first consider a case when condition no. 2 is fulfilled, but condition no. 1 is not. To take an extreme case, let us assume that an irradiated material which contains n absorbing components, is quite opaque to the radiation. This means that the value of $(\beta_1 c_1 + \beta_2 c_2 ... + \beta_n c_n + \tau)l$ in eq. (2) is so large that the value of $\exp[-(\beta_1 c_1 + \beta_2 c_2 ... + \beta_n c_n + \tau)l]$ is so small that it can be neglected. Therefore, eq. (2) transforms into

$$x_{ab} = \frac{\beta_1 c_1 x_{in}}{\beta_1 c_1 + \beta_2 c_2 ... + \beta_n c_n + \tau}. \tag{7}$$

It is assumed that the fluid was stirred during irradiation so that all molecules or particles of the material were equally exposed to the radiation.

Substituting eq. (7) into eq. (5) gives

$$\frac{dy}{dx_{in}} = \frac{dy}{dx_{ab}} \frac{\beta_1 c_1}{(\beta_1 c_1 + \beta_2 c_2 ... + \beta_n c_n + \tau)} = \frac{\beta_1 c_1 k_{ab}}{(\beta_1 c_1 + \beta_2 c_2 ... + \beta_n c_n + \tau)} f(y).$$

A comparison of this with eq. (1) shows that

$$k_{in} = \frac{\beta_1 c_1 k_{ab}}{\beta_1 c_1 + \beta_2 c_2 ... + \beta_n c_n + \tau}. \tag{8}$$

Thus, even when condition no. 2 is fulfilled, i.e. when k_{ab} is the same at all the wavelengths, the values of k_{in} at different wavelengths are not proportional to those of β_1 because they are functions not only of β_1 but also of the absorption coefficients of other components of the irradiated material $(\beta_2, \beta_3,..., \beta_n)$ and of the coefficient of turbidity (τ), all of which may vary differently as the wavelength varies. Consequently, not only the shape of the action spectrum, but also the positions of its maxima and minima may differ from those of the absorption spectrum of the relevant component. When, for example, the value of c_2 is exceptionally large, a maximum of k_{in} may be nearer to the wavelength at which β_2 is at a minimum than to the wavelength at which β_1 is at a maximum.

4. Examples of theoretically computed action spectra

Fig. 1 shows a theoretically computed fictitious example of behaviour of a material with only two absorbing components, namely of a mixture of solutions of a nucleic acid and of a protein in a non-absorbing liquid, when exposed to ultraviolet radiation as a layer of a constant thickness (1 cm) and of different concentrations but at a constant ratio of that of the nucleic acid to that of the protein (namely 1/100). The nucleic acid is the relevant component. The action spectra shown in fig. 1 were computed from the absorption coefficients of the components (assuming no scatter) shown in table 1, assuming that there was no photochemical interaction between the components and that condition no. 2 was fulfilled.

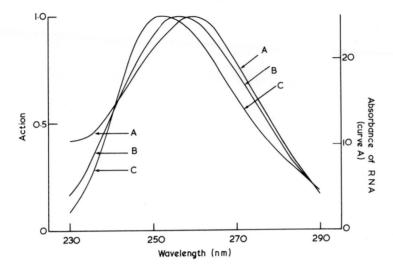

Fig. 1. Action spectra (lines A, B and C) for an effect of ultraviolet radiation on a nucleic acid when irradiated in the presence of a protein as a mixed solution in a non-absorbing liquid. Condition no. 2 is assumed to be fulfilled. The stock solution contained the nucleic acid at 0.02 g/l and the protein at 2.0 g/l. It was irradiated (with stirring) as a layer 1 cm thick at a dilution (in the same liquid) of: line A: 1/400; line B: 1/10; line C: undiluted. 'Action' is plotted in arbitrary units so adjusted for individual curves that their peaks are of the same hight. Line A is also the absorption spectrum of the nucleic acid plotted in units of 'absorbance' $\alpha = 0.4343\ \beta$, where β is the absorption coefficient.

It was also assumed for convenience of the computations that energy absorbed by the nucleic acid as a whole was involved in causing a given effect. (This need not necessarily be so. Energy absorbed not by all four bases may be involved (see, for example, Setlow [25])).

Curve A is the absorption spectrum of the nucleic acid and it is the action spectrum when the amount of the material under each unit area of the irradiated surface was so small that the irradiated layer was almost transparent to the radiation. Curve C is the action spectrum when the amount of the material under each unit area was so large that the irradiated layer was opaque to the radiation. Curves A and C are, therefore, examples of two extreme case, whereas curve B is an example of an intermediate case. It can be seen that it is only when the irradiated layer is nearly transparent to the radiation (curve A) that the action spectrum is identical in shape, and in the position of the maximum, with the absorption spectrum of the nucleic acid (i.e. of the relevant component). When the layer is not nearly transparent (curves B and C), the action spectrum may differ from the absorption spectrum of the relevant component not only in the shape, but also in the position of the maximum.

The fulfilment of conditions nos. 1 and 2 is essential for parallelism between an action spectrum and the absorption spectrum of the relevant component even when

Table 1

Absorption coefficients, at different wavelengths, of the nucleic acid and of the protein

| | | Wavelength (nm) | | | | | | |
		230	240	250	260	270	280	290
Absorption coefficient of	Nucleic acid	24.0	32.2	50.7	57.6	44.9	26.5	10.1
	Protein	13.8	2.8	1.8	2.3	2.9	3.0	2.1

Remark: The value of an absorption coefficient is 2.303 times greater than that of the corresponding specific absorbance (extinction coefficient).

the latter is the only absorbing component of an irradiated material. When there is only one absorbing component (say component no. 1) and no scatter ($\tau = 0$), eq. (2) becomes

$$x_{ab} = [1 - \exp(-\beta_1 c_1 l)] x_{in}. \tag{9}$$

When the irradiated layer is transparent to the radiation, i.e. when the value of $\beta_1 c_1 l$ is very small, and only then, eq. (9) transforms into eq. (3) the validity of which and the fulfilment of condition no. 2 are, as we have discussed above, essential for the exact parallelism between the action spectrum and the absorption of the component.

Let us consider the other extreme case, namely the results obtained when the irradiated layer is opaque to the radiation. It is opaque when the value of $\beta_1 c_1 l$ is so large that the value of $\exp(-\beta_1 c_1 l)$ is so small that it can be neglected. Consequently, eq. (9) transforms into

$$x_{ab} = x_{in}.$$

This means that k_{in} of eq. (1) is identical with k_{ab} of eq. (5). Therefore, when condition no. 2 is fulfilled, i.e. when the value of k_{ab} is the same at all the wavelengths, the value of k_{in} is also the same at all the wavelengths, and, therefore, the action spectrum is a straight line parallel to the axis of abscissae.

Fig. 2 gives a theoretically computed example of behaviour of a material with a single absorbing component when condition no. 2 is fulfilled. Action spectra are shown for an effect of ultraviolet radiation on a nucleic acid when its solution in a non-absorbing liquid was exposed to the radiation as a layer of a constant thickness (1 cm) but of different concentrations. The action spectra were computed from the absorption coefficients of the nucleic acid, shown in table 1. It was again assumed that energy absorbed by the nucleic acid as a whole was involved in causing a given effect. Line A is the absorption spectrum of the nucleic acid, and it is also the action spectrum obtained when the nucleic acid was so dilute that it was nearly transparent to the radiation at all tested wavelengths. Line E is the other

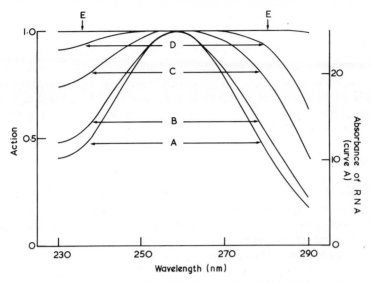

Fig. 2. Action spectra (lines A, B, C, D and E) for an effect of ultraviolet radiation on a nucleic acid when irradiated as a pure solution in a non-absorbing liquid. Condition no. 2 is assumed fulfilled. The stock solution contained the nucleic acid at 0.5 g/l. The stock solution was irradiated (with stirring) as a layer 1 cm thick at a dilution (in the same liquid) of: line A: 1/1000; line B: 1/50; line C: 1/10, line D: 1/5; line E: undiluted. 'Action' and 'absorbance' are plotted as in fig. 1. Line A is also the absorption spectrum of the nucleic acid.

extreme. It is the action spectrum obtained when the nucleic acid was so concentrated that it was opaque to the radiation. Lines B, C and D are action spectra obtained with intermediate, but progressively increasing concentrations of the nucleic acid. As the concentration increases the curves become progressively broader, but the position of the peak, as long as it is discernible, remains the same.

Although both figs. 1 and 2 show action spectra for an effect of ultraviolet radiation on the same nucleic acid, the curves of the action spectra differ, the cause of the difference being the presence or absence of a protein in the irradiated solution of the nucleic acid. Thus, the mere presence of a protein can affect an action spectrum for an effect of the radiation on a nucleic acid.

Whether the protein is present or not, the action spectra are not parallel to the absorption spectrum of the nucleic acid when the irradiated material is not transparent to the radiation. The deviation from parallelism is a result of mutual shading between molecules of the irradiated material. In other words, it is a result of a competition between the molecules for supplied radiation energy. The way in which the shape of the action spectrum deviates from that of the absorption spectrum depends on whether the competition occurs only between molecules of the nucleic acid, or whether those of the protein are also involved.

Only when the irradiated material is nearly transparent to the radiation, there is no mutual shading and, consequently, no competition for supplied radiation energy between molecules of the material. Each molecule absorbs radiation energy quite independently of any other molecule, as though the other molecules did not exist, and as though there were no scatter of radiation energy by any molecules. The amount of energy absorbed by each molecule of a given compound then depends only on the amount of supplied energy per unit area of irradiated surface, and on the absorption coefficient of the compound not including any contribution from scatter (see eq. (10) below).

5. Ways of fulfilling condition no. 1

An irradiated material may often be made nearly transparent to radiation, so that condition no. 1 is fulfilled directly, either by sufficiently diluting it in a non-absorbing medium, or by irradiating a sufficiently thin layer of it. When, however, for some technical reasons neither of these things can be done, condition no. 1 may often be fulfilled indirectly by measuring the effect of irradiating the material as it is (i.e. when it is not transparent), and then computing what the effect would have been, had the material been so dilute that it became transparent. This can be done when specific optical density and the concentration of the material, and also the kinetics of the effect, are known, in the following way.

Let E be the amount of energy absorbed per unit mass of the relevant component. If the material were transparent the value of E would, according to eq. (3), be

$$E = \frac{x_{ab}}{Ac_1 l} = \beta_1 \frac{x_{in}}{A},\tag{10}$$

(where A is the area of the irradiated surface), whereas when the material is not transparent, the value of E, according to eq. (2), is

$$E = \frac{x_{ab}}{Ac_1 l} = \frac{\{1 - \exp[-(\beta_1 c_1 + \beta_2 c_2 ... + \beta_n c_n + \tau)l]\}\, \beta_1 (x_{in}/A)}{(\beta_1 c_1 + \beta_2 c_2 ... + \beta_n c_n + \tau)l}.\tag{11}$$

Let h be a measure of the effect of irradiation, and let h be a function (say θ) of E, so that

$$h = \theta(E).$$

Then E can also be expressed as a function (say F) of h, so that

$$E = F(h).\tag{12}$$

Now, if h is the effect that was actually observed, and h' the effect that would have been observed, had the material been transparent, the combination of eqs. (10), (11) and (12) gives

$$F(h') = \frac{(\beta_1 c_1 + \beta_2 c_2 ... + \beta_n c_n + \tau)lF(h)}{1 - \exp[-(\beta_1 c_1 + \beta_2 c_2 ... + \beta_n c_n + \tau)l]} . \tag{13}$$

Let α be the specific optical density of the irradiated material that includes any scatter, and c the total concentration of the material, so that $c = \Sigma_{i=1}^{n} c_i$. If the concentration is within the limits within which the turbidity coefficient is proportional to it, then

$$\alpha cl = 2.303(\beta_1 c_1 + \beta_2 c_2 ... + \beta_n c_n + \tau)l.$$

Eq. (13) can then be transformed into

$$F(h') = \frac{2.303\alpha clF(h)}{1 - \exp(-2.303\alpha cl)}. \tag{14}$$

When an affect of irradiation is a first order reaction, $F(h) = K\log(1/h)$ where h is now the proportion of the original amount of a substance (or of activity) that has not undergone a reaction, or has not become destroyed (the surviving proportion), and K is a constant. Eq. (14) then transforms into

$$\log h' = \frac{2.303\alpha cl\log h}{1 - \exp(-2.303\alpha cl)}, \tag{15}$$

which is a particular case of the more general eq. (14).

6. Condition no. 2 and some complicating factors

When condition no. 1 is fulfilled either directly, (i.e. when an irradiated layer is transparent to the radiation) or indirectly (through application of eq. (14)), still the shape of the action spectrum may differ from that of the absorption spectrum of the relevant component when condition no. 2 is not fulfilled (i.e. when susceptibility of the component to absorbed radiation energy varies with the wavelength). This is obvious from eq. (6) which shows that k_{in} is not proportional to β_1 when k_{ab} varies with the wavelength. Whether it varies or not, is an inherent feature of the relevant component, and it may also depend on a photochemical interaction of the component with another component of the irradiated material. Not enough experimental data are yet available to show whether the fulfilment or the non-fulfilment of condition no. 2 is the more usual. Fig. 4 gives an example (which will be discussed below) of a complete lack of similarity between an action spectrum and the absorption spectrum of the relevant component (a nucleic acid) because of the non-fulfilment of condition no. 2 caused by an interaction between the component and another component (a protein) of the irradiated material.

An additional factor affecting the shape of an action spectrum may result from alterations in absorption coefficients of compounds due to structural changes caused by irradiation. As a result of this a material initially transparent to the radiation may become not transparent, and a material that was intially not transparent, may alter its behaviour because of changes in absorption coefficients of its components. When either of these things is known to occur, it should be allowed for by a procedure, suitable to a given set of circumstances, that would modify eq. (2). A discussion of such procedures is beyond the scope of this article. The problem does not usually arise when we are concerned with inactivation of specific activities of biological materials or with lethal effects of radiations, either because the effects occur before any appreciable proportions of irradiated substances are so altered that their absorption coefficients are changed, or because concentrations of the compounds with altered absorption coefficients are so small that they do not absorb any appreciable proportion of supplied radiation energy. For example, according to McLaren and Takahashi [22] inactivation by ultraviolet radiation of 99.99% of infectivity of the RNA isolated from tobacco mosaic virus is accompanied by a decrease in optical density by only 0.7%.

Finally, it should be mentioned that the problem of a comparison between an action spectrum and an absorption spectrum may be complicated by the fact that the absorption spectrum of a substance may alter as a result of an interaction with another substance in vitro or in vivo, or as a result of a change of solvent. For example, absorption spectra of extracts of chlorophyll and of carotenoids may differ from those of the same substances when inside the intact cells [8, 26].

We have discussed above the behaviour of irradiated materials that contain only one relevant component. A very similar, though somewhat more complicated, reasoning applies to the behaviour of materials that contain several relevant components. Modifications that would have to be applied to eqs. (2) and (10) and to those derived from them, are obvious and will not be discussed here.

7. Examples of action spectra with solutions or suspensions

To illustrate on examples the way action spectra may be interpreted, we shall now consider action spectra for inactivation by ultraviolet radiation of the infectivity of two plant viruses, namely tobacco mosaic virus and tobacco necrosis virus, and of their free nucleic acids, all as solution in a non-absorbing medium, namely in 0.067 M phosphate buffer at pH 7.0. Tobacco mosaic virus has rigid rod-shaped particles about 300 nm long and 15 nm wide, and tobacco necrosis virus has spherical particles with a diameter of about 28 nm. They are both RNA viruses whose contents of RNA are about 5% and 18% respectively, the remainder being protein.

Before we come to the action spectra, we must first consider and discuss two kinds of circumstances that may complicate the interpretation of the action spectra.

One is the repair in the host plant of some kinds of damage caused in a virus particle by irradiation. As the residual infectivity of an irradiated preparation of a virus is assayed by inoculating it into a host plant and counting the number of local lesions, we are immediately confronted with this problem.

One kind of repair of a reversible kind of damage may occur when the host plant, inoculated with an irradiated virus, is exposed to daylight. This is called *photoreactivation*. Most of plant viruses so far tested, but not all, show this phenomenon.

Another kind of repair of a reversible kind of damage does not need light, and so it can be called *dark reactivation*. With bacteria and bacteriophages the problem of dark reactivation received much attention. With plant viruses the problem was not studied at all until quite recently, and even now it has only just been touched. Evidence for dark reactivation has so far been obtained only with one plant virus, namely tobacco necrosis virus, and only in one species of host plant, namely in *Chenopodium amaranticolor*. In this case it seems that dark reactivation repairs the same kind of damage that would otherwise be repaired by photoreactivation [17].

The other kind of circumstances that may complicate the interpretation of the action spectra, is some kind of photochemical interaction that may occur between the nucleic acid and the protein of a virus and influence the effects of the radiation on the nucleic acid. The interaction may be evident from qualitative and quantitative differences between inactivation of the nucleic acid when inside the virus and when free.

Table 2 shows the amounts of energy in joules that must be absorbed by 1 mg of nucleic acid of either of two viruses to reduce infectivity to 50% of the original. All the irradiations were done at a wavelength of 254 nm, and the residual infectivities were assayed in plants in which there was no evidence of any dark reactivation. The word 'dark' means that the inoculated assay plants were kept in darkness to prevent photoreactivation of lost infectivity, and the word 'light' means that the plants were exposed to daylight immediately after inoculation to allow photoreactivation to occur.

All the data given in table 2 refer to the materials irradiated as solutions in 0.067 M phosphate buffer at pH 7.0. Susceptibility of TMV-RNA to inactivation by absorbed radiation energy increases when salt concentration of the medium is decreased [10]. This was also found to apply to the RNA isolated from potato virus X (Govier and Kleczkowski, unpublished), and so it may apply to TNV-RNA.

The amount of energy needed to inactivate half of the infectivity of tobacco mosaic virus (TMV) was the same irrespective of whether the test plant was kept in light or in darkness. This shows that irradiated tobacco mosaic virus was not photoreactivated at all. Now, the inactivate half of the infectivity of free RNA isolated from TMV (TMV–RNA), twice as much energy was needed when the test plant was kept in daylight, that when it was kept in darkness. The phenomenon of photoreactivation was, therefore, very well pronounced. About half of the absorbed radiation energy caused the kind of damage that could be reversed by photoreactivation, and the other half caused the kind of damage that could not be repaired in this way.

It can also be seen that when the nucleic acid was inside the virus, very much more radiation energy was needed to inactivate infectivity than when the nucleic acid was free. Thus, when inside the virus, the nucleic acid was protected to a very large extent by the virus protein from damage caused by the radiation. The extent of the protection would probably have appeared even greater, had the salt concen-

Table 2
Radiation energy (at λ = 254 nm) absorbed by virus RNA to reduce infectivity to 50% of the original

		J/mg
TMV	Dark	2.2
TMV	Light	2.2
TMV-RNA	Dark	0.2
TMV-RNA	Light	0.4
TNV	Dark	0.3
TNV	Light	0.4
TNV-RNA	Dark	0.3
TNV-RNA	Light	0.4

TMV: tobacco mosaic virus.
TMV-RNA: free RNA isolated from TMV.
TNV: tobacco necrosis virus.
TNV-RNA: free RNA isolated from TNV.
The amount of energy are average values computed from data of Bawden and Kleczkowski (1955, 1959), Kleczkowski (1963) and Kassanis and Kleczkowski (1965).

tration in the medium been smaller. The complete absence of photoreactivation of the whole irradiated tobacco virus can be explained by assuming that when the nucleic acid is inside the virus it is completely protected from the photoreactivable kind of damage, so that only non-photoreactivable kind of damage can occur, and even this at a much slower rate than in the free nucleic acid

The behaviour of tobacco necrosis virus (TNV) was quite different from that of tobacco mosaic virus. Exactly the same amounts of energy must be absorbed by the nucleic acid of this virus to reduce infectivity to 50%, irrespective of whether the nucleic acid was inside the virus or free. The extent of photoreactivation was also the same. Thus, when the nucleic acid of this virus was inside the virus, it was not protected at all from any kind of damage by the radiation. It is possible, however, that some degree of protection could occur in a different environment, such as, for example, a medium with a lower concentration of salt or salt-free.

We have, therefore, an example of a striking difference in photochemical behaviour of two viruses: a strong interaction between the nucleic acid and the protein in tobacco mosaic virus, and no such interaction at all in tobacco necrosis virus. The difference is presumably a result of a difference in the way the nucleic acid and the protein are combined.

In spite of all the complicating circumstances due to the presence or absence of repair of some damage caused by UV radiation in a virus particle, or to the presence or absence of a photochemical interaction between virus RNA and virus protein, in any particular set of circumstances, when everything is kept constant except the dose of *supplied* radiation energy, the rate of inactivation of infectivity follows the

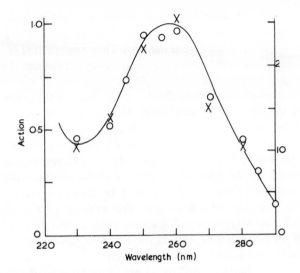

Fig. 3. Action spectrum for inactivation of the infectivity of tobacco necrosis virus and of the free RNA isolated from the virus. ○ and x are experimentally obtained points of the action spectrum for inactivation of the virus and of the free RNA, respectively. 'Action' is in arbitrary units proportional to the values of r of eq. (16). The solid line is the absorption spectrum of the RNA plotted in units of 'absorbance' $\alpha = 0.4343 \, \beta$, where β is the absorption coefficient of the RNA.

first order kinetics quite closely. This means that the proportion of the original infectivity that remains after an exposure of a virus preparation to a dose of radiation is

$$p = e^{-rD}, \tag{16}$$

where D is the dose of *supplied* radiation energy, and r is a constant which will be called the constant of the rate inactivation. (The constant r is identical with k_{in} of eq. (1) when $f(y) = -y$ because the reaction proceeds according to the first order kinetics and p is identical with h of eq. (15)). Of course the value of r is constant only in a given set of circumstances, and it varies when the circumstances vary. For example, the value of r will be smaller in the presence of photoreactivation than in its absence. When the radiation is monochromatic, the value of r will also depend on the wavelength of the radiation. Plotting the values of r corresponding to different wavelengths (or quantities proportional to the values of r) against the wavelengths will give an action spectrum for inactivation of infectivity.

All the action spectra that are discussed in this section are concerned with results of irradiating dilute preparations of viruses or nucleic acids, which were nearly transparent to the radiation. This means that condition no. 1 was fulfilled.

Fig. 3 shows the result obtained with tobacco necrosis virus and with the free

RNA isolated from this virus [15]. The circles are the experimentally obtained points of the action spectrum for inactivation of the whole virus, and the crosses are those for free RNA isolated from the virus. They were plotted in units that are arbitrary (proportional to the values of r of eq. (16)), but the same for both sets of results.

It is obvious that the two sets of results are identical for the circles and the crosses follow the course of the same curve. Thus, exactly the same action spectrum was obtained for inactivation of the whole virus and for inactivation of free nucleic acid isolated from the virus. We can conclude, therefore, that the loss of infectivity of the virus was entirely due to the radiation affecting its nucleic acid, and that the protein component of the virus did not affect at all the result or irradiation.

The solid line is the absorption spectrum of the RNA of the virus. It can be seen that by suitably adjusting the scales it is possible to make the absorption spectrum coincide with the action spectrum. The action spectrum and the absorption spectrum of the RNA are, therefore, exactly parallel. This means that the irradiated material was such that condition no. 2 was fulfilled.

We can conclude, therefore, that the extent of inactivation of infectivity depended entirely on the amount of radiation energy absorbed by the nucleic acid irrespective of whether the nucleic acid was inside the virus or free, and irrespective of the wavelength. The possibility has already been mentioned, however, that in some other environmental conditions susceptibility of the nucleic acid to inactivation by absorbed radiation energy may depend on whether the nucleic acid is free or inside the virus.

The results shown in fig. 3 were obtained in non-photoreactivating conditions. Those obtained in photoreactivating conditions would give a curve parallel to the curve shown in fig. 3 because the extent of photoreactivation of the nucleic acid of this virus does not depend appreciably on the wavelength of inactivating radiation between 230 and 290 nm, and is the same irrespective of whether the nucleic acid is inside the virus or free.

An action spectrum for inactivation of the infectivity of free RNA isolated from tobacco mosaic virus was exactly parallel to its absorption spectrum [26], but that for inactivation of intact tobacco mosaic virus, first obtained by Hollaender and Duggar [13] and shown in fig. 4, was quite different. It did not resemble at all the absorption spectrum of the RNA. We have, therefore, here an example of an action spectrum diverging entirely from the shape of the absorption spectrum of the relevant component which is the nucleic acid, and this is so in spite of the fact that condition no. 1 for parallelism between the two was fulfilled, which means that the irradiated preparation was nearly transparent to the radiation.

It had been suggested that the reason for the divergence is that the loss of infectivity of a virus particle may occur not only when the radiation damages its nucleic acid, but also when the radiation damages its protein envelope. However, it was shown later that this is not so. When the nucleic acid was isolated from an irradiated virus preparation, the relative infectivity of the nucleic acid appeared the same as

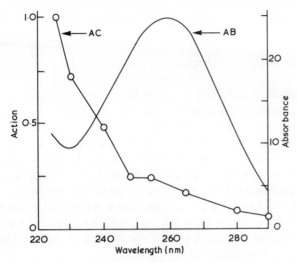

Fig. 4. Action spectrum for inactivation of the infectivity of tobacco mosaic virus. AC is the action spectrum (according to Hollaender and Duggar, [13] ; AB is the absorption spectrum of the RNA of the virus. Units of 'action' and 'absorbance' as in fig. 3.

that of the irradiated virus. Thus, as regards the relative infectivity, the removal of the protein envelope from the nucleic acid of the irradiated virus does not seem to make any difference, and so any damage caused by the radiation in the protein envelope is obviously irrelevant. Consequently, the loss of infectivity of the virus is due entirely to damage caused by the radiation in the nucleic acid of the virus [18] .

Therefore, to explain the divergence between the action spectrum for inactivation of the virus, and the absorption spectrum of the RNA, we are forced to assume that when the RNA is inside the virus its susceptibility to damage by absorbed radiation energy varies with the wavelength of the radiation, increasing as the wavelength decreases. This, of course, means that condition no. 2 for parallelism between an action spectrum and the absorption spectrum of the relevant component is not fulfilled. The reason for this must be some kind of interaction between the virus RNA and the virus protein, because as we have seen, susceptibility of free nucleic acid to absorbed radiation energy is about the same at all the tested wavelengths.

The mechanism of the interaction can at the moment be only a matter of speculation, but we have already concluded for another reason that there must be an interaction, namely to account for the partial protection of the nucleic acid by protein from damage by the radiation. We also concluded that there is no such interaction at all in tobacco necrosis virus. We shall now have to assume that the degree of protection in tobacco mosaic virus varies with the wavelength, increasing as the wavelength increases. As a result of this, more energy must be absorbed per unit mass of the nucleic acid at a longer wavelength than at a shorter wavelength to achieve the same degree of inactivation. For example, about 24 times more energy must be absorbed when the wavelength is 280 nm than when the wavelength is 230 nm [18].

8. Action spectra obtained with macro-organisms

When surfaces of macro-organisms are irradiated, a mathematical treatment of the results is as simple as that of results obtained with solutions or suspensions in liquid media only when the layer of tissue, in which a given result of irradiation occurs, is transparent to the radiation, i.e. when condition no. 1 is fulfilled. This may be so when the irradiated organism is so thin that it is nearly transparent. However, even when an organism is large and bulky, the layer of tissue that is directly involved in the effect may still be thin enough to be nearly transparent, and it may not be covered with another layer that functions as a filter. When this is so, the action spectrum for the effect will parallel the absorption spectrum of the relevant substance, provided the condition no. 2 is also fulfilled, because eq. (3) will then be applicable just as it is with solutions or suspensions.

When, however, the layer of tissue that is directly involved in the effect, is not transparent to the radiation, i.e. when condition no. 1 is not fulfilled, the shape of the action spectrum will differ from that of the absorption spectrum of the relevant substance more or less depending on circumstances even if condition no. 2 is fulfilled, and in extreme circumstances there may be no similarity at all between the action spectrum and the absorption spectrum.

The causes of the divergence that apply to solution and suspensions (as discussed above) also apply to layers of tissues, but there may also be other causes and these may make any mathematical treatment of the results rather artificial and complicated. The simple eq. (2) may not apply to results of irradiating layers of tissue for various reasons, such as an uneven distribution of the relevant substance and of other absorbing substances, uneven exposure of various parts of the layer because the equalising effect of stirring cannot be applied, the presence of another layer that may act as a selective filter, and so on. (An example of such a selective filter in the ultraviolet region is the layer of dead horny cells (*Stratum corneum*) at the surface of human skin.) The layer is usually 20 to 80 μm thick. Where it is 75 μm thick, it allows less than 1% of incident radiation energy at 254 nm to reach the living cells of the skin. The absorption spectrum of the layer at any particular point resembles that of a typical protein, with a maximum near 280 nm and a minimum near 250 nm). In addition to all this the thickness of the layer of tissue which is involved in a given effect of irradiation, may vary or be unknown, and there may even not be any definite limits of the layer. A further complication may arise from a part of the layer nearest to the surface being so much affected by the radiation that its absorption of radiation energy may alter, whereas parts of the layer further away from the surface may be hardly affected at all.

It is obvious, therefore, that eq. (14) cannot usually be applied to results of irradiating the surface of a macro-organism to compute what the result would be if

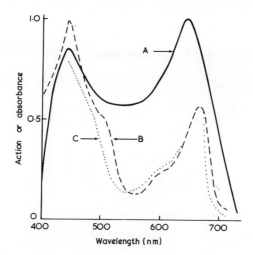

Fig. 5. Action spectrum for photosynthesis in the leaves of wheat plants (curve A from Hoover, [14] compared with the absorption spectrum (including light scatter) of a suspension of chloroplasts from spinach (curve B from Chen, [5]) and with an action spectrum for photosynthesis in a thin suspension of *Chlorella* cells (curve C from Emerson and Lewis, [8]). All the curves are plotted in arbitrary units. The original curves B and C were transformed to correspond to transparent suspensions.

the layer that is involved in the effect were transparent. Some simplifying assumptions may, however, be made in some circumstances to overcome any of the complications that were mentioned above, and to apply some kind of mathematical treatment, but this is beyond the scope of this article.

We shall discuss only one example of an action spectrum for an effect of irradiating a macro-organism, namely the action spectrum for assimilation of CO_2 in wheat plants (obtained by Hoover [14]) shown by curve A of fig. 5.

The most obvious conclusion from this action spectrum is that the entire visible spectrum is effective in producing photosynthesis. The curve has, however, two distinct peaks, one in the blue and the other in the red region of the spectrum. Their positions coincide roughly with those of the two peaks of the absorption spectrum of chlorophyll (see fig. 6), although otherwise the shape of the action spectrum is quite different from that of the absorption spectrum. Thus, apart from the obvious conclusion that the blue and the red regions are particularly effective in photosynthesis, one can also conclude that chlorophyll is probably involved as an absorber of radiation energy for photosynthesis.

If chlorophyll were the only absorber of radiation energy for photosynthesis, the action spectrum for photosynthesis would be expected to parallel the absorption spectrum of chlorophyll if conditions nos. 1 and 2 were fulfilled. However, when a wheat plant is exposed to light, condition no. 1 is not fulfilled because the

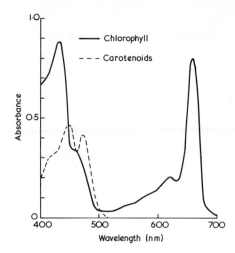

Fig. 6. Absorption spectra (plotted in arbitrary units) in an ethanol solution of pigments extracted from *Chorella* cells (from Emerson and Lewis [8]).

tissue layer, that is involved in photosynthesis, is not transparent to light. This may be the main reason, or possibly the only reason, for the action spectrum being different from the absorption spectrum. If condition no. 2 were also not fulfilled, this would also influence the shape of the action spectrum.

As absorption spectra of extracts of plant pigments may differ from those of the same substances when inside intact chloroplasts, it may be more pertinent to compare the action spectrum for photosynthesis (curve A in fig. 5) with an absorption spectrum of chloroplasts. Curve B in fig. 5 shows an absorption spectrum of a thin suspension of chloroplasts from spinach, obtained by Chen [5]. A reference to fig. 6 will show that at wavelengths longer than 520 nm absorption of light energy by chloroplasts is probably due to chlorophyll only, whereas at the wavelengths shorter than 520 nm carotenoids also contribute to the absorption.

Curve A differs from curve B (fig. 5) just about as much as it does from the absorption spectrum of a solution of chlorophyll shown in fig. 6. The positions of two peaks of each curve roughly coincide, but otherwise the shapes of the curve are quite different. To see whether the difference could be due to the non-fulfilment of condition no. 1 in wheat plants, curve B is also compared in fig. 5 with curve C which is an action spectrum for photosynthesis in a thin layer of *Chlorella* cells that was nearly transparent to light. It should be added that curve B is almost identical in shape with the absorption spectrum of a suspension of *Chlorella* cells (as obtained by Emerson and Lewis [8]), so that comparing the action spectrum for photosynthesis in *Chlorella* cells (curve C) with curve B is equivalent to comparing it with the absorption spectrum of the same cells.

The fact that curve C, obtained when condition no. 1 was fulfilled, deviates from curve B only slightly, whereas curve A, obtained when condition no. 1 was not fulfilled, deviates very considerably, suggests that the non-fulfilment of condition no. 1 may indeed be the cause of the considerable deviation of curve A. This conclusion is supported by the fact that action spectra for photosynthesis in thin suspensions of other monocellular green algae, and even in multicellular semi-transparent thin plants such as *Ulva* or *Elodea densa* (i.e. when condition no. 1 was not far from being fulfilled), are very similar to curve C [11, 12]. Further support comes from the fact that the action spectrum for the Hill reaction (which depends on a part of the chain of redox reactions involved in photosynthesis) in a thin suspension of spinach chloroplasts is also similar to curve C [5].

The problem still remains of the size and the mode of the slight deviation of action spectra for photosynthesis when condition no. 1 is fulfilled (such as curve C), from the absorption spectrum of chloroplasts (curve B) in view of the fact that the latter includes scatter of light an absorption by carotenoids. This is, however, a rather complicated and still a controversial problem that is outside the scope of this article.

References

[1] M.B. Allen, Absorption spectra, spectrophotometry and action spectra. Photophysiology Vol. 1, ed. A.C. Giese (Academic Press, New York and London, 1964) 83–110.

[2] F.C. Bawden and A. Kleczkowski, Studies on the ability of light to counteract the inactivating action of ultraviolet radiation on plant viruses. J. Gen. Microbiol. *13* (1955) 370–382.

[3] F.C. Bawden and A. Kleczkowski, Photoreactivation of nucleic acid from tobacco mosaic virus. Nature *183* (1959) 503–504.

[4] H.F. Blum, Action spectra and absorption spectra. In: Biophysical Research Methods, ed. F.M. Uber (Interscience, New York, 1950) 417–450.

[5] S.L. Chen, The action spectrum for the photochemical evolution of oxygen by isolated chloroplasts. Plant Physiol. *27* (1952) 35–48.

[6] C. Daubeny, On the action of light upon plants and plants upon the atmosphere. Phil. Trans. Roy. Soc. London (1836) 149–179.

[7] J.W. Draper, Note on decomposition of carbonic acid by the leaves of plants under the influence of yellow light. Phil Mag., Ser 3, *25* (1884) 159–173.

[8] R. Emerson and C.M. Lewis, The dependence of the quantum yield of Chlorella photosynthesis on wavelength of light. Amer. J. Bot. *30* (1943) 165–178.

[9] T.W. Engelmann, Ueber Sauerstoffausscheidung von Pflanzenzellen im Microspectrum. Bot. Zeit. *40* (1882) 419–426.

[10] N.A. Evans and A.D. McLaren, Quantum yields for inactivation of tobacco mosaic virus nucleic acid by ultraviolet radiation (254 nm). J. Gen. Virol. *4* (1969) 55–63.

[11] F.T. Haxo, The wavelength dependence of photosynthesis and the role of accessory pigments. In: Comparative Biochemistry of Photoreactive Systems, ed. Mary B. Allen (Academic Press, New York and London, 1960) pp. 339–359.

[12] F.T. Haxo and L.R. Blinks, Photosynthetic action spectra of marine algae. J. Gen. Physiol. *33* (1950) 389–422.

[13] A. Hollaender and B.M. Duggar, Irradiation of plant viruses and of micro-organisms with monochromatic light. III. Resistance of the virus of typical tobacco mosaic and *Escherichia coli* to radiations from λ3000 to λ2250 Å. Proc. Natl. Acad. Sci. US *22* (1936) 19–24.

[14] W.H. Hoover, The dependence of carbon dioxide assimilation in a higher plant on wave length of radiation. Smithsonian Misc. Coll. *95*, No. 21 (1937) 1–13.

[15] B. Kassanis and A. Kleczkowski, Inactivation of a strain of tobacco necrosis virus and of the RNA isolated from it, by ultraviolet radiation of different wavelengths. Photochem. Photobiol. *4* (1965) 209–214.

[16] A. Kleczkowski, The inactivation of ribonucleic acid from tobacco mosaic virus by ultraviolet radiation at different wavelengths. Photochem. Photobiol. *2* (1963) 497–501.

[17] A. Kleczkowski, Dark reactivation of ultraviolet-irradiated tobacco necrosis virus. J. Gen. Virol. *3* (1968) 19–24.

[18] A. Kleczkowski and A.D. McLaren, Inactivation of infectivity of RNA of tobacco mosaic virus during ultraviolet-irradiation of the whole virus at two wavelengths. J. Gen. Virol. *1* (1967) 441–448.

[19] F. Kubowitz and E. Haas, Über das Zerstörungsspektrum der Urease. Biochem. Z. *257* (1933) 337–343.

[20] A.D. McLaren and I. Moring-Claesson, An action spectrum for the inactivation of infectious nucleic acid from tobacco mosaic virus by ultraviolet light. Proc. 3rd Int. Photobiol. Congr. (Elsevier Publishing Co, Amsterdam, 1961) 573.

[21] A.D. McLaren and D. Shugar, Photochemistry of Proteins and Nucleic Acids (Pergamon Press, London, 1964) pp. 320–334.

[22] A.D. McLaren and W.N. Takahashi, Inactivation of infectious nucleic acid from tobacco mosaic virus by ultraviolet light (2537 Å). Radiat. Res. *6* (1957) 532–542.

[23] J. Senebier, Expériences sur l'action de la lumière solaire dans la végétation. Génève (1788).

[24] R. Setlow, Action spectroscopy. Advan. Biol. Med. Phys. *5* (1957) 37–74.

[25] R. Setlow, The use of action spectra to determine the physical state of DNA in vivo. Biochim. Biophys. Acta *39* (1960) 180–181.

[26] T. Tanada, The photosynthetic efficiency of carotenoid pigments in Navicula minima. Amer. J. Bot. *38* (1951) 276–283.

[27] C. Timiriazeff, Recherches sur la décomposition de l'acide carbonique dans le spectre solaire par les parties vertes des végétaux. Ann. Chim. Phys., 5e Sér., *12* (1877) 355–396.

[28] O. Warburg and E. Negelein, Über den Einfluss der Wellenlänge auf den Energieumsatz bei der Kohlensäureassimilation. Physik. Chem. *106* (1923) 191–218.

[29] O. Warburg and E. Negelein, Über die photochemische Dissotiation bei intermittierender Belichtung und das absolute Absorptionsspektrum des Atmungsferments. Biochem. Z. *202* (1928) 202–228.

CHAPTER 5

ANALYSIS OF IRRADIATED MATERIAL

R.B. SETLOW

Biology Division, Oak Ridge National Laboratory,
Oak Ridge, Tennessee, USA

1. Introduction

Analysis of the effects of radiation on molecules and on living systems is of interest in its own right, but is perhaps of more importance because the methodology used in such analyses points the way toward a detailed understanding of the effect of other environmental agents on living systems. There are many different types of light-mediated effects on biological systems — direct, sensitized, and photo-oxidized. Our problem is to relate the photochemical changes and their effects to observed photobiological changes such as killing and mutation. It is not sufficient to show that a biological effect increases with dose when a particular photochemical change also increases with dose, unless we know something about changes in other photo-products and have some estimate of the effects of the various products on biochemical and biological reactions. We must be able to relate numbers of photochemical changes with numbers of affected biological systems. The problem is not at all simple because (a) many photochemical products are not stable, (b) some products may be removed or altered by various in vivo repair systems and (c) the biological effect may not be observed until a long time after the initial photochemical change [25].

Just as a biological system involves interactions among its many constituent parts and exhibits reactions and behaviors that are more complex than those of its individual subunits, so are the photochemical reactions of polymers not equal to the sum of the reactions of their subunits. Thus, although we are always tempted to look at simple systems, the analysis of such systems may be totally incomplete in describing the effects on more complicated ones. Two examples are the formation of cyclobutyl pyrimidine dimers between adjacent pyrimidines in polynucleotides [50, 53] and the formation of DNA-protein cross-links in irradiated bacteria [62]. The first reaction is a minor one for bases such as thymine in aqueous solution, but is an important one for thymine in frozen solution [3] (conditions under which the reaction was discovered) and for thymine in polynucleotides. The kinetics of photo-dimerization are very sensitive functions of the conformation of the polynucleotide, because the two pyrimidines must be in appropriate positions to react. Thus it is possible, by changing the conformation of DNA, to change its photochemical reac-

71

tivity and photobiological effects and so to arrive at a correlation between these two quantities. Similarly, DNA–protein cross-links depend critically on the intimate spatial relationships between the DNA and protein in cells [65].

In this article I attempt to point out some of the particular difficulties, as well as some of the successful approaches, in relating the photochemical reactions of poly-nucleotides to photobiological effects. I concentrate on polynucleotides rather than proteins because nucleic acids are involved in the many steps that give specificity to protein synthesis, and moreover the subject of proteins and their photochemical reactions are covered in other chapters of this volume. I make no pretense about being complete in my treatment of the discussion of the various types of photo-products; for these, the reader is referred to a number of reviews [47, 48, 53, 63, 64, 76]. The topics to be covered will be (1) methods of irradiation, (2) analysis of the photoproducts in polynucleotides, (3) ways and tricks of correlating such pro-ducts with biological changes, and (4) the analysis of isotopic incorporation data.

2. Methodology of irradiation

If one is to make quantitative correlations between photoproducts and photo-biological effects, it is important that reasonable attention be paid to quantitative dosimetry. It is best if both photochemical and photobiological determinations are made on the same irradiated material, but even in this case it is important that all parts of the sample are exposed to the same average fluence of light. This point has been extensively discussed by Hiatt in chapter 2. Fig. 1 is a hypothetical example illustrating how one could be led completely astray by inattention to these dosi-metry problems. A bacterial culture is being irradiated. In order to get enough material for analysis the contents of each chamber is irradiated without stirring for a different period of time. After irradiation the contents are mixed, and analyzed for colony-forming ability and the photoproduct A and the photoproduct B. Let us further assume that photoproduct B can only be made in a limited amount and it saturates at very low fluences. The real survival curve of the organism is exponential (fig. 1a). However, if the irradiation vessels have small recesses that effectively shield 10% of the sample, the result in fig. 1b is obtained. There is a big change in the shape of the survival curve but the amounts of photoproducts change by a hard-ly detectable amount ($\sim$ 10%). A naive analysis of fig. 1b would result in a 1 : 1 correlation between photoproduct B and inactivation of the cells. Fig. 1a and a knowledge of the irradiation geometry tells us that this conclusion is absurd.

Most of us would not make the type of mistake illustrated in fig. 1. More subtle variations of this mistake, those that involve absorbing, unstirred samples, are al-most as bad. Even extensive stirring with such samples may not suffice to distribute all parts of it uniformly through the light beam. The moral: try to use transparent samples. If you cannot, give the radiation dosimetry extra special throught and make corrections for the absorption of light passing through the sample [21, 22, 37].

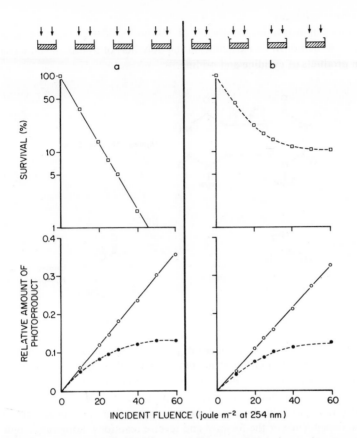

Fig. 1. A hypothetical example illustrating the necessity of good dosimetry in attempts to correlate photobiology (the survival curves above) with photochemical products (photoproducts A
and B below). (a) Good irradiation conditions. (b) 10% of the samples are shielded from the incident light beam. This has a big effect on the shape of the survival curve but a very small effect
on the yield of photoproducts.

3. Analysis of photoproducts in polynucleotides

There are four general ways of detecting photoproducts in polynucleotides. These
involve measurements of: (a) optical changes, (b) immunological reactions, (c) the
photoproducts isolated from hydrolysates of the irradiated polymer, and (d) macromolecular changes. The first method is probably the least sensitive but the most important and aesthetically satisfying, because it involves detection of changes in irradiated material under conditions in which the detection method does not alter the
photoproducts. Thus, if one wishes to demonstrate the existence of a particular product in a polynucleotide as distinct from its presence in an acid hydrolysate of the
polynucleotide, one should look at absorbance changes in the polymer.

3.1. *Absorbance changes*

The changes in ultraviolet absorbance of polynucleotides as a result of irradiation may be used to demonstrate existence, in the polymer, of pyrimidine dimers and of hydration products of cytidine and uridine.

Thymine – Thymine Dimer

Cytidine Hydrate

Uridine Hydrate

The formation of dimers between adjacent pyrimidines results in a loss of ultraviolet absorbance because of saturation of the two 5–6 double bonds [3]. Moreover, from studies on the irradiation of thymine in frozen solutions and in model polymers, it is known that dimer formation is a reversible process; because of the different absorption coefficients of the dimer and the monomers, there is a big difference in wavelength dependence of the forward and reverse reactions. Longwavelength irradiation favors the formation of dimers and leads to a decrease in ultraviolet absorbance, whereas short-wavelength irradiation shifts the equilibrium far to the side of monomers and, if given subsequent to the long wavelength, would result in the monomerization of dimers and an increase in ultraviolet absorption. This behavior, predicted on the basis of the photochemical reactions of simple model compounds [7, 23, 49], is just what is observed for the case of irradiated DNA [54] (fig. 2) and synthetic polymers such as polythymidylic acid [7] and polyuridylic acid [71]. The kinetics of these absorbance changes are also those expected from the reactions of simple model polymers, and the magnitudes of the absorbance changes also are in rough agreement with the changes expected from the known sequences of adjacent pyrimidines in DNAs. Such data are direct evidence for the existence of cyclobutyl pyrimidine dimers in irradiated DNA.

The irradiation of uridine or cytidine in solution leads to hydration of the 5–6 double bond [35]. The cytidine hydrates have reduced absorbance at 270 nm and increased absorbance at 240 nm. They are unstable and revert rather rapidly by loss of water to the original compound. Such transient heat-labile photoproducts may

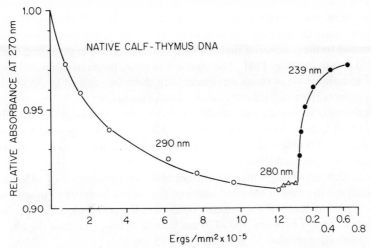

Fig. 2. Absorbance changes in native calf thymus DNA that result from irradiation with 290 nm followed by 280 nm and 239 nm (Setlow and Carrier [54]). The decrease in absorbance observed at 270 nm is a measure of the formation of cyclobutyl pyrimidine dimers.

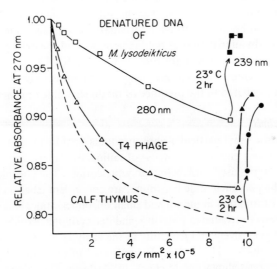

Fig. 3. Absorbance changes in several denatured DNAs as a result of irradiation at the indicated wavelengths. The increases in absorbance at 23°C for *M. lysodeikticus* and calf thymus DNAs represent the dehydration of cytidine hydrates. There is no such increase observed for the DNA of T4 phage. The increase resulting from 239 nm represents monomerization of dimers (Setlow and Carrier [54]).

be observed in irradiated, denatured DNA by the change in ultraviolet absorbance that decays relatively quickly, even at room temperature [54] (fig. 3). These transient absorbance changes are observed in denatured DNA and not in native DNA, thus leading to the conclusion that few hydrates are formed in organized double-stranded polynucleotides [54]. The numbers of photoproducts estimated by these optical techniques are in rough agreement with those determined by isolation of photoproducts from radioactively labeled DNA (see below).

The deamination product of an adduct of cytosine and thymine [74] is characterized by an increase in absorbance at 315–320 nm. During the irradiation of DNA such an increase is observed [8, 75], supplying evidence that such adducts may actually exist in irradiated DNA.

The absorbance changes that result from ultraviolet irradiation of tRNA have been analyzed in terms of the production of dimers and hydrates of uracil and the production of hydrates of the cytosine residues in the polynucleotide [14, 36]; absorbance changes between 310–370 nm have been used to follow the alteration of the thiouridine residues by irradiation at 335 nm [12].

3.2. *Immunological reactions*

It is possible to prepare antibodies against irradiated and photo-oxidized DNA and, by measuring the inhibiting effects of irradiated model compounds on the antibody–antigen reaction, to show that the antibodies have specificity for particular products. Antibodies to ultraviolet-irradiated DNA have a high specificity toward cyclobutyl thymine dimers in polynucleotides and the existence of this immunological reaction is evidence for this product in DNA [31, 46]. Immunological techniques have been used to demonstrate a specific product of guanine photo-oxidation in DNA irradiated in the presence of methylene blue or other dyes [32, 45].

3.3. *Isotopic labels*

If one is satisfied that particular photoproducts exist in a polynucleotide, then the products are best enumerated by use of radioactive labels. Table 1 lists four of the common isotopes used for labeling polynucleotides. The phosphorous isotopes, being part of the polynucleotide backbone, are convenient labels in the investigation of product formation in mono- and small oligonucleotides and in the analysis of enzymic digests of irradiated polyribo- and deoxyribonucleotides. Such labels have been used extensively and fruitfully in the analysis of the photochemical reactions of model polymers [1, 13, 19]. The existence of two easily distinguishable phosphorous isotopes allows one to check, by cochromatographic techniques, the identity of products from irradiated polynucleotides and model compounds.

Two widely used radioactive labels are ^{3}H- and ^{14}C-labeled thymidine. Such compounds are readily incorporated into the DNA of many organisms and, because thymidine is specific for DNA, photoproduct formation may often be followed by the hydrolysis of the whole organism (see below) [4, 52]. Both ^{3}H and ^{14}C labels are stable, and the two labels give the same amount of pyrimidine dimer formation

Table 1
Some isotopic labels that are useful for photoproduct analysis

A. Characteristics

Isotope	Max. β-ray energy (keV)	Half-life	Specific activity (Ci/milliatom)
^3H	18	12.5 yr	29
^{14}C	155	5700 yr	0.062
^{32}P	1710	14.3 day	9120
^{33}P	248	25 day	5200

B. Typical compounds

(within less than 1%) [52]. Labeled cytidine is also a useful radioactive precursor. However, it is not incorporated exclusively into DNA and so, if one wishes to observe photoproducts in this polymer, the DNA itself must be isolated. The 5-^3H label in cytidine is stable, but photoproducts of this material may not be. Saturation of the 5–6 double bond labilizes the 5-hydrogen [16, 17]. Presumably, the 5-^3H exchanges with the medium via the amino group so that in H_2O one observes reactions of the form [17]:

The lability of the 5-hydrogen means that photoproduct determination with such an isotopic label may give an answer completely different from one found by using ^{14}C-labeled cytidine (see below). Moreover, saturation of the 5—6 double bond also labilizes the amino group [1, 13, 58]. Hence the original cytosine photoproducts may appear in the final analysis as uracil photoproducts.

$$CH_2O \rightarrow UH_2O,$$

$$\widehat{CT} \rightarrow \widehat{UT},$$

$$\widehat{CC} \rightarrow \widehat{UU}.$$

3.4. Acid hydrolysis

DNA may be hydrolyzed by acid (formic acid at 175°C, trifluoroacetic acid at 155°C, or perchloric acid at 100°C) to the free bases plus any acid-stable photoproducts that involve these bases [4, 76]. The sugars and phosphates are lost in this analysis, and the only extensively used labels are those on thymine or cytosine. The hydrolysate may be separated into individual bases and photoproducts by paper, thin-layer, or column chromatography [4, 8, 55, 69], and the separated components may be measured by the counting of radioactivity. A typical radiochromatogram of irradiated T4 phage, labeled with ^{3}H-thymidine, is shown in fig. 4 [56]. There is only one major peak other than that of thymine. This material, amounting to 1.1% of the total radioactivity, results from the irradiation and has the chromatographic mobility of a thymine—thymine cyclobutyl dimer. The lower part of the figure indicates that, when this radioactive peak is eluted from the chromatogram and irradiated with a short wavelength (a treatment that would monomerize any thymine dimers), the photoproduct disappears and thymine is produced as measured in three different chromatographic solvents. Thus the irradiation of T4 phage DNA makes only one detectable acid-stable photoproduct a thymine—thymine dimer. There are no detectable photoproducts that involve combinations of any of the other bases,

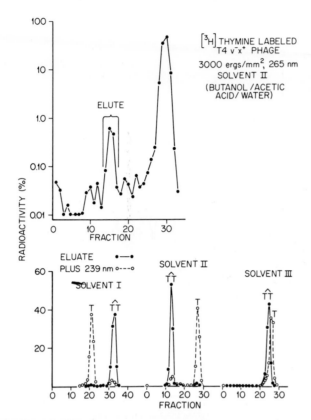

Fig. 4. Radiochromatograms, on Whatman No. 1 paper, of a formic acid hydrolysate of irradiated, ^{3}H-thymine-labeled T4 phage (3000 erg/mm^2, 265 nm). Upper section: 1.1% of the radioactivity was in the photoproduct peak that was eluted for reirradiation. Lower section: chromatograms, in three solvents, of the eluate and the reirradiated eluate ($\sim 10^5$ erg/mm^2, 239 nm) that indicate that most of the eluate is thymine–thymine dimer. Solvent I: saturated ammonium sulfate/M sodium acetate/isopropanol; 49/9/1. Solvent II: n-butanol/acetic acid/H$_2$O; 80/12/30. Solvent III: tert-butanol/methylethylketone/H$_2$O/NH$_4$OH; 40/30/20/10. Unirradiated DNA has < 0.001% radioactivity in dimers (Setlow and Carrier [56]).

such as hydroxymethylcytosine, with thymine. These data show that hydroxymethylcytosine does not participate readily in dimer formation, and fig. 3 indicates that the base also does not participate in hydrate formation the way cytidine itself does.

Different DNAs and different irradiation conditions results in the formation of different types and numbers of photoproducts [41, 55, 64]. Fig. 5 shows an example. An *E. coli* DNA labeled with ^{3}H-thymidine was irradiated at low temperature [41]. Three major peaks of radioactivity induced by ultraviolet radiation are apparent: The peak marked ÛT (the deamination product of a cytosine–thymine

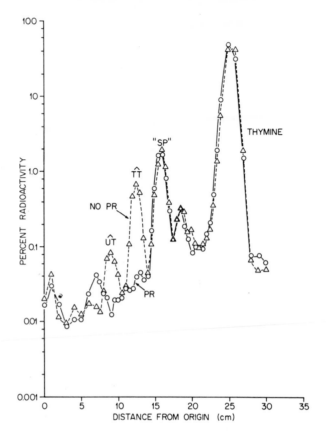

Fig. 5. The curve designated "no PR" is a chromatogram of an acid hydrolysate of an *E. coli*
DNA labeled with ^{3}H-thymine that was dissolved in ethyleneglycol : water and irradiated (inter-
ruptedly) with 2×10^5 erg/mm^2, 280 nm, at $-196°C$. The curve designated "PR" represents
the hydrolysate of the same irradiated DNA after exposure of the DNA to "black light" in the
presence of photoreactivating enzyme. Solvent II of fig. 4 (Rahn and Hosszu [41]).

cyclobutyl dimer), the thymine—thymine dimer and the peak labeled SP (a photoproduct also observed in irradiated bacterial spores) [11]. Fig. 5 also demonstrates
clearly one of the reactivation mechanisms — the enzymic monomerization of
cyclobutyl-pyrimidine dimers by treatment with photoreactivating enzyme. Such
treatment only affects cyclobutyl dimers and provides a way of demonstrating the
importance of these photoproducts in biological effects. Enzymic photoreactivation
does not affect the spore photoproduct.

The ^{3}H label in cytosine photoproducts may readily exchange with the medium
under appropriate conditions [56]. Fig. 6 illustrates the results of such an exchange
on the analysis of hydrolyzed DNA. A DNA labeled with ^{3}H- and ^{14}C-cytosine was

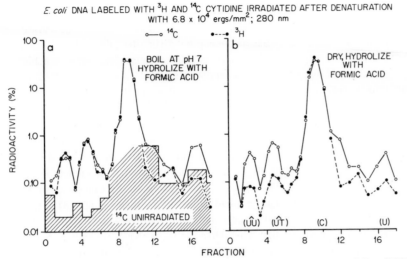

Fig. 6. Chromatograms of formic acid hydrolysates of cytidine-labeled denatured *E. coli* DNA treated in different fashions after irradiation with 6.8 × 10[4] erg/mm[2] at 280 nm. After irradiation, the DNA labeled with both 2-^{14}C and 5-^{3}H was (a) boiled at pH 7 before hydrolysis or (b) dried and hydrolyzed. Solvent II of fig. 4 (Setlow and Carrier [56]).

denatured by heat and irradiated. When the DNA was dried and then hydrolyzed with formic acid, the results in fig. 6b were obtained. It is obvious that many fewer photoproducts are observed with the ^{3}H label than with the ^{14}C label. An explanation is that at low pH there can be rapid exchange between the ionized amino group and the 5-^{3}H in a photoproduct, resulting in the loss of the ^{3}H to the medium. This exchange cannot take place if the cytidines are first deaminated to uracils by heating at neutral pH. The chromatogram of the irradiated DNA that was boiled before hydrolysis is given in fig. 6a. Both ^{3}H and ^{14}C labels give the same amounts of products and the ^{14}C label, as expected, gives the same result as in fig. 6b. Fig..6 is an example of the precept that it is more important to obtain a result by a different experiment than to repeat the experiment. The results for the ^{3}H label in fig. 6b are reproducible, but wrong. The lability of the 5-^{3}H atom at acid pH led to the erroneous conclusions that cytosine–cytosine dimers were unstable and were destroyed by hydrolysis [58], whereas uracil–uracil ones were not. The proper conclusion is that 5-^{3}H in cytosine–cytosine dimers may be labile. It is not labile in a uracil–uracil dimer.

The lability of particular atoms in a photoproduct must be distinguished from the lability of the photoproduct itself – for example, the deamination and dehydration of cytidine hydrates and the acid-lability of the cyclobutyl dimers that have the head-to-tail configuration [20]. (In DNA the dimers are head-to-head [2].)

3.5. Enzymic hydrolysis

The identification of photoproducts in polynucleotides from the observation of products in acid hydrolysates has its conceptual difficulties, difficulties associated with the instability of photoproducts. A much more gentle hydrolysis is achieved by enzymic action. For example, venom diesterase hydrolyzes polynucleotides to 5'-mononucleotides and, if the polymer is labeled with ^{32}P or ^{33}P, these mononucleotides are easy to detect in small amounts. The hydrolysate of an irradiated polymer that has been digested to the limit by venom diesterase includes not only mononucleotides but a number of enzyme-resistant sequences [9, 57]. Fig. 7 shows a radiochromatogram of an enzymic digest of unirradiated and irradiated ^{32}P-labeled DNA. The spots designated products are the enzyme-resistant sequences that result from the irradiation of the DNA. The products shown on this one-dimension chromatogram may be chromatographed in a second dimension (fig. 8) in which case the original product-spots break up into subspots, some of which (spots 1, 2, 3 and 4) are relatively easy to identify. They have the form pNpPypPy, where N represents any nucleoside, pN a 5' nucleotide, and Np a 3' nucleotide; at the high fluxes used in this experiment they contain mostly TT dimers. There is appreciable radioactivity in a number of yet unidentified products in the limit digests of irradiated native DNA. Many of these products probably do not appear in acid hydrolysates.

Photosensitized reactions make a much more limited class of products in DNA than does direct irradiation, as judged by analyses of acid hydrolysates [29, 30]. It would be useful to investigate the products in these photosensitized DNAs by enzymic hydrolysis.

Enzymic hydrolysis has been used extensively to study the photochemical reactions of irradiated ^{32}P-labeled polyuridylic acid [39, 40]. A combination of ribonuclease digestion of the irradiated polymer plus sensitive chromatographic techniques has permitted the detection in the limit digest of the enzyme-resistant sequences UpUpUp, UpUpU(hydrate)p, and U(hydrate)p, as well as clusters of dimers. Similar products are formed in poly U · poly A, but the yield of hydrate is much lower and only beings to rise rapidly when some dimers have been formed — a result indicating once more the importance of conformation in photochemical reactivity.

The methods of enzymic hydrolysis of irradiated polynucleotides have been used to determine the numbers of hydrates and dimers of uracil in tobacco mosaic virus RNA [61] and to show that guanine residues are altered in the methylene blue photosensitized reaction of poly(U, G) [60].

3.6. Macromolecular techniques

The photoproducts we have mentioned — pyrimidine dimers, spore photoproducts, and hydrates — are relatively easy to detect because they may be labeled with radioactive precursors and they are produced with high yields. With physical techniques, on the other hand, it is possible to measure very small numbers of photoproducts that involve large molecules, such as chain breaks, cross-links be-

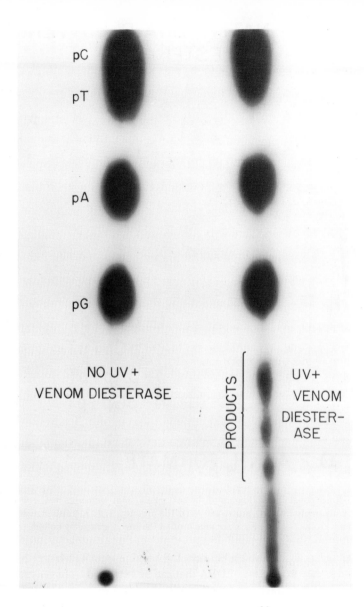

Fig. 7. A radioautogram of a chromatogram of a limit digest of ^{32}P-labeled denatured *E. coli* DNA. Left streak, unirradiated; right streak, irradiated (1.5×10^5 erg/mm^2, 280 nm) (Setlow, Carrier and Bollum [57]).

tween DNA strands, and DNA–protein cross-links. The latter product of irradiation of bacteria, a product that seems to be formed in larger amounts at low temperature

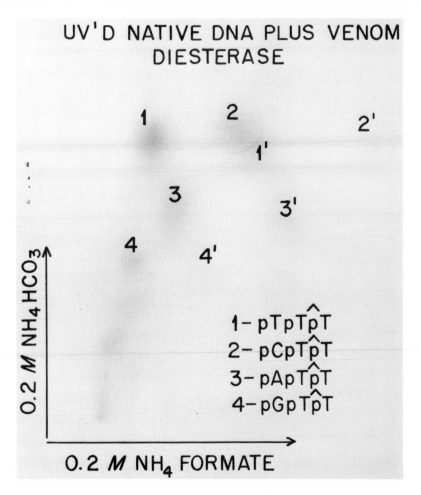

Fig. 8. A two-dimensional chromatogram of the enzyme-resistant sequences in *E. coli* DNA irradiated in the native state (10^5 erg/mm^2, 280 nm). The products 1, 2, 3, 4 also include some UT dimers. Products 1′, 2′, 3′, 4′ are unidentified (Setlow, Carrier and Bollum [57]).

[66], is detected as an association between DNA and protein which may be measured by the decrease in amount of DNA that can be extracted from cells following irradiation [62, 65]. A fraction of the DNA is associated with precipitated protein and the fraction increases with dose. The existence of cross-links between DNA and protein is easy to establish, but it is difficult to estimate the precise numbers of cross-links per micron of DNA.

The high sensitivity of the physical methods is illustrated by data on the sedimentation in alkali of bromuracil-substituted DNA of T3 phage [33] (fig. 9). In alkali the individual DNA strands separate and, using a radioactive label, one can

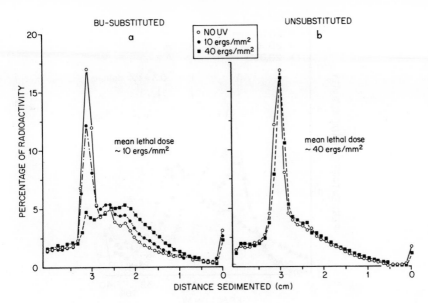

Fig. 9. Sedimentation in alkaline sucrose of the ^{3}H-labeled DNA of T3 phage [33]. The peaks near 3 cm correspond to the intact single strands of the DNA. The phage were irradiated as indicated. (a) Ninety percent of the thymine was substituted by 5-bromuracil; (b) unsubstituted phage.

measure the sedimentation of the individual strands or parts of strands that may result from the combination of irradiation and alkaline treatment. The impressive thing about the data in fig. 9 is that they permit one to detect a very clear-cut physical change in the bromuracil-substituted DNA at doses to the phage that correspond to one mean lethal dose. At these low doses the breaks that appear in alkali are a property of bromuracil-substituted material and are not observed in DNA containing only thymine. The appearance of chain breaks in bromuracil-substituted phage DNA are well correlated with phage killing but the observed breaks are not necessarily the primary photoproduct, but may arise from the action of high pH on the primary photoproduct.

Cross-links between DNA strands may be estimated either in terms of the reversible denaturation of DNA [15, 27, 28] or by the fact that in alkali, cross-linked DNA would have a higher molecular weight and would sediment more rapidly than normal, single strands if the cross-links were stable in alkali. An example illustrating the production of cross-links and of chain breaks in normal bacterial DNA irradiated in vivo is shown in fig. 10. If bacteria are lysed on the top of an alkaline gradient [34], single-stranded pieces of relatively high molecular weight are obtained (number average $\sim 1.5 \times 10^8$). Irradiation with large UV doses results in the appearance of material that sediments faster (alkaline-stable, cross-linked) and slower (alkaline-

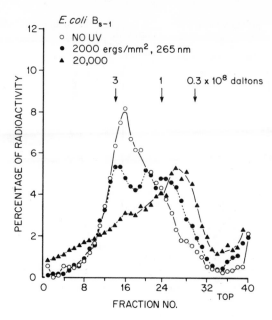

Fig. 10. Sedimentation in alkaline sucrose of the DNA of E. coli B_{s-1} immediately after large exposures in vivo to show the presence of slower (broken molecules) and faster (cross-linked strands) than normal molecules (Setlow and Carrier [56]).

labile regions) than the unirradiated material [56]. One may use such data to estimate that on the average an incident fluence of 200 erg/mm^2 will make approximately one chain break (or alkaline-labile region) and one-tenth of a cross-link per bacterial genome (2.5 × 10^{9}dalton).

4. The relation between photochemistry and photobiology

In order to find a cause-and-effect relation between photochemical products and photobiological results, some kind of trick is needed. For example, short-wavelength reversal and photoreactivation are used as indications of pyrimidine dimer involvement. A correlation between the biological effect of ultraviolet light on DNA as a function of the temperature during irradiation with the production of various types of photoproducts as a function of temperature has been used to relate pyrimidine dimers with most of the inactivation of transforming DNA at low temperature [42] and to show that the non-photoreactivable damage is partly associated with spore photoproducts. A correlation of spore photoproduct and killing of spores as a function of temperature indicates the lethality of this product for bacterial spores [10], and a correlation between the production of the spore photoproducts by a given UV dose as a function of germination time and the killing of germinating spores has

also been used to implicate this product in the killing of spores [68] . The killing of
E. coli at temperatures below freezing has been correlated with an enhanced appearance of DNA–protein cross-links [66] .

In most systems, it is apparent that several photoproducts contribute to the
photobiological effect. For example, the photoreactivable sector for *E. coli* irradiated in acqeous medium is approximately 0.8, a result indicating that 80% of the
damage may be ascribed to pyrimidine dimers, leaving another 20% unaccounted
for. For bacteria irradiated at low temperature the photoreactivable sector is very
small [66] , but protein–DNA cross-links do not seem to be the only lethal damage,
since pyrimidine dimers and spore photoproduct are still formed in large numbers
under these conditions. Thus the solution to the problem of the relation between
photochemistry and photobiology is not necessarily straightforward or simple, and
satisfactory approaches and conclusions involve knowing a great deal about the
photochemistry and photobiology of the molecules that make up living systems.

Damage to DNA may be repaired in the dark by in vivo enzymic systems. The
existence of such repair systems for the simpler biological entities such as bacteria
(and the repair-defective bacterial mutants that simplify analysis) leads one to suppose that the existence of such repair systems in more complicated cells is evidence
for the importance of the lesion that is being repaired. Such a very indirect chain
of evidence has been used to implicate pyrimidine dimers in the production of
some skin cancer. Normal human fibroblasts are able to excise cyclobutyl pyrimidine dimers from their DNA [43] . (The lesions are removed in a complicated
series of events that are described more fully in another chapter). Fibroblasts from
individuals suffering from the disease Xeroderma pigmentosum – a disease characterized by extreme sensitivity to sunlight and in many cases the eventual production
of numerous skin cancers – are unable to excise pyrimidine dimers and hence unable to do the other steps in repair [5, 59] . Fig. 11 shows such excision data. The

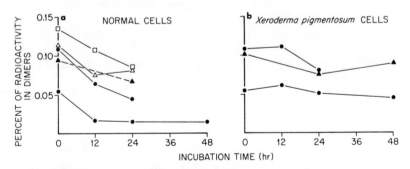

Fig. 11. The fraction of radioactivity in DNA that is in thymine-containing dimers at various
times after irradiation with different doses. At 265 nm 0.05% corresponds to an incident fluence of ~ 100 erg/mm^2 and 0.10% to ~ 200 erg/mm^2. (a) Normal human fibroblasts showing
dimer excision; (b) Xeroderma pigmentosum fibroblasts showing no excision [59] .

lack of excision in Xeroderma pigmentosum cells has been hypothesized to result in the accumulation of large numbers of dimers in such cells, and the production of skin cancer as a result of the biochemical upset produced in the translation and transcription of such altered DNAs.

5. The analysis of isotopic incorporation

Because the growth of cells is intimately related to the synthesis of macromolecules and to the interactions among them, it is fruitful to examine the effects of radiation on synthesis of the three types of macromolecules about which a great deal is known – proteins, RNA and DNA. DNA may be the most affected because its molecular weight is so much larger than the others (see table 2). Because the synthesis of DNA is approximately sequential, a single photoproduct could conceivably affect all the synthesis beyond it. A much larger number of products would be necessary to inhibit the synthesis of RNA from a DNA template, because RNA synthesis has many starting points on the DNA.

Measurements of macromolecular synthesis, especially by the incorporation of radioactive precursors, are complicated by the effects of pools of precursors. Thus the incorporation of a radioisotope from the medium into macromolecules goes throught the schematic sequence of steps illustrated in fig. 12. In many cases there may be a rapid turnover of macromolecules; they break down into subunits, and the subunits go into the pool from which they are reincorporated into new macromolecules. Many of the intermediate pools are of fixed size and do *not* equilibrate with the external medium, and breakdown products are preferentially incorporated. In such a situation, the data on incorporation of radioactivity from the medium tell nothing about the molecules that turn over and only represent net macromolecular synthesis. For example, although messenger RNA in bacteria turns over very rapidly, the rate of incorporation of RNA precursors from the medium is determined only by the synthesis rate of stable RNA, although much of the radioactivity may initially go into unstable RNA [38].

Table 2

Molecular weights and approximate numbers of molecules in a growing cell of *E. coli*

Type	Molecular weight	Numbers*
Proteins	$\sim 3 \times 10^4$	4×10^6
RNA		
transfer	2.5×10^4	4×10^5
ribosomal	$0.5 \times 10^6, 1 \times 10^6$	3×10^4
messenger	$\sim 10^6$	10^3
DNA	2.5×10^9	2

* Actual values depend on conditions of growth [6, 24, 26].

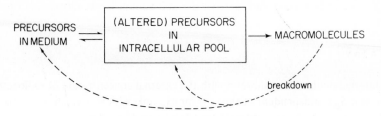

Fig. 12. A schematic diagram indicating the flow of precursors into and out of intracellular pools.

Let us consider the mathematical analysis of the incorporation of radioactivity into an exponentially growing culture. (The synthesis of macromolecules in a synchronized culture may be easier to analyze, but it involves the unproven assumption that the synchronization does not affect the sequence of macromolecular events or that the combined effect of synchronization and some radiation injury will not affect these sequences. Thus what may seem to be a mathematical simplification is an unknown biological complication.) In a unit volume of an exponentially growing culture the number of molecules N, or the number of cells, increase at an average rate proportional to the number of molecules, or cells, that are present at that time:

$$\frac{dN}{dt} = \alpha N, \tag{1}$$

where α is the growth-rate constant. If there are N_0 molecules at time $t = 0$, eq. (1) gives

$$N = N_0 e^{\alpha t} \text{ or } N = N_0 2^{t/\tau}, \tag{2}$$

where τ is the average doubling time of the molecule (or cells) and is related to α by

$$\alpha = \frac{\ln 2}{\tau}.$$

The rate of incorporation per unit volume of a radioactive precursor into macromolecules, proteins or nucleic acids, is given by

$$\frac{dR}{dt} = S \frac{dN}{dt}, \tag{3}$$

where R is the radioactivity incorporated per unit volume into N molecules or cells per unit volume and S is a proportionality factor. S may be thought of as made up of two parts — a part S_m proportional to the fraction of the macromolecular composition represented by the precursor (nonradioactive), and a fraction S_p representing the specific activity of the precursor in the internal pool of the cell:

$$S = S_m S_p. \tag{4}$$

The time rate of increase of radioactivity per unit volume is given by eqs. (1) and (3) as

$$\frac{dR}{dt} = S\alpha N. \tag{5}$$

If the internal pool is in equilibrium with the external concentration of radioactivity, that is if S_p is independent of time (as for example with very small pools that equilibrate rapidly), then substituting eq. (2) into eq. (5) yields

$$R - R_0 = S\alpha N_0 \int_0^t e^{\alpha t} dt,$$

or

$$R - R_0 = SN_0(e^{\alpha t} - 1) = SN_0(2^{t/\tau} - 1). \tag{6}$$

In many experiments zero time is that at which radioactivity is introduced into the solution. In this case $R_0 = 0$ and the incorporated radioactivity per unit volume is given by

$$R = SN_0(e^{\alpha t} - 1) = SN_0(2^{t/\tau} - 1). \tag{6'}$$

Thus a sensitive test of exponential growth and the constancy of the specific activity is obtained by plotting the incorporated radioactivity R versus $(2^{t/\tau} - 1)$. A straight line through the origin indicates the correctness of eq. (6'). Fig. 13 shows an example of such data [73], which indicate that the assumptions are correct over a range of 10^4 for *E. coli* incorporating ^{3}H-thymidine into DNA. The straight line represented by a graph of R versus $(2^{t/\tau} - 1)$ has a slope proportional to both S and N_0 — the specific activity of the precursor pool and the initial number of macro-molecules. It is obvious that a graph of R versus t is not a straight line, nor are graphs of $\log R$ versus t or R on a log scale versus t straight lines (see below).

    If a biological system is irradiated at time zero, we may think of the experiment as a classic pulse—relaxation one in which there has been a sudden change in the system and the system adapts to meet this change. Fig. 14, for example, shows several ways in which bacterial cultures of different genetic backgrounds adapt to short pulses (< 1 min) of radiation given at the same time that ^{3}H-tymidine is introduced into the system [51]. We may suppose that radiation affects one of a number of parameters in eq. (6'), such as the number of synthesizing molecules, the specific activity of the pool, or the doubling time of the molecules in question. Obviously, there is not going to be one simple, single explanation for the different types of incorporation curves shown in fig. 14. Its up to the investigator to be ingenious and to attempt to construct physical models that fit these data and to relate the models to photochemical changes in DNA or other macromolecular structures. For example, we might consider two extreme hypotheses to explain the data in fig. 14c: (a) radiation slows drastically the DNA synthesis of all bacteria for a fixed interval of time, after which they all begin synthesizing at their normal rates,

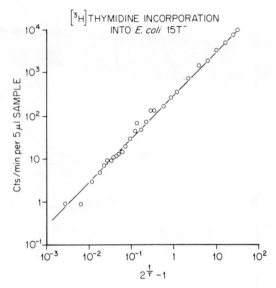

Fig. 13. Data for ^{3}H-thymidine incorporation by *E. coli* 15 T$^-$ that verify eq. (6'). The straight line on this log-log plots has a slope 1. At zero time there were 10^7 cells/ml and they grew at 23°C with τ = 126 min (Van Tubergen [73]).

or (b) some of the bacteria stop synthesis forever and the remainder of them continue at their normal rates [72]. One easily can construct incorporation curves for these two models, and typical ones are shown in fig. 15. Fig. 15a shows such manufactured data on linear plots and fig. 15b shows the same data, plotting radioactivity versus $2^{t/\tau} - 1$, where the parameter τ (as indicated above) would be the normal generation time of these molecules. It is apparent that fig. 15b allows a more accurate averaging of all the data, but obviously one could reach similar conclusions from both methods of plotting. Of the two hypotheses above, the data in fig. 14c fit better the one in which all bacteria stop synthesis and resume at a later time. A still better fit to the data is obtained if some of the bacteria do not resume synthesis or resume it at a less than normal rate [72].

Such models do not apply to *E. coli* B_{s-1} (fig. 14a), in which uptake of ^{3}H-thymidine is inhibited by small UV fluences. For example, at 2 erg/mm^2 there are only ~ 10 pyrimidine dimers per bacterial chromosome [44], and it would take the growing point of DNA moving at its normal rate about 5 min to get from one dimer to the next. If dimers inhibit synthesis (and the photoreactivation data indicate that they do), then we expect that along an individual chromosome there are fits and starts of synthesis and that the average rate depends critically on the time it takes for synthesis to go past a dimer, or to initiate on the other side of a dimer [44]. The second round of synthesis, on the model, should go at a faster rate since the dimers per chromosome are diluted out by a factor of two. Thus as time goes on

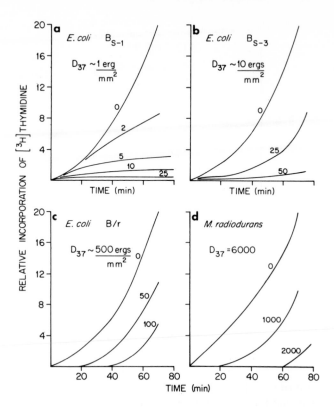

Fig. 14. The effect of ultraviolet radiation (265 nm) on the incorporation of ^{3}H-thymidine by different bacterial strains. The isotope was added immediately after irradiation at zero time with the incident fluences in erg/mm^2 indicated by the numbers next to the curves. The D_{37} is the fluence that leaves 37% of the cells able to form colonies. Doubling times for (a), (b), and (c) were 35 min and for (d) was 90 min. The sources of the original data are given in ref. [51].

the molecular population passes through a maximum of heterogeneity until finally the viable, dividing cells with normal chromosomes outnumber all the others. A further complication in the case of fig. 14a is the breakdown of irradiated DNA [70] (see below) masking much of the increased incorporation that is apparent in fig. 14b for a typical ultraviolet-sensitive straint that is defective in dimer excision The model described above should also apply to radiation-resistant strains of bacteria, but its kinetic implications are masked by the rapid recovery of synthesis that is correlated with the excision of dimers in these strains. In these resistant strains there should be much less molecular heterogeneity because the lesions are removed, not bypassed.

The introduction of radioactive precursors at the time of irradiation permits one to measure the net incorporation of precursors into macromolecules. There are cases in which this information is useful but not sufficient. We cite two examples:

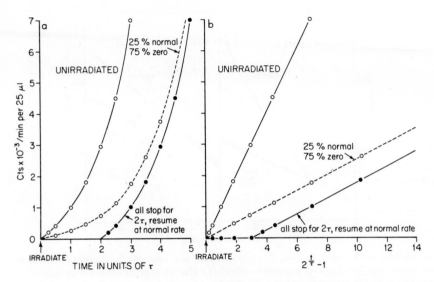

Fig. 15. Calculated, hypothetical curves that represent the incorporation, R, of a radioisotope (^{3}H-thymidine, for example) introduced into culture immediately after irradiation. The curves for the irradiated cultures have been calculated for two different models. In one all the cells stop for 2 division times (2τ) and then resume at their original rates, and in the other 25% of the cells synthesize at their normal rates and the remainder never synthesize. (a) The data plotted as R versus t; (b) the same data as R versus $2^{t/\tau} - 1$.

If there is an appreciable intracellular pool, then the simple kinetics given by eq. (6′) for unirradiated cells do not hold because S is a function of time. In this case it may be easier to find a simple interpretation if the cells are labeled for an appreciable time before the irradiation. Since $2^{3.5} = 11$, if radioactivity is allowed to incorporate for a time greater than 3.5τ, the incorporation is given within 10% by an expression of the form

$$R = SN_0 e^{\alpha t}, \, t > 3.5\tau. \tag{6″}$$

This is a simple exponential function and, when plotted as $\log R$ versus t or on semilog paper, gives a straight line of slope α and intercept at $t = 0$ of SN_0. In such experiments the cells have been in the presence of label for a long time and radiation is then given at a time t_r much later than zero. The radiation perturbs the steady state and the radioactivity measures the net number of molecules — new plus old — at times t_r and $t_r + \Delta t$ and does not measure, as in the above examples, only the newly synthesized material. An example of data obtained and treated in this way is shown in fig. 16, which shows the effects of ultraviolet radiation on the incorporation of ^{35}S into exponentially growing bacterial cultures [18]. The data are plotted on two different scales, one linear and one logarithmic, to illustrate that the unirradiated population incorporates isotopes at an exponentially increasing rate whereas

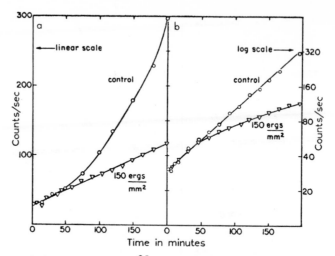

Fig. 16. Data on the incorporation, R, of ^{35}S into the acid-insoluble fractions of *E. coli* B. In this experiment the cells were exposed to label for a number of generations before irradiation at time zero with 150 erg/mm², 265 nm. (a) The data plotted as R versus t; (b) the same data plotted as $\log R$ versus t (Hanawalt and Setlow [18]).

the irradiated population synthesizes at a constant rate. This is a clear demonstration of why it is important to plot data in several ways if one is to get a feel for the type of function that the data represent. There is no preferred way to plot data.

Measurements based on eq. (6″) give the total amount of a given molecular species. In addition to eliminating problems of pools, they facilitate easy comparison of macromolecular synthesis over many division times. (The value of R in eq. (6′) is exceedingly sensitive to the choice of the parameter τ.) A perturbation of the kinetics by irradiation at time t_r may be observed by changes from the straight-line relationship between $\log R$ and t (fig. 17). The method, based on eq. (6″), has two disadvantages that are often ignored [67] : (1) Small changes that may take place near zero time — changes induced by a pulse of radiation — are much more difficult to measure experimentally, because these changes represent the difference between two large numbers. In the formulation alluded to in fig. 15, the small number is measured directly. (2) Since eq. (6″) measures the total amount of material, it cannot distinguish between the cessation of synthesis and synthesis equal to degradation in a culture.

Obviously, it is the wise investigator that makes use of all possible techniques in analyzing a problem. With the ready availability of modern scintillation counters, for example, it is comparatively easy to label cells continuously with a ^{14}C-labeled precursor, give a pulse of radiation, add the same precursor labeled with ^{3}H (keeping the ^{14}C label there), and so in one sample measure the total amount of a particular molecular type and the net synthesis of this type as a function of time. In

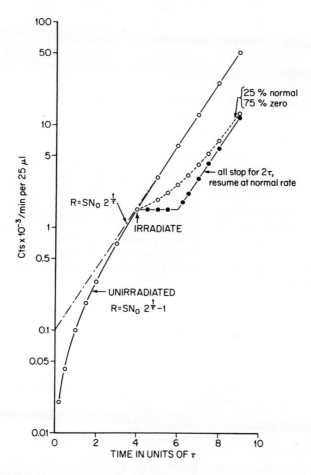

Fig. 17. The calculated incorporation of a radioisotope by an exponentially growing culture, eq. (6'), plotted as $\log R$ versus t (R on a logarithmic scale versus t). Label is introduced at zero time and the culture irradiated after 4 doubling times. The cell concentration at zero time is assumed to be one-tenth that of the cultures of fig. 15. The unirradiated culture approaches $R = SN2^{t/\tau}$ (eq. (6")) as a limit. The curves for the irradiated culture are calculated, as indicated, for the two models described in fig. 15.

principle, if there are not elaborate pool complications, one should be able to separate new synthesis, net synthesis, and breakdown.

6. Conclusion

There are many ways of analyzing photochemical products in polynucleotides and the photobiological consequences of such products. The correlation between the effects of these products and photobiology is difficult to make. The most important advice to a new investigator in the field is: know what you are doing, try to do experiments in several ways, and analyze your data in several ways.

Acknowledgements

Research sponsored by the United States Atomic Energy Commission under contract with the Union Carbide Corporation.

I thank J.E. Donnellan, Jr. and Jane K. Setlow for helpful comments on the manuscript.

References

[1] H. Becker, J.C. LeBlanc and H.E. Johns, The UV photochemistry of cytidylic acid. Photochem. Photobiol. 6 (1967) 733–743.

[2] E. Ben-Hur and R. Ben-Ishai, Trans-syn thymine dimers in ultraviolet-irradiated denatured DNA: Identification and photoreactivability. Biochim. Biophys. Acta 166 (1968) 9–15.

[3] R. Beukers and W. Berends, Isolation and identification of the irradiation product of thymine. Biochim. Biophys. Acta 41 (1960) 550–551.

[4] W.L. Carrier and R.B. Setlow (1970) The excision of pyrimidine dimers. Methods in Enzymology, Vol. XXI, part D, eds. L. Grossman and K. Moldave (Academic Press, New York, 1970) 230–237.

[5] J.E. Cleaver, Defective repair replication of DNA in Xeroderma pigmentosum. Nature, 218 (1968) 652–656.

[6] S. Cooper and C.E. Helmstetter, Chromosome replication and the division cycle of Escherichia coli B/r. J. Mol. Biol. 31 (1968) 519–540.

[7] R.A. Deering and R.B. Setlow, Effects of ultraviolet light on thymidine dinucleotide and polynucleotide. Biochim. Biophys. Acta 68 (1963) 526–534.

[8] H. Dellweg and A. Wacker, Strahlenchemische Veränderung von Thymin und Cytosin in der DNS durch UV-Licht. Z. Naturforschung, 17b (1962) 827–834.

[9] H. Dellweg and A. Wacker, The enzymatic degradation of ultraviolet-irradiated deoxyribonucleic acid. Photochem. Photobiol. 5 (1966) 119–126.

[10] J.E. Donnellan, Jr., J.L. Hosszu, R.O. Rahn and R.S. Stafford, Effect of temperature on the photobiology and photochemistry of bacterial spores. Nature 219 (1968) 964–965.

[11] J.E. Donnellan Jr. and R.B. Setlow, Thymine photoproducts but not thymine dimers found in ultraviolet-irradiated bacterial spores. Science 149 (1965) 308–310.

[12] A. Favre, M. Yaniv and A.M. Michelson, The photochemistry of 4-thiouridine in Escherichia coli t-RNA Val. Biochem. Biophys. Res. Commun. 37 (1969) 266–271.

[13] K.B. Freeman, P.V. Hariharan and H.E. Johns, The ultraviolet photochemistry of cytidylyl-(3′–5′) cytidine. J. Mol. Biol. 13 (1965) 833–848.

[14] H. Fukutome, A. Wada and Y. Kawade, A photometric analysis of ultraviolet photoreaction kinetics of transfer RNA. Radiat. Res. 37 (1969) 599–616.

[15] V.R. Glisin and P. Doty, The cross linking of DNA by ultraviolet radiation. Biochim. Biophys. Acta 142 (1967) 314–327.

[16] L. Grossman, Studies on mutagenesis induced in vitro. Photochem. Photobiol. 7 (1968) 727–735.

[17] L. Grossman and E. Rodgers, Evidence for the presence of cytosine photohydrates in UV irradiated nucleic acids. Biochem. Biophys. Res. Commun. 33 (1968) 975–983.

[18] P. Hanawalt and R. Setlow, Effect of monochromatic ultraviolet light on macromolecular synthesis in *Escherichia coli*. Biochim. Biophys. Acta *41* (1960) 298–294.

[19] P.V. Hariharen and H.E. Johns, Dimer photoproducts in cytidylyl-(3′–5′)cytidine. Photochem. Photobiol. *8* (1968) 11–22.

[20] M.A. Herbert, J.C. LeBlanc, D. Weinblum and H.E. Johns, Properties of thymine dimers. Photochem. Photobiol. *9* (1969) 33–43.

[21] J. Jagger, Introduction to Research in Ultraviolet Photobiology (Prentice-Hall, Englewood Cliffs, New Jersey, 1967).

[22] H.E. Johns, Dosimetry in photochemistry, Photochem. Photobiol., *8* (1968) 547–563.

[23] H.E. Johns, S.A. Rapaport and M. Delbrück, Photochemistry of thymine dimers. J. Mol. Biol. *4* (1962) 104–114.

[24] D. Kennell, Titration of the gene sites on DNA by DNA-RNA hybridization. II. The *Escherichia coli* chromosome. J. Mol. Biol. *34* (1968) 85–103.

[25] R.F. Kimball, Repair of premutational damage. Advan. Radiat. Biol. *2* (1966) 135–166.

[26] N.O. Kjeldgaard and C.G. Kurland, The distribution of soluble and ribosomal RNA as a function of growth rate. J. Mol. Biol. *6* (1963) 341–348.

[27] K.W. Kohn, C.L. Spears and P. Doty, Inter-strand crosslinking of DNA by nitrogen mustard. J. Mol. Biol. *19* (1966) 266–288.

[28] K.W. Kohn, N.H. Steigbigel and C.L. Spears, Cross-linking and repair of DNA in sensitive and resistant strains of *E. coli* treated with nitrogen mustard. Proc. Nat. Acad. Sci. US *53* (1965) 1154–1161.

[29] A.A. Lamola, Specific formation of thymine dimers in DNA. Photochem. Photobiol. *9* (1969) 291–294.

[30] A.A. Lamola and T. Yamane, Sensitized photodimerization of thymine in DNA. Proc. Nat. Acad. Sci. US *58* (1967) 443–446.

[31] L. Levine, E. Seaman, E. Hamerschlag and H. Van Vunakis, Antibodies to photoproducts of deoxyribonucleic acids irradiated with ultraviolet light. Science *153* (1966) 1666–1667.

[32] L. Levine, E. Seaman and H. Van Vunakis, Immunological evidence for the identity of a photoproduct formed during photooxidation of DNA with methylene blue, rose bengal, thionin, and acridine orange. Nucleic Acids in Immunology, eds.. D. Plecia and W. Braun (Springer-Verlag, New York, 1968) pp. 165–173.

[33] M.B. Lion, Search for a mechansim for the increased sensitivity of 5-bromouracil-substituted DNA to ultraviolet radiation. II. Single strand breaks in the DNA of irradiated 5-bromouracil-substituted T3 coliphage. Biochim. Biophys. Acta *209* (1970) 24–33.

[34] R.A. McGrath and R.W. Williams, Reconstruction in vivo of irradiated *Escherichia coli* deoxyribonucleic acid; the rejoining of broken pieces. Nature *212* (1966) 534–535.

[35] A.D. MacLaren and D. Shugar, Photochemistry of purine and pyrimidine derivatives. Photochemistry of Proteins and Nucleic Acids. (Pergamon Press, London, 1964) pp. 162–220.

[36] A. Matsukage, Y. Kawade and H. Fukutome, Kinetics of formation and splitting of uracil dimer in transfer RNA irradiated by ultraviolet. Radiat. Res. *37* (1969) 617–626.

[37] H.J. Morowitz, Absorption effects in volume irradiation of microorganisms. Science *111* (1950) 229–230.

[38] D.P. Nierlich, Radioisotope uptake as a measure of synthesis of messenger RNA. Science *158* (1967) 1186–1188.

[39] M. Pearson and H.E. Johns, Excision of dimers and hydrates from ultraviolet-irradiated poly U by pancreatic ribonuclease. J. Mol. Biol. *19* (1966) 303–319.

[40] M. Pearson and H.E. Johns, Suppression of hydrate and dimer formation in ultraviolet-irradiated poly(A + U) relative to poly U. J. Mol. Biol. *20* (1966) 215–229.

[41] R.O. Rahn and J.L. Hosszu, Photoproduct formation in DNA at low temperatures. Photochem. Photobiol. *8* (1968) 53–63.

[42] R.O. Rahn, J.K. Setlow and J.L. Hosszu, Ultraviolet inactivation and photoproducts of transforming DNA irradiated at low temperatures. Biophys. J. *9* (1969) 510–517.

[43] J.D. Regan, J.E. Trosko and W.L. Carrier, Evidence for excision of ultraviolet-induced pyrimidine dimers from the DNA of human cells in vitro. Biophys. J. *8* (1968) 319–325.

[44] W.D. Rupp and P. Howard-Flanders, Discontinuities in the DNA synthesized in an excision-defective strain of *Escherichia coli* following ultraviolet irradiation. J. Mol. Biol. *31* (1968) 291–304.

[45] E. Seaman, L. Levine and H. Van Vunakis, Antibodies to the methylene blue sensitized photooxidation product in deoxyribonucleic acid. Biochemistry *5* (1966) 1216–1223.

[46] E. Seaman, L. Levine and H. Van Vunakis, Immunological evidence for the existence of thymine dimers in ultraviolet irradiated DNA. Nucleic Acids in Immunology, eds. O. Plescia and W. Braun (Springer-Verlag, New York, 1968) pp. 157–164.

[47] J.K. Setlow, The molecular basis of biological effects of ultraviolet radiation and photo-reactivation. Current Topics in Radiation Research, Vol. II, eds. M. Ebert and A. Howard (North-Holland, Amsterdam, 1966) pp. 195–248.

[48] J.K. Setlow, The effects of ultraviolet radiation and photoreactivation. Comprehensive Biochemistry, Vol. 27, eds. M. Florkin and E.H. Stotz (Elsevier, Amsterdam, 1967) pp. 157–209.

[49] R. Setlow, The action spectrum for the reversal of dimerization of thymine induced by ultraviolet light. Biochim. Biophys. Acta *49* (1961) 237–238.

[50] R.B. Setlow, Cyclobutane-type pyrimidine dimers in polynucleotides. Science *153* (1966) 379–386.

[51] R.B. Setlow, Repair of DNA. Regulation of Nucleic Acid and Protein Biosynthesis, eds. V.V. Koningsberger and L. Bosch (Elsevier, Amsterdam, 1967) pp. 51–62.

[52] R.B. Setlow, Photoproducts in DNA irradiated in vivo. Photochem. Photobiol. *7* (1968) 643–649.

[53] R.B. Setlow, The photochemistry, photobiology, and repair of polynucleotides. Advances in Nucleic Acid Research and Molecular Biology, Vol. 8, eds. J.N. Davidson and W.E. Cohn (Academic Press, New York, 1968) pp. 257–295.

[54] R.B. Setlow and W.L. Carrier, Identification of ultraviolet-induced thymine dimers in DNA by absorbance measurements. Photochem. Photobiol. *2* (1963) 49–57.

[55] R.B. Setlow and W.L. Carrier, Pyrimidine dimers in ultraviolet-irradiated DNAs. J. Mol. Biol. *17* (1966) 237–254.

[56] R.B. Setlow and W.L. Carrier, unpublished experiments.

[57] R.B. Setlow, W.L. Carrier and F.J. Bollum, Nuclease-resistant sequences in ultraviolet-irradiated deoxyribonucleic acid. Biochim. Biophys. Acta *91* (1964) 446–461.

[58] R.B. Setlow, W.L. Carrier and F.J. Bollum, Pyrimidine dimers in UV-irradiated poly dI : dC. Proc. Nat. Acad. Sci. US *53* (1965) 1111–1118.

[59] R.B. Setlow, J.D. Regan, J. German and W.L. Carrier, Evidence that Xeroderma pigmentosum cells do not perform the first step in the repair of ultraviolet damage to their DNA. Proc. Nat. Acad. Sci. US *64* (1969) 1035–1041.

[60] M.I. Simon, L. Grossman and H. Van Vunakis, Photosensitized reactions of polyribonucleotides. I. Effects on their susceptibility to enzyme digestion and their ability to act as synthetic messengers. J. Mol. Biol. *12* (1965) 50–59.

[61] G.D. Small, M. Tao and M.P. Gordon, Pyrimidine hydrates and dimers in ultraviolet-irradiated tobacco mosaic virus ribonucleic acid. J. Mol. Biol. *38* (1968) 75–87.

[62] K.C. Smith, Dose-dependent decrease in extractability of DNA from bacteria following irradiation with ultraviolet light or with visible light plus dye. Biochem. Biophys. Res. Commun. *8* (1962) 157–163.

[63] K.C. Smith, Photochemistry of the nucleic acids. Photophysiology, Vol. II, ed. A.C. Giese (Academic Press, New York, 1964) pp. 329–388.

[64] K.C. Smith, Biologically important damage to DNA by photoproducts other than cyclobutane-type thymine dimers. Radiation Research 1966, ed. G. Silini (North-Holland, Amsterdam, 1967) pp. 756–770.

[65] K.C. Smith, B. Hodgkins and M.E. O'Leary, The biological importance of ultraviolet light induced DNA-protein cross-links in *Escherichia coli* 15 TAU. Biochim. Biophys. Acta *114* (1966) 1–15.

[66] K.C. Smith and M.E. O'Leary, Photoinduced DNA protein cross-links and bacterial killing: A correlation at low temperatures. Science *155* (1967) 1024–1026.

[67] K.C. Smith and M.E. O'Leary, The pitfalls of measuring DNA synthesis kinetics as exemplified in ultraviolet radiation studies. Biochim. Biophys. Acta *169* (1968) 430–438.

[68] R.S. Stafford and J.E. Donnellan Jr., Photochemical evidence for conformation changes in DNA during germination of bacterial spores. Proc. Nat. Acad. Sci. US *59* (1968) 822–829.

[69] B.M. Sutherland and J.C. Sutherland, Mechanisms of inhibition of pyrimidine dimer formation in deoxyribonucleic acid by acridine dyes. Biophys. J. *9* (1969) 292–302.

[70] K. Suzuki, E. Moriguchi and Z. Horii, Stability of DNA in *Escherichia coli* B/r and B_{s-1} irradiated with ultraviolet light. Nature *212* (1966) 1265–1267.

[71] P.A. Swenson and R.B. Setlow, Kinetics of dimer formation and photohydration in ultraviolet-irradiated polyuridylic acid. Photochem. Photobiol. *2* (1963) 419–434.

[72] P.A. Swenson and R.B. Setlow, Effects of ultraviolet radiation on macromolecular synthesis in *Escherichia coli*. J. Mol. Biol. *15* (1966) 201–219.

[73] R.P. Van Tubergen, Radio-autographic studies on *Escherichia coli*. Ph. D. thesis, Yale University (1959).

[74] A.J. Varghese and M.H. Patrick, Cytosine derived heteroadduct formation in ultraviolet-irradiated DNA. Nature *223* (1969) 299–300.

[75] A.J. Varghese and S.Y. Wang, Ultraviolet irradiation of DNA *in vitro* and *in vivo* produces a third thymine-derived photoproduct. Science *156* (1967) 955–957.

[76] A. Wacker, Molecular mechanisms of radiation effects. Progress in Nucleic Acid Research, Vol. 1, eds. J.N. Davidson and W.E. Cohn (Academic Press, New York, 1963) pp. 369–399.

CHAPTER 6

CHEMI- AND BIOLUMINESCENCE

E.J. BOWEN

Physical Chemistry Laboratory,
South Parks Road, Oxford, England

Visible radiation can be emitted, as fluorescence or phosphorescence, only from an electronically excited molecule. All strong examples of chemiluminescence must have a fluorescent emitter, and one of the tasks in investigating the subject is to identify it. The emitter may be a newly formed product molecule or may be excited by energy transfer from some energised molecule itself incapable of radiating; in that case the presence of other fluorescent molecules in the system, as impurities or deliberatly added, or even the reactant, if fluorescent, may allow of chemiluminescence.

The over-all light yield of chemiluminescence is a product of several factors, the fluorescence quantum yield of the emitter, a factor reducing this because of the presence of oxygen or other fluorescence quenchers, an energy transfer factor if such is needed, and the fraction of molecules chemcially acquiring excitation energy. A further complication is that since light yields are small, and sometimes very small, there is the possibility that the emission may not be associated with the main reaction but come from some side reaction. Most chemiluminescent reactions are vigorous oxidations with complex intermediate stages difficult to sort out. The additional fact that fluorescence spectra rarely provide adequate information for unambiguous identification reinforces the obvious conclusion that the solution of chemiluminescence problems faces formidable difficulties.

Modern photomultipliers are so sensitive that they will detect one photon per 10^{12} or more molecules reacting. Very weak chemiluminescent emissions of obscure nature are sometimes observed. At this level of measurement it is very difficult to avoid effects due to impurities; for example, luminescence from what is presumably the autoxidation of grease can be detected from heated glass-ware contaminated by finger prints. This field therefore is one where great caution is required in interpreting the results of observations.

Some examples of types of chemiluminescent reactions are given below.

(1) Recombination of radical ions gives luminescence if the products are fluorescent.
Anthracene (A) electrolysed in dimethylsulphoxide solution gives A^+ and A^-, and these have enough energy to recombine to A + singlet excited A, (A*), or to the excimer $A:A^*$, followed by A* or $A:A^* \to h\nu$ [1].

(2) Lucigenine (NN′dimethyl biacridylium nitrate) is oxidised by H_2O_2 with emission from the methyl acridone product:

Dication   N methyl acridone

One of the product molecules appears in its electronically excited state, and being strongly fluorescent, emits luminescence [2]. The formation of excited ketone molecules seems to be a basic reaction in many examples of chemi- and bioluminescence.

(3) Tetrakis (dimethylamino) ethylene readily takes up oxygen from the air and emits a strong green luminescence of long duration. The substance is proposed for use in producing 'chemical light' for rescue work in the dark, etc. The emission is apparently identical with the fluorescence of the reactant itself. Whether this is because of recombination of half-molecule radicals, or, as seems more likely, energy transfer occurs from the product dimethylamino urea in an excited state, is not settled. The oxidation mechanism is complex and involves many stages and side branches [3].

(4) Warm ethyl benzene (RH) autoxidises by a chain reaction initiated by radicals R−, and continued by peroxy-radicals ROO−. Ending of reaction chains by the process:

$$2 \text{ ROO}- \rightarrow \text{RHOH} + O_2 + R = O$$

leaves the product R = O (acetophenone) in an electronically excited state. This molecule has poor radiative powers, but if the fluorescent molecule diphenyl anthracene (stable against oxidation) is added to the reacting system energy transfer to it occurs and gives rise to its characteristic emission [4].

(5) Base-catalysed oxidations of esters or breakdown of acyl peroxides often show luminescence. Electronically excited ketonic products appear to be formed, and these may either radiate or transfer their energy to some fluorescent additive [5].

excited ketone

A reaction of this type is that between oxalyl chloride and hydrogen peroxide, with the over-all change:

$$\begin{matrix} COCl \\ | \\ COCl \end{matrix} + H_2O_2 \rightarrow 2\,HCl + 2\,CO_2$$

If diphenyl anthracene is present a strong luminescence occurs. The final product are most unlikely to be concerned in the energy transfer; intermediate peroxides or peroxy-radicals must be involved, but it is difficult to understand the effect in terms of recognisable electronic states of these [6].

(6) Certain reactions liberate oxygen in excited singlet states, from which radiational return to the ground state is 'forbidden'. Association in pairs of such excited molecules takes place, and radiation to the ground state by emission of a double-sized quantum then becomes 'allowed', the effect appearing as a red glow. Energy transfer from such pairs to other molecules is conceivable, but does not seem to be a common cause of chemiluminescence in organic systems [7].

(7) In the field of bioluminescence the problems of characterisation of the emitter and the elucidation of the reaction chemistry become more difficult. The reactions are enzymic oxidations with all the complexities of co-enzymes, presence of certain inorganic ions, etc. usual in biochemical changes. Recent work has isolated fluorescent emitters from two systems, showing that (a) such emitters have quite complicated chemical structures which are entirely different for different organisms, (b) the emitters readily undergo reversible chemical changes which are not simple oxidation-reductions, and (c) they are sensitive to changes in pH and to binding effects on the enzyme sites so that the spectral emissions show variable features [8].

Three examples have been intensively studied.

Firely: A substance of structure:

which complexes with an enzyme, and in presence of ATP, Mg^{++} and oxygen is oxidised and decarboxylated to:

This fluorescent product appears in its electronically excited state and radiates. The over-all light yield of the process is believed to approach 100%, far exceeding that of any chemiluminescent reaction in solution, and a very delicate test for ATP can be devised from it. By liberating ATP the animal can control its light flashes.

Cypridina: Here the emitter has the formula $C_{22}H_{27}O_2N_7$, and the structure:

where the R's are chains or rings containing nitrogen. The enzymic oxidation has a light yield of about 0.3. The substance can also be oxidised in dimethylsulphoxide solution, giving an emission somewhat different in spectrum, the difference being attributable to enzyme complex formation in the biochemical reaction.

Bacteria: The emitter is a flavin derivative. Reduced flavin mononucleotide ($FMNH_2$) reacts with the enzyme and with oxygen, and a bound form of FMN if produced in its electronically excited state. The enzyme reaction is greatly enhanced by the presence of an aldehyde which enters the reaction at its later stages, and by oxidation provides extra chemical energy for excitation.

It is interesting to speculate on the use of bioluminescence to an organism. It may be a way of disposing of surplus energy, or may just be the accidental result of fluorescent substances becoming involved in oxidative processes.

References

[1] C.A. Parker and G.D. Short, Trans. Farad. Soc. *63* (1967) 2618.
[2] J.R. Totter and G.E. Philbrook, Photochem. Photobiol. *5* (1966) 177.
[3] J.P. Paris, W.H. Urry and J. Sheeto, Photochem. and Photobiol. *4* (1965) 1059, 1067.
[4] R.F. Vassil'ev, Progress in Reaction Kinetics *4* (1967) 305.
[5] F. McCapra et al., Quart. Rev. *20* (1966) 485; Chem. Commun. (1966) 522, (1967) 1011.
[6] M.M. Rauhut et al., Photochem. Photobiol. *4* (1965) 1097.
[7] A.U. Khan and M. Kasha, J. Amer. Chem. Soc. *88* (1966) 1574; D.R. Kearns and A.U. Khan, Photochem. Photobiol. *10* (1969) 193.
[8] J.W. Hastings, Ann. Rev. Biochem. *37* (1968) 597; W.D. McElroy, H.H. Seliger and E.H. White, Photochem. Photobiol. *10* (1969) 153. For a complete general account see: K.D. Gundermann, Chemiluminescence (Springer, Berlin, 1968).

SECTION II

MECHANISMS OF PHOTODYNAMIC ACTION

CHAPTER 7

MOLECULAR MECHANISMS OF PHOTODYNAMIC COMPOUNDS

Adolf WACKER

Institut für Therapeutische Biochemie der
Universität Frankfurt a. M., West Germany

1. Introduction

The scientific basis for the action of light was recognized at the end of last century. This began with the pioneering work of Finsen [9, 10], whose therapeutic use of light, and other photobiological studies aroused widespread interest among biologists and physicians. Raab [37] soon reported that light in the presence of a sensitizer damaged living organisms. This destructive action of light was termed by him and his teacher von Tappeiner as 'photodynamische Erscheinung', or photodynamic action. Thus the action of light has two aspects: its usefulness and its destructive action. Recent developments in molecular genetics have made use of light as a 'repair' agent for UV damages. This field has progressed astonishingly in the last ten years. The process by which light repairs UV damages is known as 'photoreactivation'. Our recent results show that in addition to UV damages, photosensitized damages can also be repaired by light under similar conditions.

Photodynamic reactions, by definition, involve the participation of a sensitizer which absorbs the light energy involved in the reaction. Thus the chemical nature of the sensitizer plays a vital role in the effectiveness of the photodynamic action. During the last years, a number of compounds have been studied which in the presence of visible light or near UV light act as sensitizers. These compounds can be classified on the basis of their chemical nature into specific groups. The most commonly investigated are cancerogenic hydrocarbons [18, 24, 40, 44, 53], acridine derivatives [19, 57, 58], phenothiazine derivatives [56, 47, 52, 53, 57, 58], porphyrins [39, 41, 42], furocoumarins [5, 6, 13, 26, 30, 31, 51, 53] and ketones [4, 8, 27, 51, 55]. Besides the literature cited here, there have been various reports on these, and other types of sensitizers; but space does not permit all of them to be mentioned here.

Molecular specificity exists among sensitizers belonging to the same group, for example, substitution at C atoms or N atoms in the ring or elsewhere may completely inactivate the photodynamic response of a sensitizer. According to the specific reaction velocities reported by Yamamoto [57], the following order of increasing effectiveness as photosensitizers against bacteriophages becomes apparent in the anthracene-like compounds with meso atoms as indicated: $C,C < C,O \leq N,N < N,O \leq C,N < N,S$ [12]. In the case of psoralens, substitution of nitro or amino group at the 5 position results in loss of its tanning activity [30]. A similar loss of activity is observed in bacterial systems [13, 53] and cell-free systems [5, 6]. The studies with furocoumarins will be described in detail.

Among the dyes a structural specificity has been observed between methylene blue, thiopyronin and pyronin. The photodynamic effectiveness of these compounds was first compared by Herzberg [21] in vaccinia virus inactivation. The effects were repeated on bacterial systems [53] and on macromolecules [5].

The effect of structural specificity on the photodynamic action of ketones has been studied by Wilucki, Matthäus and Krauch [55]. These authors have compared the photodynamic activities of acetone, propiophenone, acetophenone and benzophenone on the formation of cyclic dimers of thymine. Among these sensitizers benzophenone was found to be the weakest.

2. Photodynamic action of dyes

Several studies have been made of the dye-sensitized photoautooxidation of biopolymers and their components. The importance of proteins as possible sites of dye-sensitized damages was recognized as early as 1904 by Jodlbauer and von Tappeiner [23]. They observed that the enzymes diastase, invertase and papain are inactivated on illumination in the presence of eosin. The fact that this photodynamic damage to proteins results from the destruction of amino acids was shown by Gaffron [14], Harris [20], Carter [3], and others. These authors found that most of the aromatic amino acids are quite susceptible to photodynamic action, whereas the aliphatic amino acids are oxidized very slowly, if at all (see Vodrazka [49]). Later investigations, e.g. by Goda [16], Galston [15], Shugar [46], Weil et al. [54], Isaka and Kato [22] and Wacker and Türck (unpublished results) showed that histidine, methionine, tryptophan, and tyrosine are photooxidized very rapidly with both methylene blue, and riboflavin (see Spikes and Straight [48]). The photodynamic behaviour of thiopyronin is similar to that of methylene blue, except that under the same conditions the former is more potent [53].

In general, one could say that the photodynamic sensitivity of proteins to dyes is dependent on the molecular nature of certain amino acids. These amino acids, due to their structural characteristics, are very susceptible to photosensitized oxidation by dyes. A protein consisting of such amino acids, will evidently, be more susceptible to the photodynamic action of dyes. Very similar to this phenomenon is the dye-sensitized destruction of nucleic acids and their structural components. These studies are described in the succeeding pages. From a number of dyes, that we have investigated, we will compare, below, the photodynamic activities of three selected dyes with similar structures.

Replacement of the ring nitrogen in the methylene blue molecule by carbon gives the structure of thiopyronin (fig. 1). The latter is an even more potent sensi-

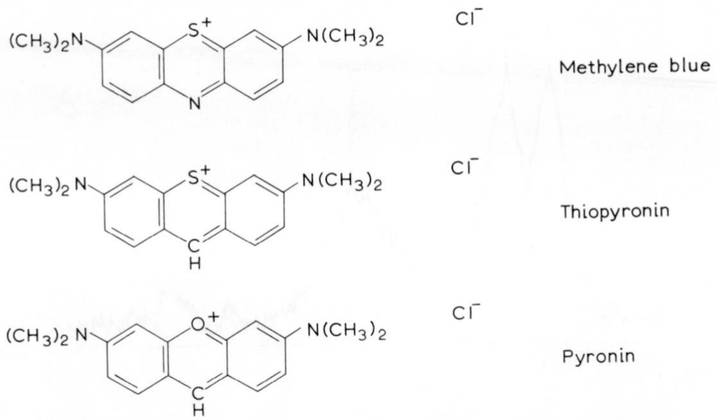

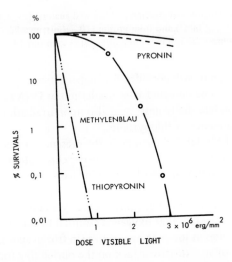

(CH₃)₂N⟶[phenothiazine ring with S⁺ and N]⟶N(CH₃)₂   Cl⁻   Methylene blue

(CH₃)₂N⟶[ring with S⁺ and C-H]⟶N(CH₃)₂   Cl⁻   Thiopyronin

(CH₃)₂N⟶[ring with O⁺ and C-H]⟶N(CH₃)₂   Cl⁻   Pyronin

Fig. 1. Chemical structures of methylene blue, pyronin and thiopyronin.

tizer than methylene blue for the inactivation of bacteria (fig. 2) and viruses by visible light. Pyronin, whose structure can be derived from thiopyronin by exchange of sulfur for an oxygen atom, shows less photodynamic activity. This photodynamic action can be correlated to chemical changes in nucleic acids produced by visible light in the presence of the dyes.

On irradiation of a DNA solution containing methylene blue with a daylight source in the presence of oxygen guanine is preferentially destroyed, whereas the other bases are practically unaffected [47]. Thiopyronin in equal weight concen-

Fig. 2. Photodynamic inactivation of *E. coli* 15T⁻ by dyes. Cells were irradiated at a conc. 1×10^5 cells/ml saline in the presence of dyes (5 μg/ml).

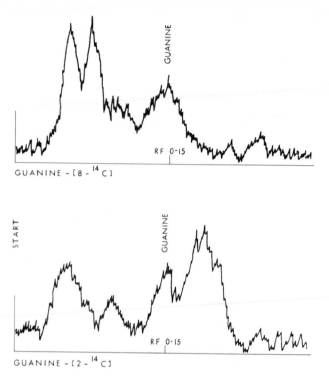

Fig. 3. Radio-paperchromatograms of guanine-[2-^{14}C] and guanine-[8-^{14}C] after irradiation with visible light (3 × 10^7 erg/mm^2) and thiopyronin (10 μg/ml); guanine conc. 1 μg/ml. Chromatograms were run in n-butanol/water system (86:14, v/v).

tration reacts similarly, but with considerably higher efficiency. In contrast no destruction of any base can be detected after irradiation of DNA in the presence of pyronin. Guanine is preferentially destroyed also upon irradiation of nucleosides or the free bases in the presence of thiopyronin.

After irradiation of ^{14}C-labeled guanine in the presence of thiopyronin several radioactive photoproducts can be found using paper chromatography, some of which are different depending on whether guanine has been labeled at 2-C or 8-C position (fig. 3).

The photochemical alteration of guanine occurs by a radical mechanism. This can be concluded from the fact that oxygen is essential to this reaction, and that the reaction rate decreases at lower temperatures (in frozen state). The photodynamic action consists of an oxidative attack on the purine ring followed by a further degradation. This degradation depends very strongly on the kind of substitutions in the purine structure.

Table 1

Photodynamic decomposition of various purines by thiopyronin

Compound	Decomposition (%)	Substituent in position	
		2	6
Guanine	80	$-NH_2$	$-OH$
2-Aminopurine	78	$-NH_2$	$-H$
2,6-Diaminopurine	64	$-NH_2$	$-NH_2$
Isoguanine	44	$-OH$	$-NH_2$
Xanthine	30	$-OH$	$-OH$
Hypoxanthine	3	$-H$	$-OH$
Adenine	4	$-H$	$-NH_2$
8-Azaguanine	1	$-NH_2$	$-OH$

Dose of visible light: 7×10^6 erg/mm^2. Thiopyronin concentration: 10 μg/ml. Purine concentration: 10 μg/ml. Per cent decomposition determined by percent decrease of UV absorption.

Table 1 presents the results of irradiation experiments on a series of purine compounds in the presence of thiopyronin. In 8-azaguanine, where the carbon atom in 8 position is replaced by nitrogen, no photochemical oxidation occurs. Furthermore, substitution at the 2-C position is of great importance. An amino group there is more effective than a hydroxy group, almost no reaction takes place without some substituent at this position. Accordingly, adenine and hypoxanthine are affected only to a very small extent.

This differential degradation of bases occurs also in bacteria. Bacteria containing radioactively labeled purine and pyrimidine were irradiated, in the presence of thiopyronin, with visible light at 10^6 erg/mm^2. Following perchloric acid hydrolysis only guanine degradation could be detected by paper chromatography. At higher doses, for example 1.5×10^7 erg/mm^2, about 8 percent of the thymine and about 4 percent of the cytosine are also destroyed, whereas adenine and uracil are unaffected. It can be concluded that the different photodynamic activities of the three dyes are reflected in their varying potencies degrading guanine in vivo and in vitro.

Thiopyronin and methylene blue in combination with visible light also exhibit a mutagenic effect on bacteria [1]. This effect was studied with two mutants of *Proteus mirabilis* strain VI/str-d-3 (streptomycin/dependent) and strain VI/99 (met⁻phen⁻). The frequency of mutation from streptomycin dependence to independence is induced by thiopyronin and by methylene blue to the same extent. However, for the reversion of phenylalanine auxotrophy to prototrophy thiopyronin is more effective than methylene blue.

In addition thiopyronin was compared to methylene blue with respect to photodynamic inactivation of enzymes. Lactate-dehydrogenase activity is reduced by 50 percent after irradiation with about 4×10^6 erg/mm^2 visible light in the presence of thiopyronin or of methylene blue (5 μg/ml). At a dye concentration of 10 μg/ml thiopyronin is a little more effective than methylene blue, and at a concentration

of 1 µg/ml or less both the dyes are inactive. We find a similar situation in the photodynamic inactivation of pyruvate kinase. Neither thiopyronin nor methylene blue
show any inactivating effect on both enzymes in the dark. The enzyme activity of
catalase is markedly lowered on irradiation even in the absence of photodynamic
agents. Irradiation in the presence of dyes causes additional inactivation. Thiopyronin is twice as active as methylene blue. Since the visible light used in our experiments has a wide range of wavelengths, and the distribution of energy is uneven
(see the chapter by Chandra), one could question the action spectrum of both the
compounds. Action spectrum is essentially the wavelength at which a maximum
effect is achieved. This wavelength usually corresponds to a maximum absorption
by the sensitizer. Spectrophotometric studies showed that at 585 nm both the dyes
have the same absorption. Repetition of our experiments using a monochromatic
light of 585 nm did not alter their potencies, i.e., thiopyronin was always more effective than methylene blue.

3. Photodynamic action of carcinogenic hydrocarbons

The problem of a possible correlation between photodynamic action and carcinogenicity by polycyclic hydrocarbons has been discussed by Santamaria [40], Graffi
[17], Mottram and Doniach [29] and others. Santamaria [40] found that the accel-
eration of polycyclic hydrocarbon carcinogenesis by light is dependent on a definite
optimum concentration of photodynamic substance and the intensity of radiation.
Using a microbial system as a model, the molecular mechanism of photodynamic
action of polycyclic hydrocarbons was studied.

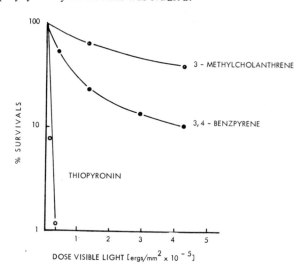

Fig. 4. Photodynamic inactivation of *E. coli* 15T⁻ by polycyclic hydrocarbons (20 µg/ml, suspension).

In fig. 4 the inactivation of bacterial cells after irradiation in the presence of methylcholanthrene, 3,4-benzpyrene and thiopyronin is presented. This photodynamic effect suggests that the polycyclic hydrocarbons destroy the nucleic acid bases as does thiopyronin. After irradiation of a DNA solution (6 mg DNA, 2.5 mg hydrocarbon, 3 ml water and 1 ml acetone) with 10^8 erg/mm^2 visible light, we find a preferential loss of guanine (70 percent) in the presence of either 1,2,5,6-dibenzanthracene or 3-methylcholanthrene, and of adenine (40%) if 3,4-benzpyrene is present. With this latter compound guanine is attacked only at higher energy doses. Pyrimidines under these conditions are not significantly affected. The photodynamic action of these hydrocarbons also requires oxygen as does that of thiopyronin. It is now well known that the polycondensed hydrocarbons have especially reactive centers, for example, irradiation of anthracene results in the addition of oxygen in the 9,10 positions, thus forming an endocyclic peroxide. It may be postulated that the hydrocarbons mentioned here also form oxygen containing intermediary products which act on purine bases by oxidative cleavage. Photodynamic activity of benzpyrene can also be detected when the hydrocarbon containing solution is first irradiated and then added to the free bases, nucleosides or bacterial cells in the dark.

4. Photodynamic action of furocoumarins

The furocoumarins are a well known group of natural substances, some of which exhibit interesting biological properties. At present the best known effect is skin photosensitization [11, 31, 32, 37] : the active furocoumarins when applied to human or guinea-pig skin which is later exposed to sunlight or long wavelength ultraviolet radiation, provoke the formation of erythemas followed by a dark pigmentation [31, 32]. Some furocoumarins have been used in the treatment of vitiligo. Since ancient times the fruits of *Ammi majus* and the seeds of *Psoralea corylifolia* have been employed in popular Egyptian and Indian practices to cure leucodermic spots. Both these vegetable drugs contain very active furocoumarins. In more recent times the treatment of vitiligo was carried out by use of pure substances. Experiments were made by Fitzpatrick et al. [11] on the use of xanthotoxin as an oral drug to increase the tolerance of human skin to sun light. Besides these skin-photosensitizing properties, furocoumarins may exert some other photosensitizing effects which have been studied in recent years, such as the lethal [13, 35, 51, 53], and mutagenic effect exerted on bacteria cultures (Mathews [28] ; Wacker and Kraft, unpublished results), the formation of giant cells by mammalian cells [7], the inactivation of some DNA viruses, and so on.

The mechanism by which furocoumarins exert their photodynamic action has been elucidated only lately [26, 34, 51]. While the photodynamic action of compounds mentioned in sections 2 and 3 is oxygen dependent, that of furocoumarins does not require oxygen. Oginsky and Fowlks (see Fowlks [12]) have observed that in the photosensitized action of 8-methoxy-psoralen on bacteria the presence of oxygen had an inhibitory effect. The photoinactivating effect on *E. coli* 15T$^-$ of psoralen is similar to that of thiopyronin (fig. 5).

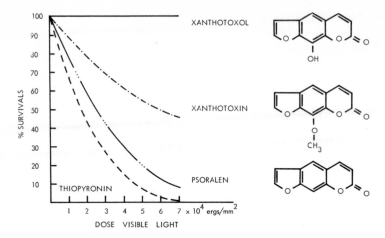

Fig. 5. Photodynamic inactivation of *E. coli* 15T⁻ by furocoumarins (5 µg/ml).

If the molecular structure of psoralen is altered, e.g. by substitution of the hydrogen in position 8-C by a methoxy group (xanthotoxin) the photodynamic activity decreases by about 50%; a hydroxy group in this position (xanthotoxol) leads to a complete loss of photodynamic activity. These different photosensitizing effects are in accordance with the tanning effects on skin [30, 37] and the template

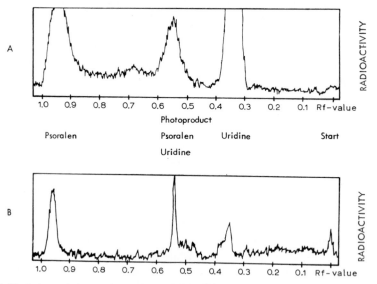

Fig. 6. Photodynamic interaction between ³H-uridine and psoralen-[4-¹⁴C] . The photoproduct (*Rf* = 0.53) was separated by paper chromatography using n-butanol/acetic acid/water system (4:1:5, v/v). (A) = irradiation at 365 nm, (B) = re-irradiation of the photoproduct at 315 nm.

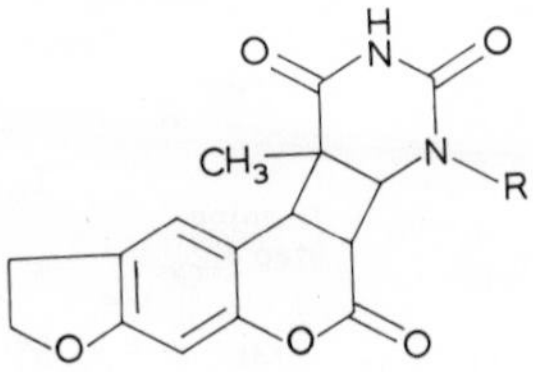

Fig. 7. Chemical structure of the photoadduct of thymine and psoralen.

activity of DNA [5] ; psoralen is the most active agent followed by xanthotoxin and xanthotoxol respectively.

The molecular mechanism of these compounds was studied by their interaction with nucleic acids in the presence of light of 365 nm. When an aqueous solution containing DNA and an active furocoumarin is irradiated at 365 nm, a photoreaction takes place with the formation of a stable chemical linkage between the two substances. These studies were carried out using psoralen-[4-^{14}C] ; the radioactivity was eluted in the DNA peak separated by gel-chromatography. To locate the structural element responsible for psoralen binding, homopolymers of adenylic acid, uridylic acid and cytidylic acid were irradiated with the sensitizer. We found that labeled psoralen binds preferentially to the uracil of polyuridylic acid.

To ascertain the nature of the photoproduct ^{3}H-uridine was irradiated with psoralen-[4-^{14}C] at 365 nm. The reaction mixture was separated by paper chromatography using n-butanol : acetic acid : water [4 : 1 : 5 v/v] system. Under these conditions the photoproduct emerges at *Rf* 0.53, having an isotopic ratio between uridine and psoralen 1 : 1. This shows that the product is 1 : 1 adduct between the two substances (fig. 6a). Re-irradiation of this adduct at 315 nm leads to its monomerization (fig. 6b). Spectrophotometric data and acid-stability suggest that it is a C-4 cycloadduct (fig. 7). Similar observations have been made by Musajo and Rodighiero [33].

5. Photodynamic action of ketones

The electronic mechanism by which pyrimidine dimers are formed when DNA is irradiated is not yet completely understood. This problem can be attacked by studying the effects of the excitation energy transfer occurring either from a sensitizer to DNA or its isolated constituents, or from DNA to a quencher. Experiments were performed where acetone, acetophenone or benzophenone have been used to sensitize the action of light of 315 nm [4, 8, 27, 51, 55]. It has been assumed that the excitation energy transfer which eventually will give rise to a photochemical lesion takes place at the triplet level [25].

Since UV light of 315 nm does not excite pyrimidines, it is obvious that the ketone molecule plays a role in transferring energy. In particular, the carbonyl group is responsible for the photosensitive action. From the fact that the light of

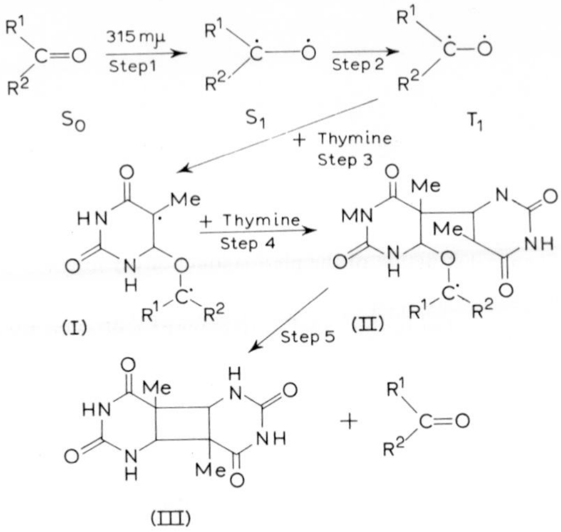

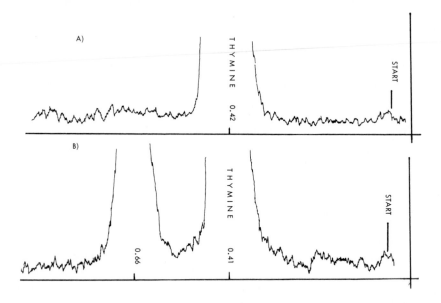

Fig. 8. Reaction scheme of the photosensitized dimerization of pyrimidines by acetone.

Fig. 9. Photosensitized dimerization of thymine in DNA. Thymine-labelled DNA was isolated from *E. coli* 15T⁻ cells. This DNA was dissolved in water (5 optical units/ml/appx. 10^6 cpm) and irradiated at 315 mμ (1.15×10^5 erg/mm²) in the presence of 10% acetone. This was applied on an IRC-50paper (Serva, Heidelberg) and the chromatogram run in an ascending manner using 0.1 M acetic acid (pH adjusted to 4.5 with NH₄OH as developer. Ordinate: radioactivity (cpm). (A) without acetone, (B) with 10% acetone.

315 nm is used in the experiments and that the efficiency of photodimerization is low, the authors [25] conclude that the n$-\pi$* transitions are those which promote carbonyl groups from the S_0 to the S_1 state (step 1, see fig. 8). Step 2 is just a rapid radiationless transition from S_1 to T_1. The triplet state rather than the singlet state takes part in the reaction of the sensitizer with the pyrimidine base. Thus the triplet state is responsible for the reaction of the excited carbonyl group with the pyrimidine molecule in its ground state. The attack can take place at only one position of the thymine ring. The remaining two electrons have parallel spins and therefore repel each other. Thus, it can be regarded as a pseudoradical reaction (step 3). Complex (I) behaves like a biradical. It reacts like a radical with another thymine molecule (step 4) forming an unstable complex (II) which is converted into a dimer (III). Besides compound (III), other stable species or oxetans have also been found [55], thus indicating that the other reactions are also possible but less probable.

The data of the sensitized reaction experiments show that the photoproducts in DNA consist almost exclusively of thymine containing dimers. These studies were carried out by irradiating thymine-labelled DNA, isolated from *E. coli* 15T⁻ cells grown on thymine-[2-^{14}C], at 315 nm in the presence of acetone. Paper chromatography of the hydrolyzed DNA shows a photoproduct, which on re-irradiation at 240 nm gives rise to thymine (fig. 9).

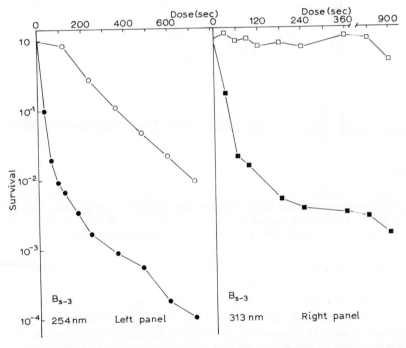

Fig. 10. Photosensitized inactivation and reactivation of *E. coli* B_{s-3}. Left panel: irradiation at 254 nm. Right panel: irradiation at 313 nm in the presence of 10% acetone. Closed symbols: without photoreactivation; open symbols: with maximum photoreactivation.

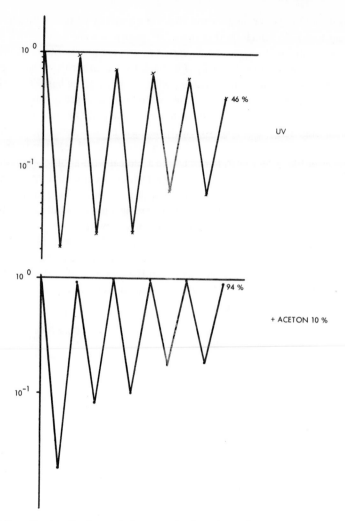

Fig. 11. Alternating inactivation and photoreactivation of *E. coli* B_{s-3}. Upper panel: irradiation at 254 nm. Lower panel: irradiation at 313 nm in the presence of 10% acetone. Closed symbols: after irradiation; open symbols: after maximum photoreactivation.

These findings suggest a new method of discovering the nature of a particular lesion and its biological significance. Pyrimidines on irradiation by UV light undergo several photochemical reactions. As a result the specific role of dimers on UV irradiation is not easy to predict. The part played by the total pyrimidine photoproducts, as opposed to thymine dimers alone, in biological damage can be studied by determining the extent to which they are split by the photoreactivating enzyme or are excised by the dark repair enzyme system. However, using the acetone-sensitized reaction one can study the biological effects of a particular lesion, namely thymine dimers, and their susceptibility to repair processes. The studies on this aspect were carried out in collaboration with Dr. H.D. Mennigmann.

The dose–response curves for the loss of colony forming ability of *E. coli* B_{s-3} are shown in fig. 10. The left panel of the curve represents inactivation at 254 nm, whereas the right panel shows photosensitized inactivation by acetone. Also included in the figure (open symbols) are the survival curves after a maximum photoreactivation (45 min at 37°C the irradiated culture was diluted to 10^{-2} before illumination). An obvious difference between the survival curves for the two types of irradiation is that with irradiation at 313 nm a plateau of about 0.5% survivors is reached. For irradiation at 254 nm no such plateau seems to exist. Maximum photoreactivation is lower after irradiation at 254 nm than at 313 nm. Cells were inactivated down to about 2 percent survivors, i.e. slightly above the extrapolation value. Then, maximum photoreactivation was allowed to take place. *E. coli* B_{s-3} tolerates 10 percent acetone for several hours without any loss in viability. But, after irradiation at 313 nm, further cells die if kept undiluted during photoreactivation at 37°C. This effect is not observed if the irradiated culture is diluted 10^{-2} or if undiluted cultures are kept at 30°C for not more than 15 min. Therefore, the latter procedure was used here.

The cells photoreactivated to a maximum level were reirradiated at the same dose. This cycling of irradiation and illumination was repeated several times. The results of such experiments are given in fig. 11 for inactivation at 254 nm and 313 nm, respectively. Closed symbols denote the inactivation after irradiation at 254 nm or 313 nm, whereas the open symbols signify the maximum photoreactivation.

The influence of irradiation lessens with increasing doses of light (which serves at the same time as photoreactivating and photoprotective light). Indeed survival after five successive doses is better than after the first irradiation. Furthermore, this effect is greater, and photoreactivation takes place more readily following irradiation at 313 nm than after 254 nm. This result is in favour of the hypothesis that survival of the fraction in question is due to dark repair.

6. Photochemical cleavage of thymine dimers

In the last section we showed that the dimerizarion of thymine, under photosensitized conditions or upon UV irradiation either of DNA or bacteria, is responsible for biological damages. Under conditions of photoreactivation cleavage of dimerized thymine has been established [43, 50, 56]. After irradiation with visible light in the presence of yeast extract the amount of thymine dimer in UV-irradiated DNA decreases about 10 percent [50].

There are also other conditions under which photochemical cleavage of thymine dimers occurs. Irradiation, with a daylight lamp, of thymine dimers in aqueous solutions in the presence of uranyl acetate, results in a conversion to thymine and some unidentified products (fig. 12). Irradiation without uranyl acetate, or with uranyl acetate in the dark are ineffective.

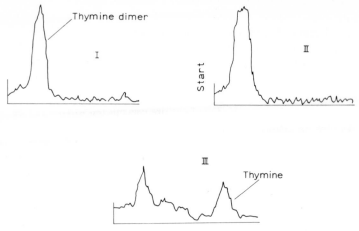

Fig. 12. Splitting of thymine dimers with uranyl acetate and visible light. I = thymine dimer treated with UAc (40 μg/ml). II = dimer irradiated with visible light. III = dimer treated with light and UAc.

The thymine dimer is not affected after addition of hydrogen peroxide and irradiation with visible light if the metal ions are bound by ethylene diamine tetra-acetate (EDTA). However, Cu^{++}, Co^{++}, or Fe^{+++} ions induce a photochemical destruction of thymine dimer with hydrogen peroxide (fig. 13) but thymine is not detected. If the energy of visible light is replaced by heat similar thymine dimer destruction is observed.

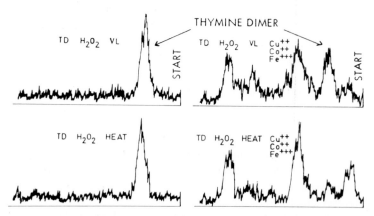

Fig. 13. Splitting of thymine dimer with hydrogen peroxide, metal ions, visible light or heat. Thymine dimer-$[2\text{-}^{14}C]$ was dissolved in water and irradiated with visible light $(1 \times 10^6$ erg/mm^2) or heat 10 min at 90°C in the presence of hydrogen peroxide (3 μg/ml) and Cu^{++}, Co^{++} and $FeCo^{++}$ and Fe^{+++} (3 μg/ml). Paperchromatogram: n-butanol/water (86:14).

Applying these results to UV-inactivated bacteria we observed that the photore-activating effect of visible light is enhanced in the presence of uranyl acetate or hy-drogenperoxide. The use of hydrogenperoxide and uranyl acetate at the cellular level, however, is limited by their toxicity to the cell.

7. Effect of caffeine and cyclic (3,5)-AMP on the dark repair of UV damage

The reversal of ultraviolet-induced biological damage by light and chemical com-pounds has been discussed in the earlier sections (sections 5 and 6). Bacterial cells under certain conditions can repair UV damage. The process, called 'dark repair', involves a number of enzymes. Bacterial cells which are highly sensitive to UV ac-tion are known to lack one, or several of these enzymes. The first dark repair sys-tem to be elucidiated at the molecular level was the so-called excision—repair sys-tem [2, 43]. It has been shown that the extracted dimers are complete, including the intervening phosphodiester bond.

It is known that the inhibition of dark repair, presumably enzymatic in nature by caffeine and theophylline can cause mutation in bacteria [45] and in mammalian

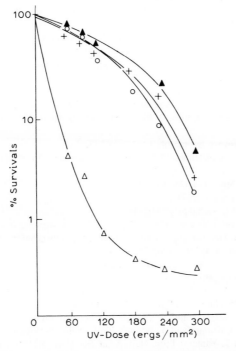

Fig. 14. Dark repair of UV-inactivated *E. coli* 15T⁻ in the presence of caffeine and cyclic AMP. Survival of *E. coli* 15T⁻ in the presence of caffeine (△————△), cyclic AMP (○————○), caffeine and cAMP (▲————▲) and the control (+————+).

cells [36]. In the mammalian system caffeine competitively inhibits phosphodi-esterase, which rapidly hydrolyzes cyclic AMP to AMP. This tempted us to examine the role of cyclic AMP in the dark recovery phenomenon of UV-inactivated *E. coli* 15T⁻ in the presence of caffeine. These studies were carried out by Dr. T.K.R. Reddy of this institute. We have found (fig. 14) that caffeine and theophyllin strongly inhibit the dark repair of UV-inactivated bacteria. This inhibition can be reversed by cAMP. The mechanism of this action is under investigation.

Acknowledgements

I thank the Deutsche Forschungsgemeinschaft and the Verband der Chemischen Industrie — Fonds der Chemie — for their financial support. The collaboration of my co-workers, cited in the literature, is gratefully acknowledged.

References

[1] H. Böhme and A. Wacker, Biochim. Biophys. Res. Commun. *12* (1963) 137.
[2] R.P. Boyce and P. Howard-Flanders, Proc. Natl. Acad. U.S. *51* (1964) 293.
[3] C.W. Carter, Biochem. J. *22* (1928) 575.
[4] P. Chandra, P. Mildner, H. Feller and A. Wacker, Europ. Symp. Photobiol., Hvar, Yugo-slavia (1967) 145.
[5] P. Chandra and A. Wacker, Z. Naturforschg. *21b* (1966) 663.
[6] P. Chandra and A. Wacker, Biophysik *3* (1966) 214.
[7] G. Colombo, A.G. Levis and V. Torlone, Progr. Biochem. Pharmacol. *I* (1965) 392.
[8] D. Elad, C. Krüger and G.M.J. Schmidt, Photochem. Photobiol. *6* (1967) 495.
[9] N.R. Finsen, Hospitalstidende *36* (1894) 1069.
[10] N.R. Finsen, Über die Anwendung des Lichtes (F.C.W. Vogel, Leipzig, 1899).
[11] T.B. Fitzpatrick, C.E. Hopkins, D. Blickenstoff and S. Swift, J. Invest. Dermatol. *25* (1955) 187.
[12] W.L. Fowlks, J. Invest. Dermatol. *32* (1959) 233.
[13] W.L. Fowlks, D.G. Griffith and E.L. Oginsky, Nature (Lond.) *181* (1958) 571.
[14] H. Gaffron, Biochem. Z. *179* (1926) 157.
[15] A.W. Galston, Science *111* (1950) 619.
[16] T. Goda, Seiri Seitai *I* (1947) 1.
[17] A. Graffi, Z. Krebsforschg. *50* (1940) 196.
[18] A. Graffi, J. Kriegel, E.J. Schreider and S.G. Sydow, Naturwiss. *40* (1953) 414.
[19] E. Graevskii, G.K. Ochinskaya and M.V. Shakk, Zhur. Obshchei. Biol. *13* (1952) 211.
[20] D.T. Harris, Biochem. J. *20* (1926) 288.
[21] K. Herzberg, Klin. Wochschr. *10* (1931) 1626.
[22] Isako and Kato, (1952).
[23] A. Jodlbauer and H. Von Tappeiner, Muench. Med. Wochschr. *26* (1904) 1139.
[24] H. Koeffler and I.L. Markert, Proc. Soc. Exp. Biol. Med. *76* (1951) 90.
[25] A. Kornhauser, J.N. Herak and N. Trinajastic, Chem. Commun. (1968) p. 1108.
[26] C.H. Krauch, D.M. Kramer and A. Wacker, Photochem. Photobiol. *6* (1967) 341.
[27] A.A. Lamola and T. Yamane, Proc. Natl. Acad. Sci. U.S. *58* (1967) 443.
[28] M.M. Mathews, J. Bacteriol. *85* (1963) 322.
[29] J.C. Mottram and J. Doniach, Nature (Lond.) *140* (1937) 933.
[30] L. Musajo, G. Rodighiero and S. Caporale, Bull. Soc. Chim. Biol. *36* (1954) 1213–1224.

[31] L. Musajo, Il Farmaco Ed. Sci. *10* (1955) 539.

[32] L. Musajo and G. Rodighiero, Experientia *18* (1962) 153.

[33] L. Musajo and G. Rodighiero, Acta Dermatol. Venereol. *47* (1967) 298.

[34] L. Musajo et al. Experientia *22* (1966) 75.

[35] E.L. Oginsky et al. J. Bacteriol. *78* (1959) 821.

[36] Ostertag, (1966).

[37] M.A. Pathak and T.B. Fitzpatrick, J. Invest. Dermatol. *32* (1959) 509.

[38] Raab, (1900)

[39] O. Reggeanini, Chem. Abstr. *34* (1940) 482.

[40] L. Santamaria, Recent. Contr. to Cancer Res. in Italy (1960) pp. 167–287.

[41] L. Santamaria, N. Fedi and A. Bartolo, Chem. Abstr. (1956) 14841b.

[42] L. Santamaria, H. Ricci and N. Fedi, 3rd Int. Cong. Biochem. Brussels (1955) p. 148.

[43] R.B. Setlow and W.L. Carrier, Proc. Nat. Acad. Sci. US *51* (1964) 226.

[44] G.O. Schenk, Naturwiss. *43* (1956) 71.

[45] D.M. Shankel, J. Bacteriol. *84* (1962) 410.

[46] D. Shugar, Bull. Soc. Chim. Biol. *33* (1951) 710.

[47] M.I. Simon and H. Van Vunakis, J. Mol. Biol. *4* (1962) 488.

[48] J.D. Spikes and R. Straight, Ann. Rev. Phys. Chem. *18* (1967) 409.

[49] Z. Vodrazka, Chem. Listy *53* (1959) 829.

[50] A. Wacker, J. Chim. Phys. *58* (1961) 1041.

[51] A. Wacker and P. Chandra, Stud. Biophys. *3* (1967) 239.

[52] A. Wacker, G. Turck and A. Gerstenberger, Naturwiss. *50* (1963) 377.

[53] A. Wacker et al. Photochem. Photobiol. *3* (1964) 369.

[54] L. Weil, W.G. Gordon and A.R. Buchert, Arch. Biochem. Biophys. *33* (1951) 90.

[55] I. Wilucki, H. Matthäus and C.H. Krauch, Photochem. Photobiol. *6* (1967) 497.

[56] D.L. Wulff and C.S. Rupert, Biochem. Biophys. Res. Commun. *7* (1962).

[57] N. Yamamoto, Virus (Osaka) *6* (1956a) 510.

[58] N. Yamamoto, Virus (Osaka) *6* (1956b) 522.

CHAPTER 8

PHOTODYNAMIC EFFECTS ON MOLECULES OF BIOLOGICAL IMPORTANCE: AMINO ACIDS, PEPTIDES AND PROTEINS *

John D. SPIKES and Martha L. MACKNIGHT

*Department of Biology, University of Utah,
Salt Lake City, Utah, USA*

1. Introduction

It appears that the first molecular-level studies of photodynamic action were carried out with proteins. In 1903, von Tappeiner [71] reported that crude preparations of the enzymes diastase, invertase, papain and trypsin could be inactivated by illumination in the presence of eosin. Shortly thereafter, Jodelbauer and von Tappeiner [33] showed that oxygen was required for the photodynamic inactivation of invertase sensitized by erythrosin; only much later were the effects of photodynamic treatment on amino acids examined [see 8, 17, 27]. The photodynamic inactivation of enzymes typically involves the consumption of several molecules of oxygen per enzyme molecule [67]. The study of photodynamic effects on proteins has been popular recently, with over 150 papers on some 60 different proteins, mainly crystalline enzymes, having been published in the last 15 years [see listing in 67].

This paper will briefly review the kinds of investigations which have been made of photodynamic action on amino acids, peptides and proteins. It will thus provide an historical introduction to the paper by Galiazzo et al. [19] which is devoted to current developments and future prospects in the field of photodynamic action on proteins, especially in terms of selective and localized effects. Space does not permit a full documentation of the field; more inclusive lists of references may be found in the reviews and other papers cited. The early work is reviewed in Blum's monograph [4], the publications to 1959 by Vodrážka [70], and the more recent work by Spikes and Livingston [67].

*The preparation of this paper was supported by the U.S. Atomic Energy Commission under contract No. AT(11-1)-875 and by National Science Foundation Graduate Fellowship stipends to M.L.M.

2. The basic photochemical mechanisms involved in photodynamic effects on amino acids and proteins

Peptide and disulfide bonds are broken on irradiation of proteins with ionizing radiation or with short wavelength ultraviolet light. With photodynamic treatment, however, there is no rupture of these bonds; changes in the protein result entirely from the photochemical alteration of certain amino acid side chains [see 67, 70]. The sensitized photooxidation of amino acid residues in proteins or of free amino acids in water solution is usually considered to involve two major types of processes which can occur separately or simultaneously. The contribution of each of the two pathways to the photooxidation depends on the amino acid being photooxidized, how the amino acid is incorporated into the peptide, the photosensitizing dye, the solvent composition, the pH, and the concentrations of the reactants.

Photodynamic reactions appear to proceed via the triplet or some other metastable state of the sensitizing dye [see 13]. In typical aqueous photodynamic systems with proteins, the solvent does not react with the excited dye, and the dye concentrations can be made sufficiently low such that dye-dye interactions are not of great importance. Under these conditions the metastable state of the dye can then react with molecular oxygen and/or proteins or amino acids. When oxygen reacts with the excited dye, energy transfer takes place which raises the ground-state oxygen to an excited state as first suggested by Kautsky [38]. The excited states involved appear to be two singlet states of oxygen, the $^1\Delta_g$ and the $^1\Sigma_g^+$ [13]. The energies of these states, 22 and 37 kcal/mole, respectively, fall in the energy range of photons of visible light. It has also been suggested that the metastable state of the dye can react with ground state oxygen to form an energy-rich sensitizer—oxygen complex [23]. In either case, the product would be a good oxidizing agent and could proceed to oxidize susceptible amino acid residues. In both reactions ground-state dye would be regenerated, so there would be no net consumption of dye. Sensitized photooxygenation reactions of this type which involve only electronically-excited states of the reactants have been termed Type II processes [24]. It should be pointed out that the suggested role of singlet oxygen in the photodynamic oxidation of amino acids and proteins is based only on analogy to studies with other organic compounds, since experimental information on interactions of such compounds with singlet oxygen has not as yet been published.

When the metastable state of the dye reacts directly with the amino acid residue, it usually does so via hydrogen- or electron-transfer processes, producing free radical forms of the amino acid capable of reacting with molecular oxygen. The reduced form of the sensitizing dye which results from this type of process then undergoes reactions, often with oxygen, to produce ground-state dye which can then participate in another photosensitizing cycle. Such reactions which involve free radical intermediates have been termed Type I processes [24]. As discussed below for specific cases, the photodynamic oxidation of an amino acid may proceed via a Type I process, a Type II process, or by both processes.

3. Photodynamic effects on free amino acids

It was shown a number of years ago that most aromatic amino acids were susceptible to photodynamic action, while the aliphatic amino acids were resistant [8, 17,
27]. Substituent groups can markedly alter the sensitivity of an amino acid. For
example, dihydroxyphenylalanine is rapidly photooxidized with methylene blue,
whereas phenylalanine is rather resistant [32]. Lysine is not photooxidized, however successive methylation of the α-amino group of lysine to give the secondary
and tertiary amines progressively increases the sensitivity to photooxidation with
methylene blue [76]. Of the amino acids which occur commonly in proteins, only
five are susceptible to photodynamic action: cysteine, histidine, methionine, tryptophan and tyrosine [see 67 and 70 for references]. The sensitivity of amino acids
to photodynamic action depends on the sensitizing dye, the pH, the solvent, etc.
Thus, by the proper choice of the reaction conditions some selectivity can be obtained in the photooxidation of certain amino acids in a mixture. For example, if
one corrects for the effects of pH on the sensitizing efficiency of dyes used, histidine is susceptible to photodynamic action only when the imidazole moiety is un-
ionized. Thus photooxidation does not take place below approximately pH 6 [2,
61, 73]. With some dyes tyrosine is not photooxidized at low pH values [see 2].
Blocking the amino and/or carboxyl groups has no effect on the pH-dependence of
the dye-sensitized photooxidation of methionine, tryptophan or tyrosine [73].
With either acetic or formic acid as the solvent, only methionine and tryptophan
are oxidized at appreciable rates with proflavine as sensitizer [3]. Of the five susceptible amino acids, only methionine is appreciably sensitive to photodynamic action with rose bengal and methylene blue in acetic acid solutions at low temperatures [34]. In addition, some dyes are highly selective sensitizers of certain amino
acids. For example, of the five susceptible amino acids, only cysteine is photooxidized with crystal violet in tris-(hydroxymethyl) aminomethane buffer at pH 9 [2;
also see [19]. There are few reports on the absolute efficiencies of the sensitized
photooxidation of amino acids. The quantum yield for the riboflavine-sensitized
photooxidation of tryptophan is 0.038 [60]. We have measured the quantum yields
of the photodynamic destruction of methionine; the values at pH 8 are 0.048 with
eosin Y, 0.13 with lumiflavine and 0.030 with methylene blue (unpublished results)

There are large gaps in our information on the intermediate and final products
resulting from the photodynamic treatment of free amino acids. In at least one case,
i.e. methionine, the stoichiometry and reaction products clearly depend on the
chemical nature of the sensitizing dye and on the reaction conditions. With dyes
such as methylene blue [76], proflavine [3] and rose bengal [34], methionine is
quantitatively oxidized to methionine sulfoxide. These dyes probably operate pre-
dominantly by a singlet oxygen pathway [13]. The oxygen stoichiometry with
methylene blue as sensitizer, however, depends on the reaction conditions. One
mole of oxygen is consumed per mole of methionine photooxidized at pH 8 giving
one mole of the sulfoxide and one mole of hydrogen peroxide [76]. We find that

only one-half mole of oxygen is consumed when the reaction is carried out at low pH; similar results were obtained with eosin Y as sensitizer [unpublished results]. In contrast, when flavine sensitizers such as lumiflavine, riboflavine and flavine mono-nucleotide (FMN) are used, less than 20% of the methionine is photooxidized to the sulfoxide, with methional appearing as the major product [11, 64]. The flavines, unlike the dyes listed above, appear to sensitize via electron-abstraction pathways [81]. The photodynamic treatment of histidine with methylene blue gives four un-identified organic products [25], while a histidine model compound, 4-methylimi-dazole, is photooxidized to acetylurea with methylene blue [51]. The methylene blue-sensitized photooxidation of tyrosine results in ring-rupture and the formation of carbon dioxide and two unidentified organic products [40, 76]. Tryptophan was reported to give rise to seven organic products, including dioxindolylalanine, on photooxidation [26]. The proflavine-sensitized photooxidation of tryptophan leads to two classes of products, melanines and kynurenines [3]. With methylene blue at pH 8, the photodynamic treatment of tryptophan gives N-formyl-kynurenine; a tryptophan model compound, skatole, is photooxidized to o-acetylformanilide [51]. An interesting result is obtained when the photodynamic treatment is carried out in acetic acid, for in this case solvent is incorporated into the reaction products giving β-carbolines [37].

4. Photodynamic effects on amino acid derivatives and small peptides

Relatively little work has been done on the effects of photodynamic treatment on simple derivatives of amino acids or on the amino acid residues in small peptides. Such studies may be useful, however, in understanding the photooxidation of amino acid residues in proteins. The rates and the chemistry of the photooxidation of amino acid residues in peptides might be expected to be different from that of free amino acids in some cases because of the blockage of the amino and/or carboxyl groups. In one of the first experiments [20], it was found that histidine was still sensitive to photodynamic treatment after incorporation into carnosine (β-alanyl-histidine). The glycyl dipeptides of histidine, methionine and tryptophan are photo-oxidized with methylene blue at about the same rate and with the same pH-depen-dence as the free amino acids [73]. Glutathione is reported to be converted to oxi-dized gluthatione on photodynamic treatment with riboflavine [62]. Methionyl residues in small peptides, like free methionine, are oxidized to the sulfoxide with proflavine [3] hematoporphyrin [36] and, as we have found, with eosin Y [un-published results]. If the α-amino group of tryptophan is blocked, photooxidation with proflavine gives only kynurenine-type products [3]. When tryptophan is in-corporated into small peptides, it is no longer susceptible to photooxidation with hematoporphyrin as sensitizer [36]. We found [unpublished work] that the methio-nyl residues in glycylmethionine and glycylmethionylglycine are oxidized more slowly than free methionine, while in methionylglycine the methionyl residue was oxidized at about the same rate. When these peptides are treated photodynamically with lumiflavine as sensitizer, most of the methionyl residues are oxidized to the sulfoxide; this is in sharp contrast to free methionine, as described above, which is photooxidized largely to methional with flavine sensitizers. In some cases, then, the incorporation of the susceptible amino acid into a peptide chain can markedly alter its sensitivity to photodynamic treatment as well as the chemistry involved in the photooxidation process.

5. Photodynamic effects on proteins

There is an extensive literature on the kinetics of photodynamic effects on proteins [see 67 and 70 for literature reviews]. It is difficult to make quantitative comparisons among much of the kinetic data in the literature, since most investigators have merely made rate measurements under some arbitrary set of conditions, rather than quantum yield determination. In the case of the photodynamic inactivation of enzymes, quantum yields are typically in the order of magnitude of a few thousandths [22, 28, 29, 44], i.e., about an order of magnitude smaller than the yields for the dye-sensitized photooxidation of free amino acids as described above. The kinetics of the photodynamic alteration of proteins are almost always first-order in protein concentration [67, 70]. Quantum yields for the photodynamic inactivation of enzymes are strongly affected by the concentrations of the reactants in the system. For example, yields typically increase as the oxygen concentration is increased from 0 to approximately 10—20%; above this concentration the yields remain rather constant [28]. In general, the quantum yields for enzyme inactivation are small at low pH values, increase rapidly with increasing pH at approximately neutrality, and then level off at high pH [22, 44]. Quantum yields also increase as the concentration of the sensitizing dye increases from zero, but then go through maxima and decline at higher concentration [22, 44]. The experimental activation energies for the photodynamic inactivation of enzymes at low temperatures are only a few kcal/mole [22, 44]; the activation energies are larger with some enzymes at high temperatures where unfolding of the molecule occurs [44]. The quantum yields are independent of light intensity, at least in the intensity range obtainable with ordinary light sources [22].

Many kinds of dyes have been examined for their ability to sensitize photodynamic reactions [14, 55]. We have studied over 200 different dyes as possible sensitizers for the photodynamic inactivation of trypsin [55, 63, 66]. In general, there is a good correlation between dye structure and sensitizing ability, i.e., certain classes of dyes sensitize, while others do not. Among the principle effective sensitizer are most of the acridine dyes examined (acridine orange, proflavine, etc.), the anthraquinones (anthraquinone-2,6-disulfonate, Cibantine golden yellow, Soledon brilliant orange, etc.), the flavine dyes (lumiflavine, riboflavine, FMN, but not flavine adenine dinucleotide), the phthalein subgroup of the xanthene dyes (fluoresceins, including eosin Y, rose bengal, mercurochrome, etc.), some of the porphyrins (hematoporphyrin, sodium magnesium chlorophyllin), the safranine subgroup of the azine dyes (rhoduline violet, safranin O), and the thiazine dyes (the azures, methylene blue, thionin, etc.). Inactive dyes for trypsin photooxidation include most of the azo dyes, the indophenol dyes, the methine dyes, the nitro dyes, the nitroso dyes, the oxazine dyes, the thiazole dyes and the triarylmethane dyes. Substituents on a dye molecule can have a large effect on the sensitizing efficiency. For example, fluorescein dyes show an increased quantum yield for the photodynamic inactivation of trypsin with increased halogenation: 0.00017 with fluorescein, 0.0016 with dibromofluorescein, 0.0022 with tetrabromofluorescein (eosin Y), 0.0030 with tetraiodofluorescein, and 0.0037 with tetrachlorotetraiodofluorescein (rose bengal).

As we have pointed out elsewhere [63, 66], one relatively little-used kinetic approach toward understanding the mechanism of photodynamic reactions is the use of chemical 'inhibitors' or 'protective' agents, i.e., chemicals which prevent the photodynamic oxidation of the compound being studied. One can envision, for example, that different protective agents might act at different points in the reaction sequence of photodynamic processes; e.g., some might interfere with excitation of the dye, some might quench the metastable state of the dye, some might quench singlet oxygen more effectively than the substrate compound, etc. We have examined the effects of a wide variety of compounds on the quantum yield of the photodynamic inactivation of enzymes using different sensitizing dyes [21, 63]. Serotonin creatine sulfate and the anti-oxidant, n-propylgallate, give about 50% protection with eosin Y, FMN and methylene blue as sensitizers. Iodide ion and tyrosine, on the other hand, were effective protective agents with FMN but not with eosin Y and methylene blue. By using flash photolysis techniques we are presently continuing our study of the mechanisms by which different protective agents function.

As might be expected, proteins often exhibit changes in their physicochemical properties as a result of photodynamic treatment. In the case of lysozome, photodynamic treatment with methylene blue results in a decrease in optical density at approximately 280 nm as a result of the destruction of tryptophyl and tyrosyl residues [18, 77]. In contrast, with ribonuclease, where the aromatic amino acid residues are destroyed only slowly, there is little change in absorption at 280 nm [75]. In cases where there is some selectivity of residue destruction with different dyes, the absorption changes during the photodynamic oxidation of a given protein depend on the sensitizing dye used [10]. Some heme-proteins such as catalase show a sharp reduction in absorbance in the Soret band on photodynamic treatment [47]. Some proteins, such as serum albumin, show an increase and then a decrease in optical rotation during photodynamic treatment [57], others, such as lysozyme, show marked changes in one direction [35], while some show no change [see 67 for references]. Many proteins show characteristic changes in electrophoretic mobility on photodynamic treatment [see 54, 67 and 70 for references]. Although the ultracentrifuge behavior of some proteins, such as ovalbumin, is not altered by photodynamic treatment [1], a number of proteins do show changes in sedimentation coefficients and degree of heterogeneity [see 67 for references]. Some proteins, including ovalbumin [1], serum albumin and lysozome, become more insoluble as a result of photodynamic treatment, while others, e.g. chymotrypsin, show little change [see 67 for references]. Most native proteins are resistant to digestion by proteases. Photodynamic treatment of glyceraldehyde-3-phosphate dehydrogenase [16] and lysozome [30, 35] markedly increases their digestibility. With some enzymes, which have been partially inactivated by photodynamic treatment, a fraction of the remaining active enzyme molecules have been altered to the extent that they are more susceptible to inactivation by heat or high concentrations of urea than the native enzyme [see 30 and 67 for references]. In the case of lysozome, the enzyme undergoes conformational changes during photodynamic treatment which parallel the loss of activity [30]. Other physicochemical alterations observed in proteins during photodynamic treatment include changes in metal-binding properties [79], heme-binding properties [7], coenzyme-binding properties [49], polarographic behavior [12, 54], diffusion properties [6], light-scattering properties [74], viscosity [56] and surface tension [58].

The photodynamic treatment of proteins usually alters their chemical and physical properties such that they do not function in their normal biological role. Such photodynamic alterations of function have been studied with many different kinds of proteins. For example, the effects of photodynamic treatment on the catalytic function of a large number of enzymes of all categories has been studied [see 67, table II, for references], including hydrolases (alkaline phosphatase, amylase, carboxypeptidase, cathepsin, chymotrypsin, fibrinolysin, hyaluronidase, lipase, lysozome, myosin, papain, pepsin, rennin, subtilisin, trypsin, urease), isomerases (phosphoglucose isomerase), ligases (aminoacyl RNA synthetase), oxidoreductases (alcohol dehydrogenase, d-amino acid oxidase, catalase, glyceraldehyde dehydrogenase, lactic dehydrogenase, peroxidase, tyrosinase), and transferases (aspartate transaminase, levansucrase, phosphoglucomutase, phosphorylase, pyruvate kinase, ribonuclease, transketolase). In general, enzymes are rapidly inactivated by photodynamic treatment. Of those listed above, only radish peroxidase [29, 46] and bacterial alkaline phosphatase [49] are resistant. Photodynamic treatment with methylene blue decreases the peptidase activity of carboxypeptidase A but increases its apparent esterase activity [69]; similar treatment initially increases the adenosinetriphosphatase activity of myosin [68]. Enzyme activity can be lost during photodynamic treatment not only as a result of the destruction of amino acid residues located in the active site, but also by the destruction of amino acid residues elsewhere which are necessary for the maintenance of the catalytically-active conformation of the enzyme [see 30].

The photodynamic inactivation of two peptide hormones has been demonstrated. The octapeptide, angiotensinamide, loses biological activity on photodynamic treatment at the same rate as the single histidine residue in the molecule is destroyed [48]. Insulin also loses hormonal activity on photodynamic treatment with methylene blue at a rate parallel to the photooxidation of histidine residues [see 52 and 78 for references]. Photodynamic treatment can destroy the antigenicity of proteins as well as the ability of proteins to react with antibodies [see 67 for references] . The toxin, ricin, from castor beans [71] , as well as bacterial toxins such as tetanus toxin [42] , botulinum toxin [5] , etc. [see 67 for additional references] can also be inactivated by photodynamic treatment. Similarly, photodynamic treatment destroys the toxicity of snake venoms [see references in 41 and 67] . The rate of clotting of fibrinogen is markedly decreased by photodynamic treatment [82] . Further, the mechanical properties of structural proteins such as collagen can be altered by photodynamic action; treated collagen is easier to stretch and shows less stress-relaxation than the native protein [65] .

It is possible to produce certain selective effects in the photooxidation of proteins, e.g., the destruction of one type amino acid residue or of a small selected group of residues due to the different rates of photooxidation of different residues which can be obtained by using various dyes and/or reaction conditions. These selective photodynamic reactions promise to serve as a useful tool in the study of proteins as discussed in greater detail by Galiazzo et al. [19] in their paper in this volume.

One means for obtaining selective destruction of residues in proteins is that of controlling the ionic state of the side chains. As mentioned above, the imidazole moiety of histidyl residues must be uncharged in order for these residues to be photooxidizable. Thus histidyl residues in proteins are not photooxidized at low pH values [9, 67] , and the pH dependence for the photodynamic loss of enzymic activity often follows that of histidine photooxidation [15, 72, 80] . By working under conditions of low pH it is possible to avoid the destruction of histidyl residues and in this way to obtain the selective destruction of other residues, such as methionine [34, 35] or tryptophan [18] .

As discussed by Galiazzo et al. [19] , the photooxidizable amino acid residues are susceptible to photooxidation according to their relative accessibility to the sensitizing dye or to the oxidizing species involved in the reaction. Ray and Koshland [50] were the first to investigate the number of accessible and inaccessible photooxidizable residues in a protein. It is possible in some cases to control the number and types of susceptible residues photooxidized by unfolding proteins to different extents. For example, Weil et al. [78] selectively photooxidized the histidyl residues in insulin at low temperatures using methylene blue as a sensitizer; however, tyrosyl residues were destroyed, in addition to histidyl residues, when the protein had been unfolded thermally or with 8 M urea. Kenkare and Richards [39] studies the accessibility of histidyl residues in RNase-S using methylene blue as a sensitizer, by separating the 20-residue N-terminal S-peptide from the residual S-protein. The selective modification of residues in proteins using unfolding techniques is treated in the paper of Galiazzo et al. [19] .

Some dyes are specific for the photodynamic alteration of a certain type of amino acid residues, and highly selective effects can be obtained by the use of such sensitizers in combination with the proper reaction conditions. For example, only methionyl residues in proteins are photooxidized using hematoporphyrin as a sen-

sitizer over a wide pH range below pH 6.5 and in aqueous acetic acid solutions [36].
Methylene blue and rose bengal in acetic or formic acid solutions have also been
used to sensitize the selective photooxidation of methionyl residues in ribonuclease
A [34] and lysozome [35]. Proflavine in 98% to 100% formic or acetic acid solutions sensitizes the photooxidation of only tryptophyl and methionyl residues in
proteins [3]. By chemical reduction with mercaptoethanol or thioglycolic acid, the
resulting methionine sulfoxide residues can be converted to methionines to obtain
the selective alteration of only tryptophyl residues [18]. Some degree of selectivity
for the photooxidation of tyrosyl residues is found with FMN as a sensitizer at low
pH values [43]. Many investigators have selectively altered histidyl residues in certain proteins because of the greater reactivity and/or accessibility of these residues
compared with the other types of photooxidizable residues present [15, 31, 72].

A very high degree of selectivity in the photooxidation of amino acid side chains
in proteins can be obtained when the sensitizer is in close association with the substrate protein. For example, Rippa and Pontremoli [53] observed that two of the
12 histidyl residues and three of the eight cysteinyl residues in 6-phosphogluconate
dehydrogenase were photooxidized when the chromophore, pyridoxal 5'-phosphate,
was bound to the lysyl residue at the active site of the enzyme. In a similar study,
only one tyrosyl residue of the six present in ribonuclease was altered when the
substrate analogue 4-thiouridylic acid was complexed to the enzyme [59]. Upon
illumination of the complex of protoporphyrin IX and sperm whale apomyoglobin,
Breslow et al. [7] observed that one or two histidyl residues were selectively photooxidized. The photochemical behavior of proteins which naturally contain sensitizing moieties, and of sensitizers covalently linked to proteins, is treated in detail
by Galiazzo et al. [19] in this volume.

References

[1] J.S. Bellin and G. Entner. Dye-sensitized photooxidation of ovalbumin. Photochem. Photobiol. *5* (1966) 251–259.

[2] J.S. Bellin and C.A. Yankus, Influence of dye binding on the sensitized photooxidation of
amino acids. Arch. Biochem. Biophys. *123* (1968) 18–28.

[3] C.A. Benassi, E. Scoffone, G. Galiazzo and G. Jori, Proflavine-sensitized photooxidation of
tryptophan and related peptides. Photochem. Photobiol. *6* (1967) 857–866.

[4] H.F. Blum, Photodynamic Action and Diseases Caused by Light (Reinhold, New York,
1941; reprinted with an up-dated appendix, in 1964 by Hafner, New York).

[5] D.A. Boroff and R.R. DasGupta, Study of the toxin of *Clostridium botulinum*. VII. Relation of tryptophan to the biological activity of the toxin. J. Biol. Chem. *239* (1964)
3694–3700.

[6] J.M. Brake and F. Wold, Photo-oxidation of yeast enolase. Biochim. Biophys. Acta *40*
(1960) 171–173.

[7] E. Breslow, R. Koehler and A.W. Girotti, Properties of protoporphyrin-apomyoglobin
complexes and related compounds. J. Biol. Chem. *242* (1967) 4149–4156.

[8] C.W. Carter, The photooxidation of certain organic substances in the presence of fluorescent dyes. Biochem. J. *22* (1928) 575–582.

[9] R.R. DasGupta and D.A. Boroff, Increased selectivity of photooxidation for amino acid residues. Biochim. Biophys. Acta *97* (1965) 157–159.

[10] D. della Pietra and K. Dose, Die durch Riboflavin-5'-phosphat und Hämatoporphyrin sensibilisierte Photo-Oxydation von Hefe-Alkoholdehydrogenase. Biophysik *2* (1965) 347–353.

[11] K. Enns and W.H. Burgess, The photochemical oxidation of ethylenediaminetetraacetic acid and methionine by riboflavin. J. Amer. Chem. Soc. *87* (1965) 5766–5770.

[12] S. Fiala, Polarographic and spectrophotometric studies on photodynamic action by fluorescein dyes. Biochem. Z. *320* (1949) 10–24.

[13] C.S. Foot, Mechanisms of photosensitized oxidation. Science *162* (1968) 963–970.

[14] W.L. Fowlks. The mechanism of the photodynamic effect. J. Invest. Dermatol. *32* (1959) 233–247.

[15] K.A. Freude. Photochemical evidence for the participation of histidine in the active center of carboxypeptidase Aγ. Biochim. Biophys. Acta *167* (1968) 485–488.

[16] P. Friedrich, L. Polgár and G. Szabolesi, The existence of a histidine residue essential for glyceraldehyde-3-phosphate dehydrogenase action. Acta Physiol. Acad. Sci. Hung. *25* (1964) 217–228.

[17] H. Gaffron, Über den Mechanismus der Sauerstoffaktivierung durch belichtete Farbstoffe. Biochem. Z. *179* (1926) 157–163.

[18] G. Galiazzo, G. Jori and E. Scoffone, Selective and quantitative photochemical conversion of the tryptophyl residues to kynurenine in lysozome. Biochem. Biophys. Res. Commun. *31* (1968) 158–163.

[19] G. Galiazzo, G. Jori and E. Scoffone, Dye-sensitized photo-oxidation as a tool for elucidating the topography of proteins in solution. In: Research Progress in Organic-Biological and Medicinal Chemistry, Vol. III (North-Holland, 1971) ch. 9.

[20] A.W. Galston, Riboflavin, light and the growth of plants. Science *111* (1950) 619–624.

[21] B.W. Glad, J.D. Spikes and L.F. Kumagai, The inhibition of the dye-sensitized photoinactivation of lysozome using tyrosine and thyronine analogues. Proc. Soc. Exp. Biol. Med. *131* (1969) 1279–1280.

[22] B.W. Glad, and J.D. Spikes, Quantum yield studies of the dye-sensitized photoinactivation of trypsin. Radiation Res. *27* (1966) 237–249.

[23] K. Gollnick and G.O. Schenck, Mechanism and stereoselectivity of photosensitized oxygen transfer reactions. Pure Appl. Chem. *9* (1964) 507–525.

[24] K. Gollnick, Type II photooxygenation reactions in solution. Advan. Photochem. *6* (1968) 1–122.

[25] S. Gurnani and M. Arifuddin, Effect of visible light on amino acids. II. Histidine. Photoche . Photobiol. *5* (1966) 341–345.

[26] S. Gurnani, M. Arifuddin and K.T. Augusti, Effect of visible light on amino acids. I. Tryptophan. Photochem. Photobiol. *5* (1966) 495–505.

[27] D.T. Harris, Observations of the velocity of the photooxidation of proteins and amino acids. Biochem. J. *20* (1926) 288–293.

[28] C.F. Hodgson, E.B. McVey and J.D. Spikes, The effect of oxygen concentration on the quantum yields of the dye-sensitized photoinactivation of enzymes. Experientia *25* (1969) 1021–1022.

[29] T.R. Hopkins and J.D. Spikes, Inactivation of crystalline enzymes by photodynamic treatment and gamma radiation. Radiation Res. *37* (1969) 253–260.

[30] T.R. Hopkins and J.D. Spikes, Conformational studies of lysozome during photodynamic inactivation. Photochem. Photobiol. *12* (1970) 175–184.

[31] Y. Ichikawa and T. Yamano, Studies on the photoinactivation of D-aminoacid oxidase. Tokushima J. Exp. Med. *10* (1963) 156–169.

[32] S. Isaka and J. Kato, Photodynamic activity of riboflavin. I. general properties of the photodynamic riboflavin. J. Coll. Arts. Sci. Chiba Univ. *1* (1952) 43–50.

[33] A. Jodelbauer and H. von Tappeiner, Ueber die Beteiligung des Sauerstoffes bei der photodynamischen Wirkung fluoreszierender Stoffe. Muench. Med. Wochschr. *26* (1903) 1139–1141.

[34] G. Jori, G. Galiazzo, A. Marzotto and E. Scoffone, Dye-sensitized selective photooxidation of methionine. Biochim. Biophys. Acta *154* (1968) 1–9.

[35] G. Jori, G. Galiazzo, A. Marzotto and E. Scoffone, Selective and reversible photo-oxidation of the methionine residues in lysozome. J. Biol. Chem. *243* (1968) 4272–4278.

[36] G. Jori, G. Galiazzo and E. Scoffone, Photodynamic action of porphyrins on amino acids and proteins. I. Selective photooxidation of methionine in aqueous solution. Biochem. *8* (1968) 2868–2875.

[37] G. Jori, G. Galiazzo and G. Gennari, Photosensitized conversion of tryptophan to β-carboline derivatives. Photochem. Photobiol. *9* (1969) 179–181.

[38] H. Kautsky, J. de Bruijn, R. Neuwirth and W. Baumeister, Photo-sensibilisierte Oxydation als Wirkung eines aktiven, metastabilen Zustandes des Sauerstoff-Molekuls. Chem. Ber. *66* (1933) 1588–1600.

[39] U.W. Kenkare and F.M. Richards, The histidyl residues in ribonuclease-S. J. Biol. Chem. *241* (1966) 3197–3206.

[40] J.H. Kim, Chemical synthesis of 4-C^{14}-DL-Tyrosine and its photooxidation. Univ. Microfilms No. 65-12, 292 (University Microfilms Inc., Ann Arbor, Michigan, 1965).

[41] W. Kocholaty, Detoxification of *Crotalus atrox* venom by photooxidation in the presence of methylene blue. Toxicon *3* (1966) 175–186.

[42] K. Lippert, The photodynamic effect of methylene blue on tetanus toxin. J. Immunol. *28* (1935) 193–203.

[43] M.L. MacKnight and J.D. Spikes, The time-course of the destruction of amino acid residues during the photodynamic inactivation of ribonuclease A. Boll. Chim. Farm. *109* (1970) 659–664.

[44] M.L. MacKnight and J.D. Spikes, Investigations of the quantum yield of the dye-sensitized photoinactivation of ribonuclease. Experientia *26* (1970) 255–256.

[45] D.B.S. Millar and G.W. Schwert, Lactic dehydrogenase. IX. effect of photo-oxidation upon activity and complex formation. J. Biol. Chem. *238* (1963) 3249–3255.

[46] M. Nakatani, Studies on hemeproteins related to their activities. IV. Photooxidation of peroxidase in the presence of methylene blue. J. Biochem. (Tokyo) *48* (1960) 640–644.

[47] M. Nakatani, Studies on histidine residues in hemeproteins related to their activities. V. Photooxidation of catalase in the presence of methylene blue. J. Biochem. (Tokyo) *49* (1961) 98–102.

[48] A.C. Paiva and T.B. Paiva, The photooxidative inactivation of angiotensinamide. Biochem. Biophys. Acta *48* (1961) 412–414.

[49] S. Plotch and A. Lukton, Active site studies of bacterial alkaline phosphatase. Biochim. Biophys. Acta *99* (1965) 181–182.

[50] W.J. Ray, Jr., and D.E. Koshland, Jr., Identification of amino acids involved in phosphoglucomutase action. J. Biol. Chem. *237* (1962) 2493–2505.

[51] G. Rhodes and P.D. Gardner, unpublished results.

[52] P. Rieser, The catalytic nature of insulin. Amer. J. Med. *40* (1966) 759–764.

[53] N. Rippa and S. Pontremoli, Pyridoxal 5'-phosphate as a specific photosensitizer for histidine residue at the active site of 6-phosphogluconate dehydrogenase. Arch. Biochem. Biophys. *113* (1969) 112–118.

[54] L. Santamaria, Photodynamic action and carcinogenicity. Recent Contributions to Cancer Research in Italy, Tumori Suppl. (Casa Editrice Ambrosiani, Milan, 1960) pp. 167–288.

[55] L. Santamaria and G. Prino, The photodynamic substances and their mechanism of action. In: Research Progress in Organic-Biological and Medicinal Chemistry, Vol. I (North-Holland, Amsterdam, 1964) pp. 259-336.

[56] R. Santamaria, G.G. Corigliano and L. Santamaria, Comportamento reologico di soluzioni di miosina 'A' fotoossidata. Boll. Soc. Ital. Biol. Sperimentale *40* (1964) 1801–1805.

[57] L. Santamaria, R. Bianco and C. Torriani, Variazioni polarimetriche di siero di sangue umano, siero-albumina cristallizzata e peptidi in seguito ad effetto fotodinamico da emato-porfirina. Atti Soc. Ital. Patol. *5(2a)* (1957) 449–455.

[58] L. Santamaria, R. Santamaria and M. Ciarfuglia, Tensione superficiale di soluzioni die siero di sangue foto-ossidato mediante ematoporfirina. Atti. Soc. Ital. Patol. *5* (1957) 457–461.

[59] F. Sawada, Photosensitized modification of bovine pancreatic ribonuclease by a colored substrate analogue, 4-thiouridylic acid. J. Biochem. (Tokyo) *65* (1969) 767–776.

[60] D. Shugar, Photosensibilisation des enzymes et des dérives indoliques par la riboflavin. Bull. Soc. Chim. Biol. *33* (1954) 710–718.

[61] L.A.Ae. Sluyterman, Photooxidation, sensitized by proflavine, of a number of protein constituents. Biochim. Biophys. Acta *60* (1962) 557–561.

[62] K. Sone and T. Koyanagi, Photooxidation of glutathione sensitized by the riboflavin. Bitamin *33* (1966) 349–354.

[63] J.D. Spikes, Kinetic studies of the photodynamic inactivation of trypsin. In: Radiation Research (North-Holland, Amsterdam, 1967) pp. 823–838.

[64] J.D. Spikes, Preliminary studies of the dye-sensitized photooxidation of methionine. Biophys. J. *9* (1969) A-272.

[65] J.D. Spikes and C.A. Ghiron, Photodynamic effects in biological systems. In: Physical Processes in Radiation Biology, eds. L.G. Augenstein, R. Mason and B. Rosenberg, (Academic Press, New York, 1964) pp. 309–336.

[66] J.D. Spikes and B.W. Glad, Photodynamic action, Photochem. Photobiol. *3* (1964) 471–487.

[67] J.D. Spikes and R. Livingston, The molecular biology of photodynamic action: sensitized photoautoxidations in biological systems. Advan. Radiat. Biol. *3* (1969) 29–121.

[68] A. Stracher and P.C. Chan, Studies on the adenosinetriphosphatase active center of Myosin A. In: Biochemistry of Muscle Contraction, ed. J. Gergely (Little, Brown, Boston, Massachusetts, 1964) pp. 106–113.

[69] B.L. Vallee, J.F. Riordan and J.E. Coleman, Carboxypeptidase A: approaches to the chemical nature of the active center and the mechanisms of action. Proc. Nat. Acad. Sci. U.S. *49* (1963) 109–116.

[70] Z. Vodrážka, The photooxidation of proteins. Chem. Listy *53* (1959) 829–843.

[71] H. von Tappeiner, Ueber die Wirkung fluorescirender Substanzen auf Fermente und Toxine. Ber. Deut. Chem. Ges. *36* (1903) 3035–3038.

[72] N.E. Vorotnitskaya, G.F. Lutovinova and O.L. Polyanovsky, Functional groups of aspartate aminotransferase. In: Pyridoxal Catalysis: Enzymes and Model Systems, eds. E.E. Snell, A.E. Braunstein, E.S. Severin and Yu. M. Torchinsky, (Interscience, New York, 1968) pp. 131–142.

[73] L. Weil, On the mechanism of the photo-oxidation of amino acids sensitized by methylene blue. Arch. Biochem. Biophys. *110* (1965) 57–68.

[74] L. Weil and A.R. Buchert, Photooxidation of crystalline β-lactoglobulin in the presence of methylene blue. Arch. Biochem. Biophys. *34* (1951) 1–15.

[75] L. Weil and T.S. Seibles, Photooxidation of crystalline ribonuclease in the presence of methylene blue. Arch. Biochem. Biophys. *54* (1955) 368–377.

[76] L. Weil, W.G. Gordon and A.R. Buchert, Photooxidation of amino acids in the presence of methylene blue. Arch. Biochem. Biophys. *33* (1951) 90–109.

[77] L. Weil, A.R. Buchert and J. Maher, Photooxidation of crystalline lysozome in the presence of methylene blue and its relation to enzymatic activity. Arch. Biochem. Biophys. *40* (1952) 245–252.

[78] L. Weil, T.S. Seibles and T.T. Herskovits, Photooxidation of bovine insulin sensitized by methylene blue. Arch. Biochem. Biophys. *111* (1965) 308–320.

[79] G. Weitzel, W. Schaeg, G. Boden and B. Willms, Histidingehalt und Aktivität von Insulin. Z. Naturforsch. *206* (1965) 497.

[80] E.W. Westhead, Photooxidation with rose bengal of a critical histidine residue in yeast enolase. Biochem. *4* (1965) 2139–2144.

[81] W.F. Yang, H.S. Ku and H.K. Pratt, Photochemical production of ethylene from methionine and its analogues in the presence of flavin mononucleotide. J. Biochem. *242* (1967) 5274–5280.

[82] P.D. Zieve and H.M. Solomon, Effect of hematoporphyrin and light on human fibrinogen. Amer. J. Physiol. *210* (1966) 1391–1395.

CHAPTER 9

DYE-SENSITIZED PHOTO-OXIDATION AS A TOOL FOR ELUCIDATING THE TOPOGRAPHY OF PROTEINS IN SOLUTION

Guido GALIAZZO, Giulio IORI and Ernesto SCOFFONE

Institute of Organic Chemistry, University of Padova,
35100 Padova, Italy

1. Introduction

The dye-sensitized photochemical modification of protein yield insights on several different levels of inquiry into the protein structure. In the first place, dye-sensitized photo-oxidations allow one to modify certain amino acid residues with a high degree of selectivity and, hence, to elucidate their role in the biological function [34, 36]. This set of studies appears to be the logical development of the exploratory experiments in this field by Weil et al. [40], Galston [10] and Shugar [31].

In the early sixties, new prospects were opened by the pioneering work of Ray and Koshland [25] about the methylene blue-sensitized photo-oxidation of phosphoglucomutase: the elegant kinetic analysis developed by these authors allowed one to evaluate the type and the number of the photo-oxidizable residues which are exposed on the surface of a protein molecule or buried in the interior folds.

A novel strategy for investigating the three-dimensional structure of proteins by means of dye-sensitized photo-oxidation is being presently developed in our laboratory. Following the first attempts of Shafer et al. [30], and of Sawada [27] this method involves the irradiation of proteins containing the sensitizer bound in known position of the molecule: in this way, only the photo-oxidizable side chains which are located close to the dye are modified. Since the position of the sensitizer in the molecule is known, once the modified residues have been identified, their position in the space can be deduced.

These two topics, i.e. the determination of the buried and exposed groups and the mapping of predetermined regions of a protein molecule, will be separately discussed in the next sections.

2. Determination of the buried and exposed groups

In the three-dimensional structure of proteins, the individual amino acid residues may be present either on the surface of the molecule and accessible to the solvent or buried in the internal hydrophobic regions. However, a rigorous classification into completely buried and completely exposed residues would be an over-simplification, since many residues belong to an intermediate class, neither fully buried nor fully accessible. In any case, as a result of their location within the spatial network of a protein, the amino acid side chains are characterized by different levels of reactivity towards specific reagents. The groups on the surface resemble simple peptides in their reactivity, whereas those buried in the interior display reduced reactivities and often are not reactive at all. Accordingly, the reactivity shown by a given amino acid in model peptides may be taken as the upper reactivity limit of this amino acid toward the selected reagent. Furthermore, since any drop in reactivity is a function of the environment of the amino acids within the protein molecule, it becomes possible to probe the state of such amino acids by studying their reactivity with specific reagents [38].

The method developed by Ray and Koshland [25] for differentiating the buried from the exposed residues relies on the principles outlined above. In fig. 1, the time course of the loss of histidine upon irradiation of phosphoglucomutase is reported. If the data are plotted on a linear scale, no direct meaning can be apparently attached to the exponential decrease of the amino acid content. A relationship does immediately emerge if the curve is analyzed kinetically by a semilogarithmic plot, i.e. the disappearance of histidine is assumed to follow a first order kinetics to the

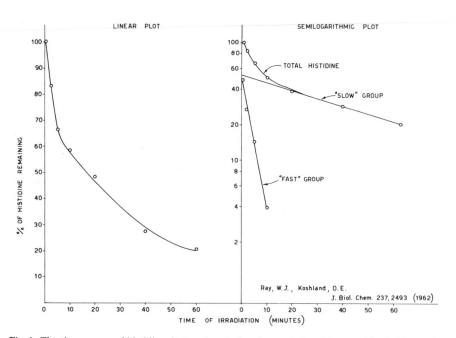

Fig. 1. The time course of histidine destruction during the methylene blue-sensitized photo-oxidation of phosphoglucomutase. The irradiations were carried out at 10°C in 0.05 M phosphate buffer, pH 7.08 [25].

expression

$$\frac{dH}{dt} = -KH. \tag{1}$$

Clearly, the plot is not linear as it would be if all the histidyl residues were oxidized at the same first order rate. On the other hand, the shape of the curve can be rationalized by assuming that some of the histidines are relatively inaccessible to the photo-oxidizing agent and the remainder readily accessible. In this case, the overall rate of histidine photo-oxidation would be the sum of the photo-oxidation rates of both the accessible and inaccessible residues, and would be given by the equation

$$\frac{H}{H_0} = f_1 e^{-K_1 t} + f_2 e^{-K_2 t}, \tag{2}$$

where H/H_0 is the amount of histidine remaining after the time t, f_1 and f_2 are the fraction of the total histidine involved in the slow and fast reacting group, and k_1 and k_2 are the first order rate constants for the photo-oxidation of the residues in the respective group. Graphical analysis of these data, as depicted in fig. 1, allows one to deduce the number of exposed and buried residues. The same procedure was successfully applied by Ray and Koshland to determine the number of the tryptophyl, tyrosyl, methionyl and cysteinyl residues which are located on the surface of the phosphoglucomutase molecule or buried in the interior.

Actually, this procedure has the favourable feature of yielding the degree of exposure of live different types of amino acids in a protein by a 'one-hit' procedure; moreover, it can often resolve the role of individual residues in contributing to enzyme photoinactivation, even when a number of such residues are simultaneously photo-oxidized. This approach was consequently applied by several authors on a wide number of proteins [1, 23, 24, 22, 41]. The same result can seldom be achieved in such a simple fashion by the presently used chemical and physicochemical techniques.

However, some limitations are inherent to this method. In brief, one can say that the full description of the state of all the residues of a given amino acid in a protein can be unequivocally achieved only if these residues can be separated into two, and no more than two, classes of reactivity. If this is not the case, additional terms must be introduced into eq. (2), i.e. one term for each class of reactivity; this results in a considerable reduction of the linear portion of the plot, while the non-linear portion displays several different shapes, which makes the extrapolation procedure rather difficult, if not impossible at all. A striking example of such difficulties has been presented in a paper by Chatterjee and Noltman [5]. As one can deduce from fig. 2, the time course of histidine and methionine degradation during the photosensitized oxidation of rabbit muscle phosphoglucose isomerase is polyphasic, so that it appears somewhat hazardous to attach a quantitative meaning to all the possible slopes. Actually, the authors were forced to limit their deductions to the initial linear part of the plot, that is they could estimate only the number of fully accessible residues.

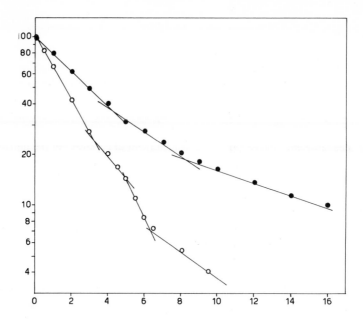

Fig. 2. The time course of methionine (●——●) and histidine (○——○) loss during the methylene blue-sensitized photo-oxidation of phosphoglucose isomerase at 10°C in 0.05 M phosphate buffer, pH 7.1 [5].

A modification of this method, which should at least in part circumvent these drawbacks, is being developed in our laboratory [15, 17]. It consists in inducing a progressive multistage unfolding of the protein molecule by the stepwise addition of a suitable denaturing agent. Both the native, the partially loosened and the extensively unfolded proteins are then subjected to dye-sensitized photo-oxidation. In this way, a gradually increasing number of residues should become susceptible to the photodynamic treatment, at first the exposed ones, then — in subsequent steps — the partially buried ones, and finally the deeply buried ones. Consequently, a concerted study of the pattern of changes produced by the photo-oxidation of the different molecules should allow one to obtain a rather detailed description of the single residues, without neglecting any classes having intermediate reactivities.

Out of the commonly used denaturing agents for proteins, the most suitable for this kind of studies were found to be the pH of the solution and the polarity of the solvent. Actually, both these parameters can be varied continuously over a sufficiently wide range, therefore offering large possibilities of selecting the proper conditions of irradiation. Furthermore, previous studies [2, 14, 33] demonstrated that these variables do not interfere with the energy transfer process between the excited sensitizer and molecular oxygen.

Of course, if the pH of the solution or the polarity of the solvent are to be changed along a given set of experiments, one must envisage the fact that the photo-sensitivity of the photo-oxidizable amino acids is strongly affected by these param-eters [14, 33, 34, 39]. In general, whereas cysteine, methionine and tryptophan are photoreactive over all the pH range, as well as in non-aqueous, solvents, such as formic or acetic acid, the amino acids histidine and tyrosine display a reactivity de-creasing with increasing acidity of the solvent and usually are no longer photo-oxi-dized at pH values lower than 6. Clearly, the result obtained in different experiments can be compared only if the pattern of the alterations introduced into the amino acid sequence of the protein is the same in all cases. Therefore, by our method, un-equivocal results can be achieved employing a sensitizer which can act selectively on a single amino acid over all the range of conditions used. The degree of the given residue within the protein matrix can then be explored.

At the present time, we succeeded in finding two dyes which mediate the photo-oxidation of a single amino acid over a large spectrum of conditions: hematopor-phyrin acts selectively on methionine both in aqueous solution of the pH range 2–6.5 and in 5 to 95% acetic acid solution [15]; on the other hand, crystal violet promotes the selective photo-oxidation of cysteine at any pH as well as in acetic acid solution [16]. Accordingly, in order to test the feasibility of our procedure for estimating the degree of exposure of amino acids, we focused our attention of the methionyl residues of ribonuclease A; this protein was chosen since its three-dimen-sional structure had already been resolved by X-ray crystallography [19], so that suitable data for comparison with our results were available.

At first we investigated the stability of ribonuclease A to the action of low pH and of acetic acid. As a probe for the conformation of both the peptide backbone and the amino acid side chains, we chose UV difference spectroscopy since this technique is known to be very sensitive to even small conformational changes of protein molecules [11]. Experiments in aqueous solution showed that the shape of the absorption spectrum of the enzyme was unaffected by lowering the pH to 2.3; on the contrary, dramatic effects appeared to occur as soon as the protein was ex-posed to concentrated acetic acid. As can be seen from fig. 3, the difference spec-trum of ribonuclease A in 80% acetic acid versus water shows a sharp negative maxi-mum at 287 nm; this clearly reflects the occurrence of a blue shift in the absorption of the tyrosyl side chains, probably owing to the disruption of the native tertiary structure and to the consequent exposure of the previously buried chromophores to the water-acetic acid environment. The observed is quantitatively analogous to that found for samples of ribonuclease A denatured by high concentrations of urea [4]. Therefore, it is reasonable to assume that the protein molecule is extensively disorganized in 80% acetic acid.

Now, as shown in fig. 4, the depression in the extinction coefficient caused by acetic acid occurs in two distinct steps. Indeed, in the plot, three plateaus can be distinguished between 0 and 5%, 10 and 55%, and over 62% acetic acid. This would suggest that at least three different conformations of ribonuclease A can exist at different levels of acetic acid concentration. On this basis, we run the hematopor-

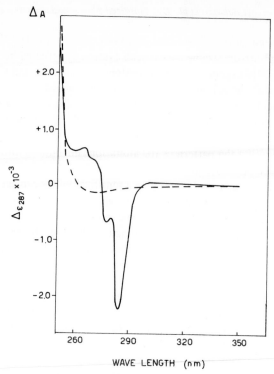

Fig. 3. Difference absorption spectrum obtained by reading a 1.8 × 10⁻⁴ M solution of RNase A in 80% acetic acid against an equimolar solution in distilled water, pH 5.95. The dotted line represents the baseline.

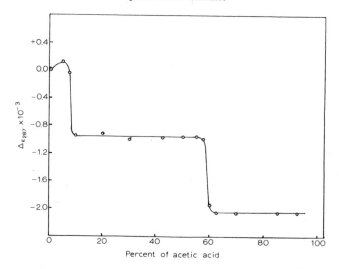

Fig. 4. The effect of the acetic acid concentration on the molar difference extinction coefficient of RNase A at 287 nm. The experimental points were obtained by reading a 1.8 × 10⁻⁴ M solution against an equimolar solution in water pH 5.95.

Table 1

Effect of acetic acid on the protoreactivity of the methyonyl residues in ribonuclease A

Percentage of acetic acid (%)	Met residues oxidized per protein molecule	Location in the amino acid sequence of the Met residues photo-oxidized
5	1	Met-29
10	2	Met-29, Met-13
50	2	Met-29, Met-13
67	4	Met-29, Met-13, Met-79, Met-30

4 ml of a 0.1 mM protein solution were irradiated in the presence of hematoporphyrin at 37°C for 5 min. No amino acids other than methionine were affected by photo-oxidation.

phyrin-sensitized photo-oxidation of ribonuclease A in different water—acetic acid mixtures. The results obtained are summarized in table 1. Apparently, one residue, Met-29, is affected when the irradiations are performed in up to 5% acetic acid. This residue is therefore likely to be exposed on the surface of the native protein. A second residue, Met-13, becomes accessible in the intermediate stage of unfolding, and it is therefore reasonable to postulate that this residue is included in a cleft which is rather easily opened by the denaturing agent. Once a general collapse of the molecule has been achieved, also Met-79 and Met-30 are photo-oxidized; accordingly, these residues must be deeply buried in the hydrophobic core of the molecule. In conclusion, our data suggest that three different degrees of reactivity, i.e. three different levels of burial, exist for the methionyl residues in ribonuclease A.

Table 2

Amino acid content of untreated and of 10-min photo-oxidized lysozome[a]

Amino acid	Theory	Untreated	Photo-oxidized in deionized water (pH 6.1)	Photo-oxidized in 84% acetic acid
Methionine sulfoxide[b]	0	0.0	0.9	1.9
Methionine[b]	2	1.8	1.2	0.0
Half-cystine	8	7.8	7.7	7.8
Tryotophan[c]	6	6.0	5.8	5.8
Tyrosine	3	2.8	2.8	2.6
Histidine	1	0.8	0.8	1.0

[a] The amino acids were determined chromatographically with a Technicon automatic analyzer after 22-hr hydrolysis in 6 N HCL at 110°C. The table includes only those amino acids which are known to be affected by photo-oxidation. No appreciable change was found in the other amino acids analyzed. The values in the table denote number of residues per molecule.
[b] Evaluated by automatic chromatographic analysis after 14-hr hydrolysis in 3.75 N NaOH at 100°C.
[c] Evaluated on the intact protein both by the method of Goodwin and Morton and by reaction with 2-nitrophenylsulfenyl chloride. The two methods agreed very well.

A similar differentiation between buried and exposed methionyl residues was detected in the case of lysozyme [15]. Actually, when this enzyme was irradiated in
the presence of hematoporphyrin in 5 to 25% acetic acid solution, only one of the
two methionyl residues was converted to the sulfoxide (see table 2); on the contrary,
in above 30% acetic acid solution, both the methionines were photo-oxidized. This
behaviour was correlated with a conformational change of lysozyme induced by
acetic acid [9]; as can be seen from fig. 5, the rotational strength of the aromatic
dichroic band, which measures the asymmetries caused by the protein conformation
in the light absorption of the tryptophyl and tyrosyl chromophores, is reduced to
zero above 30% acetic acid; this suggests that, in concentrated acetic acid, the protein conformation imposes no restrictions to the free motion of the aromatic side
chains. These data are strongly indicative that, in native lysozyme, one Met, Met-12,
is accessible to the photo-oxidizing agent, whereas Met-105 is shielded by an apolar
environment and cannot be oxidized unless the protein framework is previously destroyed.

In conclusion, this method appears to yield quite reliable data relative to the degree of burial or exposure of specific amino acids. Unfortunately, as outlined above,
this approach can be presently used only for two amino acids, that is methionine

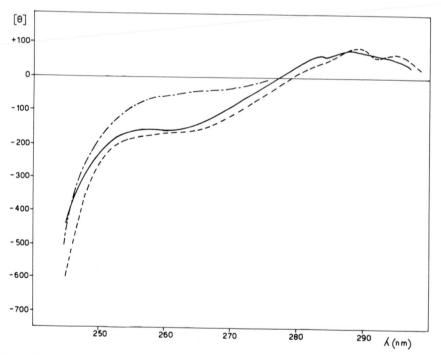

Fig. 5. The effect of acetic acid on the dichroic spectrum lysozyme: —— lysozyme in water,
––– lysozyme in up to 20% acetic acid, –.–.– lysozyme in above 30% acetic acid [9].

and cysteine. However, it may be hoped that the intensification of the studies in this area will provide new experimental conditions for the selective modification of other amino acids. It may be possible, moreover, to take advantage of the conformational transitions induced by other denaturing agents, such as temperature, as preliminary investigations from other laboratories would suggest [35]. In this way, the scope and the usefulness of the above described procedure can be appreciably enlarged.

3. Mapping of the three-dimensional topology

With the recent development of methods for the elucidation of the three-dimensional structure of proteins in the crystalline state, the detailed knowledge of the exact space coordinates of the amino acid side chains has been greatly advanced. The most promising and fruitful of these techniques proved to be X-ray crystallography [37]. However, the question of the location, in the spatial network, of specific residues in a protein molecule in solution still remains to be solved. The latter problem has been dealt with by several approaches, especially spectroscopic [38] and chemical modification [38, 32] studies. In the light of the preliminary results obtained by a novel photochemical procedure [18, 29, 26], we feel that photosensitized modification of amino acids can substantially improve the present information about the conformation of proteins in solution.

The procedure involves the irradiations of proteins containing the sensitizer covalently bound in known positions of the molecule. The sensitizer may be either naturally present in the protein, such as it occurs with the hemoproteins and the pyridoxal enzymes, or synthetically introduced by chemical reaction between it and selected functional groups of the protein. By irradiation of the protein-sensitizer complex, one should achieve the selective modification of those residues which are located in close proximity of the dye. Since the positions of the latter in the molecule is known, once the modified side chains have been identified, their spatial location can be deduced.

The basic assumption of these investigations is that the photodynamic action of the sensitizer on the substrates takes place within a well defined sphere of action. Evidently, the validity of such assumption is dependent on the mechanism of the process and, in particular, on the nature and the properties of the intermediate species involved. Unfortunately, while a great deal of researches have been carried out on the photoreactions sensitized by free dyes, very little definite work has been done on processes sensitized by a dye which has been precomplexed with the substrate. In any case, whatever the nature of the excited states involved, we ascertained that the final photo-oxidation products are identical in the two cases.

Now, if one can extrapolate the mechanistic data obtained for free sensitizers to our case, two divergent theories are to be taken into consideration. According to the first hypothesis [28], the active photo-oxidizing agents is a sensitizer–oxygen complex (photoperoxide):

$$^0Sens \xrightarrow{h\nu} {}^1Sens \xrightarrow[\text{intersystem}]{} {}^3Sens$$
$$\text{crossing}$$

$$^3Sens + O_2 \rightarrow [Sens - O_2]$$

$$[Sens - O_2] + \text{Substrate} \rightarrow Sens + \text{Oxidized substrate}.$$

An alternative theory [7, 20] postulates that the reactive species is singlet oxygen formed by energy transfer from the excited dye to the ground-state triplet oxygen:

$$^0Sens \xrightarrow{h\nu} {}^1Sens \xrightarrow[\text{intersystem}]{} {}^3Sens$$
$$\text{crossing}$$

$$^3Sens + {}^3O_2 \rightarrow {}^0Sens + {}^1O_2$$

$$^1O_2 + \text{Substrate} \rightarrow \text{Oxidized substrate}.$$

According to the first proposal, one can safely assume that the photo-oxidized measure residues are only those which can be actually contacted by the sensitizer within the protein molecule. Indeed, the overall process consists in transfer of oxygen from the excited sensitizer to the amino acid, and this evidently requires close approach (on the order of collisional diameters) of the donor and acceptor. The problem is more complex if the second path is operative. In this case, the step must occur over a very short range, probably by the exchange mechanism which requires overlap of the electron clouds, since long range energy transfer of this type has been observed only in rigid solutions [3]. However, excited $^1O_2$ possesses a rather long lifetime (10^{-6} sec) [8] and may therefore diffuse over a discrete distance before being deactivated. In any case, on a statistic ground, it appears reasonable to assume that the photo-oxidizable amino acids which are located nearest to the sensitizer are the most likely to be modified and, hence, the most frequently affected in a set of experiments. Our findings with model proteins, to be exposed hereafter, indicate that the specific photo-oxidation of the side chains close to the sensitizer takes invariably place by this method.

In order that the obtained results are absolutely reliable, some requisites must be met. First of all, the insertion of the sensitizer into the protein molecule must bring about no significant alteration of the tertiary structure. Therefore, once the protein has been labelled, a very rigorous conformational analysis must be carried out in order to ascertain that the spatial parameters of the molecular area, which is to be irradiated, are unchanged. In this connection, one can take advantage of the several refined and sensitive spectroscopic techniques presently available. Secondly, any intermolecular photoreaction must be avoided. Consequently, it would be desirable that the sensitizer is embodied into the interior of the molecule, hence shielded from any contact with the surface of other molecules. Furthermore, in order to minimize this kind of interactions, it is advisable to operate in sufficiently diluted solutions. In any case, the occurrence of intermolecular photosensitization can be easily detected by irradiating a mixture of labelled and unlabelled protein: if the

Table 3

Protein-labeling reagents acting as sensitizers for the photo-oxidation of amino acids

Reagent	Structure	Site of labeling	λ_{max} (mu)	Conditions of reaction
2,4-dinitrofluoro-benzene		α-amino group ε-amino group (lysine) thiol group (cysteine)	360 360 330	buffer pH 8.0 buffer pH 8.0 buffer pH 5.5
Carbon disulfide	CS_2	α-amino group	312	buffer pH 6.9
ω-DNP-imidic esters		ε-amino group (lysine)	365	buffer pH 8.5
Trinitrobenzenesulfonic acid		α-amino group ε-amino group (lysine)	420	buffer pH 7.5 buffer pH 7.5

Table 3 (Continued)

Reagent	Structure	Site of labeling	λ_{max} (mu)	Conditions of reaction
2-bromoacetoamide-4-nitrophenol		thioether function methionine	390	acidic media
Tetranitromethane	$C(NO_2)_4$	phenolic side chain (tyrosine)	428	buffer pH 8
2-hydroxy-5-nitrobenzyl bromide		3-position indole (tryptophan)	410	buffer pH 3-6
Methylisothiocyanate	$CH_3-N=C=S$	thiol group (cysteine)	320	buffer pH 5.5

intermolecular process does not take place, the unlabelled proteins is recovered unchanged after irradiation.

In the third place, the labelling must be performed in a highly selective degree, so that the photodynamic action of the introduced sensitizer operates over a well defined restricted area, and the results obtained can be interpreted in an unequivocal fashion. A list of reagents, which are specific for determined functional groups in proteins and, at the same time, are able to promote the photo-oxidation of amino acids, is shown in table 3. Moreover, sensitizers especial fit for a particular problem can be eventually built through the insertion of substituents. In this connection, the nitro, keto and thioketo functions proved to be the most suitable. All that is required is some synthetic organic chemistry.

Clearly, the usefulness of the method is enhanced of the sensitizer can promote the photo-oxidation of the largest possible number of amino acids. For this aim, the irradiations should be performed in slightly alkaline solutions (pH 7.0–8.5), where all the five photo-oxidizable amino acids are susceptible of being modified. We also screened the photosensitizing ability of the main chromophoric coenzymes: at pH 7–9 both porphyrins [15] and pyridoxal phosphate [16] mediate the photo-oxidation of the usual five amino acids.

As an example of the original possibilities opened by this approach, we shall briefly expose some results obtained with proteins containing both naturally or synthetically linked sensitizers.

3.1. *Horse heart cytochrome c.*

Cytochromes of the *c* type are unique in having the porphyrin prosthetic group covalently bound to the peptide chain through the sulfur atom of two cysteinyl residues; the heme group was shown by X-ray crystallography [6] to be inserted into a cleft with only one edge exposed to the solvent. Therefore, this protein appears to be suitable for this kind of investigation.

Actually, irradiation of ferricytochrome *c* at pH 5.7 (where only methionine is susceptible of photo-oxidation) and at pH 8.1 (where all the five amino acids can be modified) yielded the results reported in table 4. Sequence studies allowed us to identify the modified residues as His-18 and Met-80. Several lines of evidence indicate that these residues are located in the immediate surroundings of the heme. As shown in fig. 6, both products were chromatographically homogeneous, with R_f close to that of native cytochrome, pointing out that only limited conformational changes have occurred. Moreover, the visible absorption spectrum of both the photo-oxidized products lacked the 695 nm band (fig. 7), which is typical of the natural conformation of the heme environment [21]. Finally, His-18 is already known to be one of the two protein ligands for heme. As to the other ligand, several proposals have been made; a tryptophyal indole, a lysyl amino group, a tyrosyl oxygen, a methionyl sulfur or a second histidine. It appears suggestive to interpret our results as ruling out many of these hypotheses, since only one histidine and no tyrosine or tryptophan appear to be near the heme, and as producing additional evidence for Met-80 as a ligand.

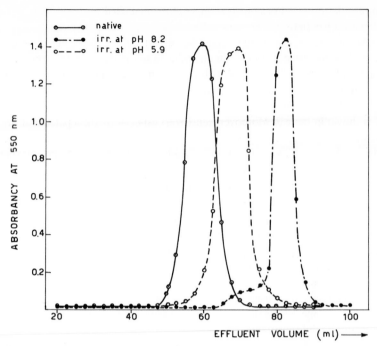

Fig. 6. Chromatographic analysis of ferricytochrome *c* on a column (0.9 × 50 cm) of Amberlite CG-50. Eluting buffer was 0.1 M phosphate, pH 8.5 [18].

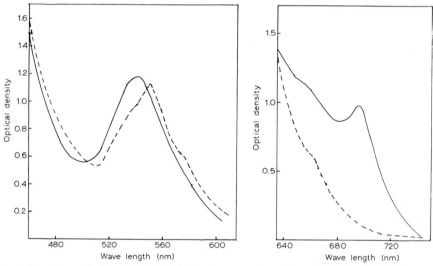

Fig. 7. The changes in the visible absorption spectrum of ferricytochrome *c* after 10 min photo-oxidation in water, pH 5.9; —— native, - - - - irradiated protein. All spectra were recorded at room temperature in aqueous solution.

Table 4

Amino acid analyses of native and photo-oxidized cytochrome *c*

Amino acid	Native	Photo-oxidized at pH 5.7	Photo-oxidized at pH 8.1
Lysine	18.8	18.7	18.7
Histidine	*2.9*	*3.0*	*1.9*
Arginine	2.0	2.1	2.0
Aspartic acid	8.1	7.9	7.8
Threonine	9.7	9.8	9.7
Serine	0.1	0.1	0.1
Glumatic acid	11.8	11.9	11.8
Proline	4.0	3.8	3.9
Glycine	12.3	12.0	11.9
Alanine	5.8	5.9	5.8
Tryptophan[a]	1.0	1.0	1.0
Valine	2.9	2.8	2.8
Isoleucine	5.8	5.7	6.0
Leucine	5.9	5.9	5.9
Tyrosine	3.9	4.0	3.9
Phenylalanine	4.0	3.8	3.9
Methionine[b]	*2.0*	*0.9*	*1.1*
Methionine sulphoxide[b]	0.0	1.0	1.0

The amino acids were determined chromatographically with a Carlo Erba 3A27 automatic analyzer after 22-hr hydrolysis in 6 N HCl at 110°C. The values in the table denote number of
residues per molecule.
[a]Determined on the intact protein by the method of Goodwin and Morton.
[b]Determined after alkaline hydrolysis in 3.75 N NaOH for 14 hr at 100°C.

This looks as an exciting example of the original prospects inherent to this
method. The possibility exists to extend these investigations to other hemoproteins,
such as the hemoglobins and many catalases and peroxidases; moreover, the coenzyme environment in the vast class of the pyridoxal enzymes and the flavoproteins
can be explored.

3.2. 41-dinitrophenyl-ribonuclease A (41-DNP-RNase A)

This enzyme was selected for a first approach since: (i) the protein can be selectively labelled at the amino group of Lys-41 by fluoro-dinitrobenzene [12]; that is
an efficient sensitizer for the photo-oxidation of amino acids; (ii) the labelling
group is hydrophobic and therefore should be sheltered in the interior of the molecule; (iii) the labelling does not induce any appreciable conformational alterations
[13]; (iv) the labelled group, Lys-41, is near the active site of RNase A.

As shown in table 5, after 10 min irradiation, one methionyl, one histidyl and
one tyrosyl residue were modified; sequence studies showed that the modified residues were Met-30, His-12 and Tyr-97. The same results were repeatedly obtained

Table 5

Amino acid analyses of 41-dinitrophenyl-ribonuclease A

Amino acid	Theory	41-DNP-RNase A irradiated	41-DNP-RNase A denatured and irradiated
Lysine	9	8.8	8.7
Histidine	4	2.8	0.0
Arginine	4	3.8	3.8
Aspartic acid	15	14.7	14.9
Threonine	10	9.7	9.7
Serine	15	15.0	14.8
Glumatic acid	12	11.8	11.7
Proline	4	4.0	3.9
Glycine	3	3.2	2.9
Alanine	12	11.7	11.9
1/2 Cystine	8	7.8	7.8
Valine	9	8.8	9.0
Isoleucine	3	2.8	2.9
Leucine	2	1.9	1.7
Tyrosine	6	4.9	0.0
Phenylalanine	3	2.8	2.9
Methionine[a]	4	3.1	0.0
Methionine sulphoxide[a]	0	1.1	3.8

The amino acids were determined with a Carlo Erba 3A27 automatic analyzer, after hydrolysis at 110°C for 22 hr. The values in the table denotes the number of residues per molecule.
[a]Determined after alkaline hydrolysis in 3.75 N NaOH at 100°C for 14 hr.

with samples irradiated for up to 120 min: this indicates that no appreciable disorientation of the protein active site took place after photo-oxidation. Moreover, the occurrence of intermolecular photosensitization was ruled out since native RNase A, irradiated in the presence of 41-DNP-RNase A, was recovered unchanged in the amino acid composition, UV-absorption spectrum, rotatory power and enzymic activity. It could also be argued that the limited number of modified amino acids is the consequence of special features inherent to the molecule of RNase A, independently of the tertiary structure. However, if the DNP-protein was denatured prior to irradiation, the total destruction of the hystidyl, tyrosyl and methionyl residues did occur (table 5). Therefore, all the available evidences point out that the selectively modified His-12, Met-30 and Tyr-97 are the only residues accessible to the sensitizer when RNase A is in its native conformation; moreover, these modified residues must be at or close to the active site.

Similar promising results are being obtained by us for the mapping of the active site of papain and chymotrypsin. It thus appears that this procedure is of general application.

It may then be possible by a concerted use of the spectroscopic techniques, of

the procedure for estimating the buried and exposed residues, and of the above outlined irradiations of proteins labelled with sensitizers of different bulkiness or specificity, to obtain a rather detailed description of the topology of a protein in solution.

References

[1] K.O. Bilitser and K.I. Kotkova, Photooxidation of fibrinogen and fibrin monomer. Ukrain.
Biochem. Zhur. *32* (1960) 3.
[2] C.A. Benassi, E. Scoffone, G. Galiazzo and A. Jori, Proflavine-sensitized photooxidation
of tryptophan and related peptides. Photochem. Photobiol. *6* (1967) 857.
[3] R.G. Bennet, Radiationless intermolecular energy transfer I. Singlet—singlet transfer. J.
Chem. Phys. *41* (1964) 3037.
[4] G.G. Bigelow, Difference spectra of ribonuclease and two ribonuclease derivatives. Compt.
Rend. Lab. Carlsberg *31* (1960) 305.
[5] G.C. Chatterjee and E.A. Noltmann, Dye-sensitized in photooxidation as a tool for the
elucidation of critical amino acid residues in phosphoglucose isomerase. European J. Biochem. *2* (1967) 9.
[6] R.E. Dickerson, M.L. Kopka, J. Weinzierl, J. Varnum, D. Eisenberg and E. Margoliash,
Location of the heme in horse heart ferricytochrome *c* by X-ray diffraction. J. Biol. Chem.
242 (1967) 3015.
[7] C.S. Foote, Mechanisms of photosensitized oxidation. Science *162* (1968) 963.
[8] C.S. Foote, W. Ando and R. Higgins, Chemistry of singlet oxygen IV oxygenations with
hypoclolorite hydrogen peroxide. J. Amer. Chem. Soc. *90* (1968) 975.
[9] G. Galiazzo, A.M. Tamburro and G. Jori, Photodynamic action of porphyrins on amino
acids and proteins. Spectral studies on the mono- and di-sulfoxide derivatives of lysozyme.
European J. Biochem. (1969) in press.
[10] A.W. Galston and R.S. Baker, Inactivation of enzymes by visible light in the presence of
riboflavin. Science *109* (1949) 485.
[11] T.T. Hershovits, Difference spectroscopy. Methods in Enzym. Vol XI, p. 748.
[12] C.H.W. Hirs, Reactions with reactive aryl halides. Methods in Enzym. Vol. XI (1967) p. 548.
[13] C.H.W. Hirs, On the structure of 41-dinitrophenyl-ribonuclease A. Solvent perturbation,
thermal transition, optical rotatory dispersion, and binding studies. Biochemistry *7* (1968)
3374.
[14] G. Jori, G. Galiazzo, A. Marzotto and E. Scoffone, Dye-sensitized selective photooxidation
of methionine. Biochim. Biopys. Acta *154* (1968) 1.
[15] G. Jori, G. Galiazzo and E. Scoffone, Photodynamic action of porphyrins on amino acids
and proteins I, Selective photooxidation of methionine in aqueous solution. Biochemistry
8 (1968) 2868.
[16] G. Jori, G. Galiazzo and E. Scoffone, Dye-sensitized selective photooxidation of cysteine.
Int. J. Protein Res. (1969) in press.
[17] G. Jori, G. Galiazzo, A.M. Tamburro and E. Scoffone, Dye-sensitized photooxidation as a
tool for determining the degree of exposure of amino acids in proteins: the methionyl
residues in ribonuclease A. J. Biol. Chem. (1969) submitted for publication.
[18] G. Jori, G. Gennari, G. Galiazzo and E. Scoffone, Photooxidation of ferricytochrome *c*.
Evidence for methionine-80 as a heme ligand. Biochem. Biophys. Res. Commun. (1969)
in press.
[19] G. Kartha, J. Bello and D. Harker, Tertiary structure of ribonuclease. Nature *213* (1967)
862.
[20] H. Kautsky, Die Wechselwirkung zwischen Sensibilisatoren und Sauerstoff in Licht. Biochem. Z. *291* (1937) 271.

[21] E. Margoliash and A. Scheiter, Cytochrome *c*. Advan. Protein Chem. *24* (1966) 113.

[22] M. Martinez-Carrion, C. Turano, F. Riva and P. Fasella, Evidence of a critical histidine residue in soluble aspartic aminotransferase. J. Biol. Chem. *242* (1967) 1426.

[23] M. Nakatani, Studies on histidine residues in hemoproteins related to their activities. Photooxidation of cytochrome *c* in the presence of methylene blue. J. Biochem. (Tokyo) *43* (1960) 633.

[24] M. Nakatani, Studies on histidine residues in hemoproteins related to their activities. J. Biochem. (Tokyo) *49* (1961) 98.

[25] W.J. Ray, Jr. and D.E. Koshland, Jr., Identification of aminoacids involved in phosphoglucomutase action. J. Biol. Chem. *237* (1962) 2493.

[26] M. Rippa and S. Pontremoli, Pyridoxal-5-phosphate as a specific photosensitizer for histidine residue at the active site of 6-phosphogluconate dehydrogenase. Arch. Biochem. Biophys. *103* (1969) 112.

[27] F. Sawada, Photosensitized modification of bovine pancreatic ribonuclease by a colored substrate analogue, 4-thiouridylic acid. J. Biochem. (Tokyo) *65* (1969) 767.

[28] G.O. Schenck, Aufgaben und Möglichkeiten der preparativen Strahlenchemie. Angew. Chem. *69* (1957) 579.

[29] E. Scoffone, G. Galiazzo and G. Jori, Dye-sensitized photooxidation as a tool for mapping certain amino acid residues in proteins. Biochem. Biophys. Res. Commun. (1969) in press.

[30] J. Shafer, P. Garonowski, R. Laursen, F.M. Finn and F.M. Westheimer, Products from the photolysis of diazoacetyl chymostrypsin. J. Biol. Chem. *241* (1966) 421.

[31] D. Shugar, Photosensibilisation des enzymes et des dérivés indoliques par la riboflavin. Bull. Soc. Chim. Biol. *33* (1951) 710.

[32] S.J. Singer, Covalent labelling of active sites. Advan. Protein Chem. *22* (1957) 1.

[33] L.A.Ae. Sluyterman, Photooxidation, sensitized by proflavine of a number of protein constituents. Biochim. Biophys. Acta *60* (1962) 557.

[34] J.D. Spikes and R. Livingston, The molecular biology of photodynamic action: sensitized photooxidations in biological systems. Advan. Radiat. Biol. *2* (1969) 28.

[35] J.D. Spikes and M.K. MacKnight, The time course of the destruction of amino acid residues during the photodynamic inactivation of ribonuclease A. Experientia (1969) in press.

[36] J.D. Spikes and R. Straight, Sensitized photochemical processes in biological systems. Ann. Rev. Phys. Chem. *18* (1967) 408.

[37] L. Stryer, Implications of X-ray crystallographic studies of protein structure. Ann. Rev. Biochem. *37* (1968) 25.

[38] S.M. Timasheff and M.J. Gorbunoff, Conformation of proteins. Ann. Rev. Biochem. *37* (1967) 13.

[39] L. Weil, On the mechanism of the photooxidation of amino acids sensitized by methylene blue. Arch. Biochem. Biophys. *110* (1965) 57.

[40] L. Weil, W.G. Gordon and A.R. Buchert, Photooxidation of amino acids in the presence of methylene blue. Arch. Biochem. Biophys. *33* (1951) 90.

[41] L. Weil, T.S. Seibles and T.T. Herskovitz, Photooxidation of bovine insulin sensitized by methylene blue. Arch. Biochem. Biophys. *111* (1965) 308.

CHAPTER 10

PHOTO-C$_4$-CYCLO-ADDITION REACTIONS TO THE NUCLEIC ACIDS

Luigi MUSAJO and Giovanni RODIGHIERO

Institute of Pharmaceutical Chemistry
of the University of Padova, Padova, Italy

The ability of some substances to give a C$_4$-cyclo-addition reaction to the nucleic acids under irradiation with long wavelength ultraviolet light has recently been discovered. Until now two groups of substances have been known to possess this property: the skin-photosensitizing furocoumarins and some aromatic polycyclic hydrocarbons, particularly benz(a)pyrene. In such a reaction an addition of the substances to the pyrimidine bases (in the case of benz(a)pyrene also to the purine bases) of the nucleic acids takes place, with the formation of a new cyclo-butane ring*.

1. Basic information on furocoumarins

Furocoumarins are a group of substances which occur in nature especially in plants of the families *Umbrelliferae* and *Rutaceae*; for a review article see [77]. Many furocoumarin derivatives have also been synthesized by several researchers [1, 16, 17, 20, 44, 45, 111]. Furocoumarins derive from the condensation of a furanic ring with the coumarine nucleus; there are many possibilities of condensation and therefore numerous isomers are possible. The isomer which has a linear structure and is known as psoralen (I) is the parent compound of the most important group of furocoumarin derivatives. The system of numeration used almost exclusively is that originally proposed by Späth [106] in his fundamental studies on the structure of naturally occurring furocoumarins. The Ring Index system by contrast is not in common use.

Among the many other isomers of psoralen we recall only iso-psoralen (II), known also as angelicin, and allo-psoralen (III), which is not found in nature, but has been obtained by synthesis.

*Kahn and Davies [43] presented at the Vth International Congress on Photobiology, Hanover, New Hampshire (USA), 1968, evidences of a binding of chloropromazine, a substance which has photosensitizing properties on the skin, with single stranded nucleic acids and only minimally with double stranded DNA under irradiation with long wavelength ultraviolet light. Moreover this substance can bind also with the pyrimidine and purine bases. The nature of this binding is, however, unknown.

I – Commonly used
numeration

I – Ring index numeration

II

III

Furocoumarins are known especially for the biologic effects which they can exert under irradiation with long wavelength ultraviolet light. The best known and most studied is the formation of erythema on human or guinea-pig skin, which manifests itself after a latent period of several hours and is followed, after some days, by a dark pigmentation [37, 51, 55]. Clinically this effect is connected with dermatitis due to contact with plants or vegetable materials containing furocoumarins [42, 51, 56]. Some of these compounds (8-methoxy-psoralen, 4,5′,8-trimethyl-psoralen) are used in the treatment of vitiligo, for obtaining the repigmentation of the leucodermic spots [33]; moreover they have been proposed to increase the tolerance of human skin to solar radiation [37].

In producing these skin-photosensitizing effects a great difference exists in the activity of the various furocoumarins; some compounds are very active, for instance psoralen and many of its methyl derivatives, others have only a weak activity, for instance angelicin, while many are completely inactive. Extensive studies have been made on the relationship between chemical structure and photosensitizing activity on the skin [19, 55, 57, 59, 61, 82, 83, 84, 89] and various tests have been used to evaluate the potency of the different substances [19, 55, 83]: the results obtained were strictly analogous and therefore at present a clear evaluation of the photosensitizing properties of furocoumarins is possible, as tables 1 and 2 show.

Various other effects have been obtained with furocoumarins, some quite recently, as indicated in table 3.

For several years the active furocoumarins have been called 'skin-photosensitizing' [61], keeping in mind their most studied effect. The most frequently used furocoumarins in photobiologic research are psoralen (I), 8-methoxy-psoralen or xanthotoxin (IV), 5-methoxy-psoralen or bergapten (V) (three naturally occurring substances) and two synthetic derivatives, 8-methyl-psoralen (VI) and 4,5′,8-tri-methyl-psoralen (VII).

Their mechanism of action remained unknown until a few years ago. When in

Table 1

Skin-photosensitizing activity of some furocoumarins. Test on human skin: substance 5 μg per cm^2; irradiation with a Philips HPW 125 lamp (365 nm) at 15 cm from the skin [55, 61]

Furocoumarins	Minimum irradiation time necessary for the appearance of erythema (min)	Relative activity (psoralen = 100)
Psoralen	6	100
4'-Methyl-psoralen	10	60
Xanthotoxin	16	37.5
4,5'-Dimethyl-xanthotoxin	18	33.3
4,4'-Dimethyl-psoralen	20	30
4-Methyl-xanthotoxin	20	30
Bergapten	22	27.5
5-Ethoxy-psoralen	25	24
4'-Methyl-xanthotoxin	25	24
5-Isopropyloxy-psoralen	35	17.1
4-Methyl-bergapten	40	15
5-Chloro-xanthotoxin	40	15
Angelicin	50	12
4',4-dimethyl-xanthotoxin	50	12
Allobergapten	50	12
4-Methyl-allobergapten	50	12
Bergapten-8-carboxylic acid methyl ester	50	12
3,4-Dimethyl-xanthotoxin	55	11
3-Methyl-xanthotoxin	60	10
Isobergapten	60	10
4-Methyl-angelicin	60	10
8-Benzyloxy-psoralen	60	10

1957, in cooperation with Santamaria [58], we studied the photosensitizing properties of furocoumarins, using some biological or non-biological tests, it appeared very clear that furocoumarins were completely lacking in photooxidative properties. By contrast these were the main characteristics of many other well known photodynamic substances, such as hematoporphyrin, chlorophyll, methylene blue, bengal rose, erythrosin, etc. Oginsky et al. [40, 80] confirmed this fact in experiments on bacteria cultures: the killing effect of xanthotoxin by irradiation at 365 nm was independent of the presence of oxygen. Analogous results were obtained also by Krauch et al. [47].

From these observations the conclusion was drawn that furocoumarins do not act as other photodynamic substances provoking a photooxidation of the substrates, but have a different mechanism of action, forming a group of substances with a unique photobiologic behaviour.

Therefore, since that time, many studies have been carried out by our group in Padua [6, 18, 38, 39, 60, 61, 62, 63, 92, 93, 96, 97] and by other researchers

[47, 52, 79, 112, 113], especially the group of Pathak et al. in Boston [85, 86, 87] to clarify this mechanism of action.

Finally, in 1964, we found that furocoumarins photoreacted with nucleic acids [65].

Table 2

Skin-photosensitizing activity of some methyl derivatives of psoralen. Test on guinea-pig skin: substance 2.5 μg per cm^2; irradiation with an Osram HWA 500 W lamp (strongly emitting at 365 nm, besides in the visible region) at a distance of 45 cm [19]

Furocoumarins	Minimum irradiation time necessary for the appearance of erythema (min)	Relative activity (psoralen = 100)
8-Methylpsoralen	5	540
5-Methylpsoralen	6	450
5',8-Dimethylpsoralen	8	337
4,8-Dimethylpsoralen	8	337
4,5',8-Trimethylpsoralen	10	270
4,4',8-Trimethylpsoralen	10	270
4',8-Dimethylpsoralen	12	225
4-Methylpsoralen	12	225
Psoralen	27	100
4'-Methylpsoralen	40	67
4,4'-Dimethylpsoralen	45	60
3,4,8-Trimethylpsoralen	50	54
3,4-Dimethylpsoralen	inactive up to 60 min	
3,4,4',8-Tetramethylpsoralen	inactive up to 60 min	

Table 3

Photobiologic effects produced by furocoumarins under irradiation at 365 nm

Skin-erythema, followed by dark pigmentation (Kuske [51])
Death of bacteria (Fowlks, Oginsky et al. [40, 80])
Formation of mutants in *Sarcina lutea* cultures (Mathews [52]); a mutagenic action was observed also on *Drosophila melanogaster* (Nicoletti and Trippa [79])
Formation of giant cells by mammalian cells adapted to in vitro growth (Colombo et al. [23])
Inactivation of DNA viruses (Musajo et al. [64])
Decrease of the template efficiency of DNA in RNA polymerase reaction (Chandra and Wacker [21])
Loss of tumor transmitting capacity of Ehrlich ascites tumor cells (Musajo et al. [74])
Formation of polynuclear cells in sea-urchin eggs fertilized with sperm irradiated in the presence of psoralen (Colombo [24])

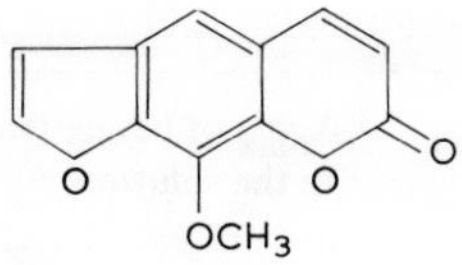

IV

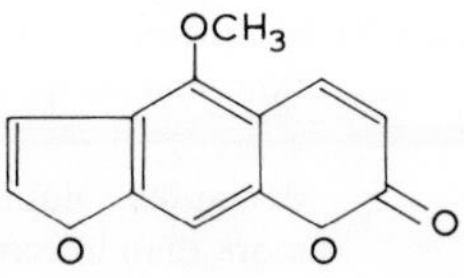

V

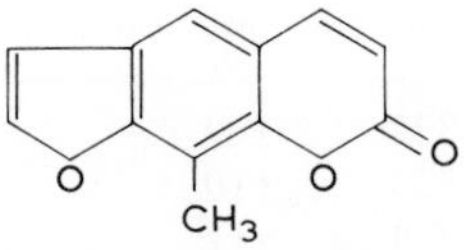

VI

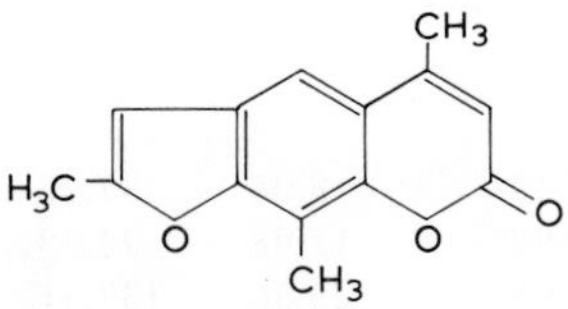

VII

2. Molecular complexes between furocoumarins and nucleic acids

When furocoumarins are added to a solution of nucleic acid without any irradiation, molecular complexes are formed in which very weak bonds are involved. These complexes have been studied by various experimental methods. For instance, it was observed [95] that the viscosity of aqueous solutions of native DNA increased sharply after addition of small amounts of furocoumarins. By contrast, after addition of coumarins, the increase was very slight. More evidence was obtained by the equilibrium dialysis method [95] which provided data indicating the amounts of furocoumarins linked to the nucleic acid.

Solubilization experiments were largely employed by several researchers for studying the interactions between nucleic acids and polycyclic aromatic hydrocarbons (see later). The same procedure was applied also to furocoumarins, especially to bergapten (5-methoxy-psoralen) which was very suitable, as it has very little solubility in water (5 μg/ml at 20°C) [26, 64].

The procedure consisted of shaking at a constant temperature an amount of solid furocoumarin with water or with an aqueous solution of nucleic acid. After the equilibrium was reached, the concentration of the furocoumarin in both the solutions was determined. The amount solubilized in excess of that which disolved in water, was considered to be bound to the nucleic acid. In table 4 are reported the results obtained in such solubilization experiments carried out with bergapten and native calf-thymus DNA [26]. We can see that the amounts of bergapten bound to DNA increased by increasing the concentration of DNA; however the ratio between the number of nucleotides present in DNA and the number of molecules of bergapten bound to it remained more or less constant in the range 1:36—1:39.

Moreover it is known that the formation of complexes between small molecules and a macromolecule is accompanied by variations of the spectrophotometric properties of the small molecules. In our case (see table 4) by increasing the concentra-

Table 4

Solubility of bergapten in aqueous solutions of native calf-thymus DNA at various concentrations and its spectrophotometric properties [26]

Concentration of DNA		Bergapten solubilized more than in water*		C_P/C_{Berg}	λ_{max} of bergapten in the solution		ϵ of bergapten at the λ_{max}*
(g%)	(μMP/l) [C_P]	(μg/ml)	(μM/l) [C_{Berg}]		(mμ)	shift with respect to aqueous solution* (mμ)	
0.05	1349	8.04	37.22	36.24	322	+ 9	9940
0.1	2699	15.98	73.98	36.48	323	+10	9050
0.2	5398	29.66	137.31	39.31	324	+11	8570
0.3	8097	44.11	204.21	39.65	325	+12	8320

*The solubility of bergapten in water at 20°C is 5 μg/ml; the λ_{max} in aqueous solution is 313 mμ and the ϵ_{313} value is 15,250.

tion of DNA we can see a progressive shift of the absorption maximum and a decrease of the absorbing power of bergapten.

As occurs in other cases, an increase in the ionic strength of the medium has a strong negative effect on the complex formation. The same negative effect is produced by the addition of small amounts of formamide [26].

Other than with native DNA, complexes are formed also with heat-denatured DNA and with RNA (ribosomal [26, 64] and transfer [98]). However, operating at room temperature (around 25°C) and using bergapten as furocoumarin, it was found that the formation of these complexes takes place at a reduced extent, while by decreasing the temperature until it reached a level of 2°C, it became much more evident (see later). By contrast with native DNA temperature has no influence on the complex formation [31].

Both active and inactive furocoumarins can form complexes with native DNA [26]. Therefore this property has no significance in connection with the explanation of the mechanism of the biological effects of furocoumarins. It is, however, a very important preliminary condition for the subsequent photoreaction, which of course takes place when the solution is irradiated.

3. The photoreaction (365 nm) between skin-photosensitizing furocoumarins and native DNA

In the early stages of our research it was not very easy to obtain evidence that a photochemical reaction occurs on irradiation of a solution containing native DNA and a furocoumarin. In fact no significant results have been obtained by examining the variation in viscosity, UV spectra and rotatory power of DNA solu-

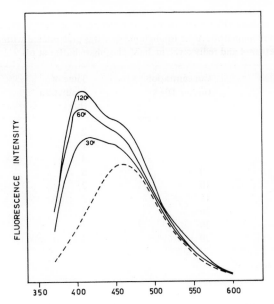

Fig. 1. Fluorescence spectra of an aqueous 0.2% DNA solution containing 20 μg/ml of psoralen before the irradiation $-----$ and after 30, 60 and 120 min of irradiation at 365 nm (Philips HPW 125 lamp at a distance of 20 cm) ————. Activating wavelength: 300 nm.

tions when irradiated at 365 nm both in the presence and in the absence of furo-coumarins.

We obtained the first indication of a photoreaction in 1964 by determining the fluorescence spectra of a solution containing native DNA and a skin-photosensitizing furocoumarin before and after irradiation at 365 nm [65]. In fig. 1 are reported, as an example, the fluorescence spectra of a solution of DNA and psoralen, before irradiation and after irradiation for increasing periods; we can see an evident and progressive modification of the spectrum: a new maximum at shorter wavelengths clearly appears, indicating the formation of a new molecular species.

Analogous modifications were observed also with other skin-active furocoumarins. In contrast in the absence of DNA or on irradiation of solutions containing inactive furocoumarins, the fluorescence spectrum showed no modifications.

The fluorescence of these solutions is due to the contained furocoumarins; therefore the results obtained indicated only that the skin-photosensitizing furo-coumarins underwent a modification after irradiation at 365 nm in the presence of native DNA.

In a subsequent study modifications were observed also in a property of DNA. In fact, the melting curves of DNA, determined before and after irradiation at 365 nm in the presence of skin-photosensitizing furocoumarins showed a sharp increase of the T_m value of DNA. These modifications did not appear on irradiation in the presence of skin-inactive furocoumarins (see table 5) [27].

Table 5

Increases in the T_m values of DNA after irradiation at 365 nm in aqueous 0.1% solution in the presence of various furocoumarins. After irradiation DNA was precipitated with ethanol, washed with 80% ethanol and redissolved in 3mM phosphate buffer at pH 7.2 [27]

Furocoumarins	Concentration (μg/mg DNA)	Time of irradiation (hr)	Increase of T_m value
None	0	2	0
Psoralen	20	1	+ 3.0
Psoralen	20	3	+ 6.1
Psoralen	20	5	+ 9.1
Psoralen	40	2	+ 7
Xanthotoxin	45	2	+ 10
Bergapten	20	2	+ 7.1
Bergaptol*	20	2	0
Isopimpinellin*	20	2	+ 1.3

*Skin-inactive furocoumarins

At this point we realized that after irradiation both furocoumarins and DNA underwent modifications, but this was not yet evidence of a combination between the two substances. That was obtained by experiments performed using labelled $-O^{14}CH_3$ bergapten [67, 68].

Aqueous solutions of DNA and labelled bergapten were irradiated at 365 nm. Then DNA was precipitated with ethyl alcohol, carefully washed, redissolved in water and its radioactivity examined. Non-irradiated samples were not radioactive, while the irradiated ones were radioactive and the radioactivity increased by increasing the period of irradiation. From the radioactivity measurements the amounts of bergapten linked to DNA were calculated. The results obtained have shown that as a consequence of the irradiation a stable covalent linkage of the furocoumarins to the macromolecule takes place. The quantum yield of this photoreaction (considered as the ratio between the number of bergapten molecules linked to DNA and the number of quanta absorbed by the solution after a given period of irradiation) was 5.2×10^{-3}. After 2 hr of irradiation (for the experimental conditions see [68]) the ratio between the number of bergapten molecules linked to DNA and the number of nucleotides present in the same DNA was 1:154. Moreover it was ascertained that this photoreaction was oxygen-independent, that is the same results were obtained operating either in the presence or in the absence of oxygen. Furthermore, in other experiments the photoreactions between native DNA and four furocoumarins, namely psoralen, bergapten, xanthotoxin and 8-methyl-psoralen, were shown to be independent of the temperature of irradiation [31].

The action spectra of two furocoumarins, xanthotoxin and bergapten, for the photoreactions with native calf-thymus DNA were also determined [30]. The following wavelengths were used to irradiate: 254, 265, 302, 312, 334, 365, 405 and 436 nm. After each irradiation with a constant number of quanta the amounts of furocoumarin linked to DNA were determined. The action spectra showed maxima at 312 nm for xanthotoxin and at 334 nm for bergapten, while the photoreactions were practically absent at 254—265 nm. With both furocoumarins the quantum yields of the photoreactions generally were low, as previously seen; however they showed their maximum values at 365—405 nm, decreasing then very rapidly at decreasing wavelengths. These results led to some conclusions:

(a) The behaviour of the quantum yield justifies the use of the 365 nm radiation, even if it does not correspond to an absorption maximum of the furocoumarins; in fact the relative highest quantum yield indicates that this radiation is utilized for the photoreaction better than others.

(b) The very low ability of the short wavelengths to provoke photoreactions, in spite of the fact that furocoumarins strongly absorb in this region, was demonstrated to be due very largely to the filter effect exerted by DNA itself on the same radiations [30].

(c) Analogies exist between these action spectra and the action spectrum of xanthotoxin for the production of erythema on human or guinea-pig skin determined by various researchers [15, 81, 88]: the active radiations is always that of the long-wave ultraviolet. This fact makes the connection between this photoreaction with native DNA and the photosensitizing effects exerted by the furocoumarins on the skin more evident.

At this point one may ask whether the insertion of a furocoumarin molecule into the macromolecule forming covalent linkages may be accompanied by a breakage of internucleotide bonds or by modifications in the conformation of the macromolecule. In fact Bellin et al. [8, 9] found that irradiation of DNA in the presence of some dyes, such as methylene blue and rose bengal, which photooxidize the guanine moieties, provoked a depolymerization of the macromolecule. In the photoreactions (365 nm) between DNA and psoralen Dall'Acqua et al. [29] by means of light-scattering measurements found that neither changes in molecular weight nor evident conformational modifications occurred in DNA.

However, one can observe that if single chain scissions occur in only a small number of sites, the macromolecule can retain its molecular weight and its conformation. This could be the case in these photoreactions: in the irradiation conditions used for the light-scattering experiments it was ascertained that 1 molecule of psoralen linked to every 110 nucleotides of DNA. Therefore if the insertion of the furocoumarin produced a breakage of the polynucleotide chain, this breakage would occur with the same frequency.

For this reason, in addition to the light-scattering measurements, the same researchers have also performed the viscosimetric 'i-assay' [41] on DNA after irradiation in the presence of psoralen, xanthotoxin and bergapten: the results indicated very clearly that no breakages of polynucleotide chains occur as a consequence of the photoreaction.

4. Photoreactions between furocoumarins and pyrimidine bases

With the aim of clarifying the reactive sites of DNA, the behaviour of the simple components of nucleic acids was examined by irradiating aqueous solutions of purine and pyrimidine bases, nucleosides and nucleotides at 365 nm in the presence of skin-photosensitizing furocoumarins [65, 66]. The results obtained showed that only pyrimidine derivatives photoreacted forming new compounds as revealed by paper or thin-layer chromatography.

Concerning the possibility of photoreaction of the irradiated compounds, it was known that pyrimidine bases, irradiated at 260 nm in frozen aqueous solutions, give photodimers reacting with their 5,6 double bond [11, 32, 104, 108]. Also by the irradiation of DNA and RNA at 260 nm, formation of such dimers occurs [105]. By contrast, when irradiated in aqueous liquid solution at 260 nm, pyrimidine bases do not give dimers but hydration products, by the addition of a molecule of water at the 5,6 double bond [53]. However, when irradiated at 365 nm both in aqueous liquid and in frozen solutions pyrimidine bases remain unchanged, because they do not absorb the radiation at this wavelength.

Furocoumarins, and also coumarins, when irradiated at 365 nm in the solid state, as well as in various organic solvent solutions or in frozen aqueous solution, give photodimers, which result from a C_4-cyclo-addition of 2 molecules at the 3,4 double bond of the α-pyrone ring [22, 48, 94, 99, 103, 109, 110]. No photodimers involving the 4′,5′ positions of the furanic ring are known. When irradiated in the presence of other substances, furocoumarins can give cyclo-addition products with these compounds: C_4-cyclo-additions to their 3,4-double bond have been obtained with some olefins, while with certain quinones furocoumarins yielded both C_4- and C_3O-cyclo-additions to their 4′,5′-double bond [36, 49].

In order to clarify the structure of the new compounds formed in the photoreac-

VIII   IX

X   XI

tions between furocoumarins and pyrimidine bases, some of these photoreactions have been worked out and studied also from a preparative point of view; small quantities of the compounds were isolated in a pure state and their structure elucidated by means of elemental analysis and of UV, IR and NMR spectroscopy [69, 70].

Now we can say that the photoreaction between a pyrimidine base and a furocoumarin leads to a C$_4$-cyclo-addition of the two substances, with the formation therefore of a new cyclo-butane ring. The pyrimidine bases react with their 5,6-double bond, while furocoumarins react either with their 3,4-double bond or with their 4',5'-double bond. Therefore two types of photoproducts may be obtained.

In the first type (VIII or IX: photoadduct is reported which involves psoralen and thymine*) the 4',5'-double bond of furocoumarin is involved. As the coumarine nucleus remains intact, compounds of this type have a brilliant violet fluorescence, when observed under long wavelength ultraviolet light.

In the second type of photoproducts (X or XI, always between psoralen and thymine) the 3,4-double bond of furocoumarin is involved. Compounds of this type are not fluorescent.

The experimental conditions under which the irradiation is performed may have a remarkable influence on determining the type of photoadduct which is formed. Experiments carried out using ^{3}H-psoralen and 2-^{14}C-thymine showed that when irradiating the substances in aqueous frozen solution both fluorescent and non-fluorescent photoproducts are formed: one fluorescent isomer and four non-fluorescent isomers were revealed. In contrast, when irradiating aqueous liquid solutions of the substances only non-fluorescent photocompounds (four isomers) were obtained (see table 6 [28]).

It has been ascertained that also in native DNA pyrimidine bases are the reactive sites, giving a photo-C$_4$-cyclo-addition reaction analogous to that which occurs using the simple compounds. The following experiment was performed [73]: a sample of native DNA was irradiated in aqueous solution in the presence of psoralen, then it was precipitated with absolute ethyl alcohol, washed with 70% ethyl alcohol and finally hydrolyzed by heating in acidic medium (70% perchloric acid or 0.4 N hydrochloric acid). Among the products of hydrolysis it was possible to isolate two fluorescent substances identical with the 4',5'-photoadducts of psoralen with thymine and with cytosine** and a non fluorescent one identical with the 3,4-photoadduct between psoralen and thymine.

Some quantitative results were also obtained using ^{3}H-psoralen: it appeared that

---

*The two forms derive from the double possibility of addition of the two substances. Moreover, we must consider that each of these structures can exist in two isomeric forms, as a consequence of the different positions of the H atoms and of the methyl group in respect to the plane determined by the furan ring. Therefore, on the whole, four isomeric forms are possible.

**The photoadducts of cytosine easily undergo a hydrolytic deamination transforming them into uracil photoadducts [69].

Table 6
Yields of most important products formed in the photoreactions between psoralen and thymine.
Aqueous solutions containing 30 μg/ml of psoralen and various amounts of thymine were irradiated in the liquid or frozen state at 365 nm for 1 hr (0.98 mW/cm^2). The reported values indicate the per cent amounts of psoralen initially present which reacted to give the compounds.
The R_f values have reference to ascending thin-layer chromatography with cellulose plates developed by water [28]

Compounds	Irradiation of the solutions at the frozen state Molecular ratio psoralen— thymine			Irradiation of the solutions at room temperature Molecular ratio psoralen— thymine		
	1:1	1:10	1:100	1:1	1:10	1:100
Fluorescent photo-adduct R_f = 0.60	trace	2.1	7.5	–	–	–
Non-fluorescent photoadduct R_f = 0.85	1	11.4	11.3	4.3	26.5	24
Non-fluorescent photoadduct R_f = 0.78	trace	4	5.8	1.25	4	5
Non-fluorescent photoadduct R_f = 0.58	trace	4.2	6.3	0.7	6	5
Non-fluorescent photoadduct R_f = 0.48	trace	trace	trace	trace	trace	trace
Dimer of psoralen	75	55.7	8	–	–	–

the amount of fluorescent photoadducts obtainable after hydrolysis of native DNA in the presence of labelled psoralen was nearly three times as great as that of non-fluorescent one [12].

5. Photoreactions between furocoumarins and denatured DNA or RNA

Furocoumarins under irradiation at 365 nm showed an ability to photoreact, other than with native DNA, with heat-denatured calf thymus DNA, ribosomal yeast RNA and some polynucleotides [31, 59, 68, 98]. However in these cases the temperature at which irradiation is performed may have marked influences of various kinds [31].

The behaviour of psoralen is different from that of other furocoumarins, such as bergapten, xanthotoxin and 8-methyl-psoralen. Therefore they must be considered separately.

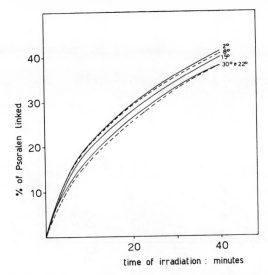

Fig. 2. Photoreactions between psoralen and heat-denatured calf-thymus DNA performed at various temperatures. Aqueous 0.1% solutions of denatured DNA containing 20 μg/ml of ^{3}H-psoralen were irradiated at various temperatures (2–30°) with 365 nm radiation (2.9 × 10^{16} hv/sec). Percentages of psoralen (referred to the amount initially present) linked to DNA as a function of the period of irradiation are reported.

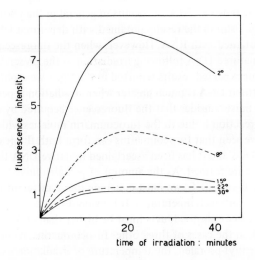

Fig. 3. Intensities of the fluorescence acquired by denatured DNA after irradiation (365 nm) in the presence of psoralen at various temperatures. Experimental conditions as indicated in fig. 2.

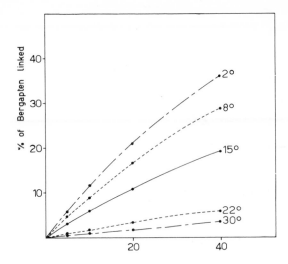

Fig. 4. Photoreactions between bergapten and denatured DNA performed at various temperatures. Aqueous 0.1% solutions of denatured DNA containing 5 μg/ml of [$-O^{14}CH_3$]-bergapten
have been irradiated at various temperatures (365 nm; 2.9×10^{16} $h\nu$/sec). Percentages of bergapten (referred to the amount initially present) linked to DNA as a function of the period of
irradiation are reported.

In the case of psoralen, the rate of the photoreactions is independent of the
temperature within the range 2–30°C, as occurs in general in the photoreactions
with native DNA. Fig. 2 shows the results obtained with denatured DNA. They are
analogous to those obtained with RNA. However, when the fluorescence acquired
by the samples of denatured DNA following irradiation in the presence of psoralen
was determined, the unexpected results reported in the fig. 3 were obtained: the
fluorescence of denatured DNA is much greater when irradiation is performed at a
low temperature. We must consider that the fluorescence acquired by denatured
DNA after the photoreaction is due to the furocoumarin moieties which are linked
to it. However we have seen that furocoumarins can form both fluorescent and non-
fluorescent photoadducts and it has been ascertained in native DNA that both
types of photoadducts are formed. As the amount of psoralen linked to denatured
DNA is the same at 2°C and 30°C, we must conclude that a different type of photo-
addition takes place at various temperatures. It is evident that at 2°C the ratio
fluorescent/non-fluorescent photoadducts must be higher than at 30°C.

On the other hand, in the cases of three other furocoumarins, namely xanthotoxin, bergapten, 8-methyl-psoralen, the temperature of irradiation exerts its influence not on the fluorescence acquired by denatured DNA and RNA, but on the
rate of the photoreactions. Fig. 4 shows, as an example, the results obtained in the
photoreaction between bergapten and denatured DNA. It can be seen that at room

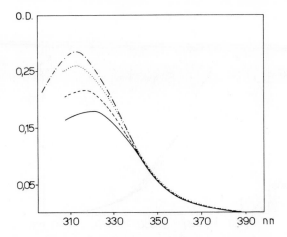

Fig. 5. Absorption UV spectra of an aqueous 0.1% solution of denatured DNA containing 4 μg/ml of bergapten at 30°C.................., at 15°C – – – – – – – and at 2°C ————; aqueous solution of bergapten (4 μg/ml) –.–.–.–.–.–.

temperature (25–30°C) the photoreaction is very slow, but, by decreasing the temperature, it becomes more and more rapid.

As the photoreactions are normally independent of temperature, a suggestion has been made that temperature may exert an influence on the preliminary formation of complexes between furocoumarins and macromolecules. In fact complexes are a very useful condition for the photoreaction [68] . Therefore, UV spectra of aqueous solutions of furocoumarins and nucleic acids have been determined at various temperatures. Figs. 5 and 6 show the UV spectra of aqueous solutions of bergapten* in the presence of denatured and native DNA respectively; they show two different situations. As we have previously seen, the formation of a complex between a small molecule and a macromolecule can be detected by observing variations in the optical properties of the small molecule. Fig. 5 shows that at 30°C there is only a small formation of the complex between bergapten and denatured DNA, but by decreasing the temperature, there is an evident decrease of the optical density of bergapten and a shift in the maximum of absorption, and this indicates an increasing formation of the complex.

In contrast, fig. 6 shows that in the presence of native DNA the complex is well formed even at 30°C and a decrease of the temperature has practically no influence on the spectrum of bergapten, that is on the formation of the complex. We recall that the photoreaction between bergapten and native DNA in independent of the temperature of irradiation.

*Here are reported only the results obtained with bergapten; with xanthotoxin and 8-methylpsoralen the results were strictly analogous.

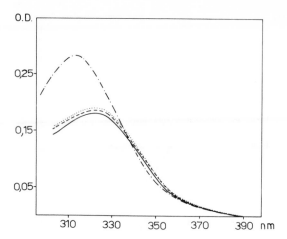

Fig. 6. Absorption UV spectra of an aqueous 0.1% solution of native DNA containing 4 μg/ml of bergapten determined at 30°C........................, at 15°C – – – – – – – and at 2°C —————; aqueous solution of bergapten (4 μg/ml) –.–.–.–.–.–.

Moreover, if we want to make a comparison between the photoreactivity of furocoumarins with native DNA and that with denatured DNA or ribosomal RNA, we can say that: (a) psoralen, independently of the temperature, photoreacts almost equally with RNA or native and denatured DNA; (b) bergapten and xanthotoxin at room temperature (25–30°C) photoreact much more easily with native than with denatured DNA or with RNA, while at 2°C the photoreactivities with the different nucleic acids become almost equal.

Finally, from all the data obtained, it appeared evident that the behaviour of a single furocoumarin in the photoreactions with the various nucleic acids may show remarkable differences from that of other furocoumarins. Therefore general conclusions concerning the whole group of furocoumarins cannot be drawn from results obtained in experiments performed with only one of these compounds.

6. Photoreactions with nucleic acids and photobiologic effects of furocoumarins

(A) After having reviewed the results obtained in vitro, the question arises, whether these photoreactions can really explain photobiologic effects. The main part of these, which are considered in table 3, are clearly connected with damage to DNA. In regard to the photobiologic effect produced on the skin, that is erythema followed by pigmentation, at present the connection between damage to DNA and the appearance of erythema is not clear; though other studies have been carried out to test this hypothesis. One of these consisted of evaluating the photoreactivities of a number of furocoumarins with native DNA by irradiating at 365 nm and comparing the results thus obtained with those previously found while determining the skin-photosensitizing activities of the same substances by irradiating also at 365 nm [100].

As we have previously said, the various furocoumarin derivatives have a very different skin-photosensitizing activity. In our early research in this field, we used a test to evaluate this activity [55, 61]: it consisted of placing 5 μg of substance per cm^2 of skin (the backs of human volunteers were used), irradiating at 365 nm under standard conditions and determining the minimum irradiation time necessary for the appearance of erythema. Considering the activity of psoralen as equal to 100, the relative activities of the other substances were calculated.

Pathak et al. [83] used another test with strictly analogous results, operating on guinea-pigs and determining the minimum amount of substance necessary to produce erythema after a constant period of irradiation.

Recently we have modified our previous test in order to make it more suitable for assaying the very active methyl-derivatives of psoralen [19]; guinea-pig skin was used and the quantity of substance applied was reduced (2.5 μg/cm^2), but the minimum irradiation time (under standard conditions) necessary to obtain erythema was always determined. The relative skin-photosensitizing activities of the various furocoumarins reported in table 7 refer to this test.

Now, using a number of tritiated furocoumarins, chosen from among those which have either a very high activity on human and guinea pig skin, or only a moderate activity, or which are inactive, we determined the rates of their photoreactions with native DNA and with ribosomal yeast-RNA by irradiation at 365 nm under standard conditions (figs. 7 and 8).

In order to obtain data which would be directly comparable with those of the photosensitizing activity on the skin, we have calculated from the results obtained the time of irradiation necessary to produce a linkage to DNA corresponding to 20% (10% in the case of RNA) of the amount of furocoumarin initially present. On the basis of the numbers so obtained, by considering the photoreactivity of psoralen as equal to 100, the relative photoreactivities of the compounds have been calculated.

The results are reported in table 7. The data on the relative photoreactivity with native DNA appear to be very close to the data on the relative skin-photosensitizing activity. Therefore, they demonstrate that a tight correlation exists between the capacity of furocoumarins to photoreact with native DNA and that of provoking skin-erythema.

In contrast, the data referring to RNA appear to be not so closely correlated to the data on skin-photosensitizing activity: in fact various evident discrepancies exist. Therefore, we must conclude that RNA, even if it can photoreact with furocoumarins, appears to be less involved than DNA in producing the photobiologic effects of these substances.

Some evidence has confirmed that photoreactions between furocoumarins and nucleic acids can occur also in vivo, that is inside the cells. In fact, Pathak and Krämer [90] have recently found that after irradiation of guinea-pig skin treated

Table 7

Relative photoreactivities with native DNA and with ribosomal RNA and relative skin-photosensitizing activities of some furocoumarins

Furocoumarins	Photoreactivity with native DNA		Photoreactivity with ribosomal RNA		Relative skin-photosensitizing activity: test on guinea-pig skin [19] (psoralen = 100)
	Irradiation time necessary for a 20% linkage (sec)	Relative photo-reactivity (psoralen=100)	Irradiation time necessary for a 10% linkage (sec)	Relative photo-reactivity (psoralen=100)	
8-Methyl-psoralen	90	373	126	190	540
5-Methyl-psoralen	144	233	360	66.6	450
Psoralen	336	100	240	100	100
Xanthotoxin	558	60	3744	6.4	71
Bergapten	1026	33	4440	5.4	61
Angelicin	2160	15	1800	13.3	12*
4',5'-Dihydro-psoralen	–	–	–	–	inactive
Xanthotoxol (5-hydroxy-psoralen)	–	–	–	–	inactive

*The low skin-photosensitizing activity of angelicin did not allow a correct determination on guinea-pig skin, which has a lesser sensitivity than human skin [19]. Therefore this datum is referred to the test on human skin [55, 61].

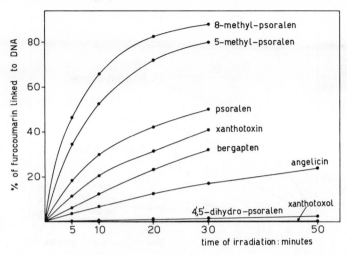

Fig. 7. Photoreactions between some furocoumarins and native DNA by irradiation at 365 nm and 22°C. Percentages of furocoumarins (referred to the amount initially present) linked to DNA as a function of the period of irradiation are reported (from Experientia [100]).

with ^{3}H-4,5',8-trimethyl-psoralen, radioactivity was present in DNA and in RNA extracted from the skin, whereas it was absent in protein fractions.

After this evidence we may conclude that the photoreactions with native DNA must really be considered as the first act which initiates the photobiologic effects of furocoumarins.

(B) The effects on viruses of irradiation at 365 nm in the presence of skin-photo-sensitizing furocoumarins have been also studied [64]. Some DNA viruses were completely inactivated while some RNA viruses were more resistant.

Recently two DNA viruses were studied in more detail, they were an adenovirus, which lacks a lipoproteic envelop, and a pseudorabies virus, belonging to the herpetic group, which has such an envelop [76a].

The suspensions of these two viruses completely lost their infectivity after irradiation for a few minutes with long-wave ultraviolet light in the presence of 8-methyl-psoralen at a concentration of 10 μg/ml.

The *adenovirus* (2×10^7 infectious unit per ml) irradiated under these conditions for 60 min exerted antigenic activity, that is it was able to produce antibodies at high titer in rabbits.

Pseudorabies herpetic virus (5×10^6 infectious units per ml) irradiated also for 60 min under the same conditions and inoculated in mice produced an antibody response of moderate intensity. However, this is a noteworthy result, considering the weak immunizing power of herpetic vaccines inactivated with other methods.

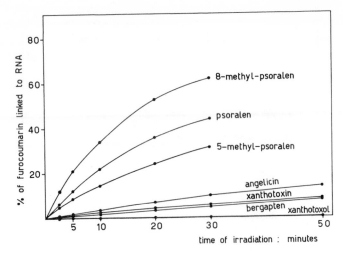

Fig. 8. Photoreactions between some furocoumarins and ribosomal yeast-RNA by irradiation at 365 nm and at 22°C. Percentages of furocoumarins (referred to the amount initially present) linked to RNA as a function of the period of irradiation are reported.

(C) In 1967 Musajo et al. [72, 74] demonstrated that the Ehrlich ascites tumor cells irradiated at 365 nm in the presence of various skin-photosensitizing furocoumarins (psoralen, xanthotoxin, bergapten) lose their tumor transmitting capacity. To obtain direct evidence that this biological effect was due to a linkage of the furocoumarins with the cellular DNA, the same researchers [75] irradiated at 365 nm a pool of Ehrlich ascites tumor cells in the presence of ^{3}H-psoralen. After this, DNA was extracted from the cells: it was found to be radioactive, but only after irradiation. Moreover, after hydrolysis of the extracted DNA a substance was identified which was identical with the fluorescent photoadduct obtained from the irradiation of psoralen and thymine [69].

Erhlich ascites tumor cells, after inactivation by irradiation in the presence of psoralen, seemed to behave like the untreated tumor cells towards Wright's liquid and trypan blue; no difference has been observed in the uptake of oxygen, as determined with the Warburg's manometric method.

Many studies have been carried out in the last five years by Musajo et al. [76b] to ascertain whether photoinactivated Ehrlich ascites tumor cells had the capacity to induce protection against the same tumor.

Various skin-photosensitizing furucoumarins were used, but the most significant results have been obtained with psoralen and 8-methyl-psoralen. Suspensions of Ehrlich ascites tumor cells have been prepared by diluting the ascitic fluid, extracted from the albino mice bearing the tumor, to a concentration of 2×10^7 cells per ml and by adding the furocoumarin at a concentration of 10 μg per ml. After irradiation at 365 nm, cells were intraperitoneally injected into albino mice at a dose of 10^7 cells per mouse, within one hour after drawing from the animals.

Various groups of mice have received one, two or four injections spaced at weekly intervals. After this, the same animals have been injected with 2×10^3 untreated cells. This dose provoked the death in 70% of the control mice, in a period of 60 days. The results of various experiments in which 568 animals on the whole have been used showed a significant decrease in mortality, that is a protection against the tumor in those animals which had received tumor cells photoinactivated by irradiation in the presence of psoralen or 8-methyl-psoralen.

7. Photoreaction (365 nm) between benz(a)pyrene and DNA

It has been known for a long time that polycyclic aromatic hydrocarbons can exert photodynamic properties on various substrates, for instance on *Paramecium caudatum* [34, 54], on yeast-cells [25], on DNA solutions producing depolymerization [46], on serum proteins [101]. These effects are explainable by their photo-oxidative properties. Santamaria [102] has carried out extensive studied on the photodynamic activities of these substances on serum proteins, finding some correlations with the carcinogenic activity of the same substances.

However, polycyclic aromatic hydrocarbons when applied to the skin and irradiated with long wavelength ultraviolet light provoke an erythemal response which is different from that provoked by photodynamic dyes such as hematoporphyrin, methylene blue, bengal rose, eosin and others, and has many analogies with that provoked by furocoumarins [58]. In fact, in the first case (after intradermical in-

Table 8

Skin-photosensitizing activity on guinea-pig of some polycyclic aromatic hydrocarbons compared with their carcinogenic activity. Substance 10 μg per cm^2; irradiation with an Osram HWA 500 W lamp (see table 2) at 30 cm from the skin [4]

Compounds	Minimum irradiation time necessary for the appearance of erythema (min)	Carcinogenic activity, according to Epstein [35]
7,12-Dimethyl-benz(a)anthracene	10	+++
7-Methyl-benz(a)anthracene	10	+++
12-Methyl-benz(a)anthracene	10	+++
6-Methyl-benz(a)anthracene	15	+
Benz(a)pyrene	20	+++
3-Methyl-cholanthrene	30	+++
Benz(c)fluoranthene	30	++
5-Methyl-benz(c)phenanthrene	40	++
3-Methyl-pyrene	40	−
Phenanthrene	45	−
Dibenz(a,h)anthracene	inactive up to 60	+++
Pyrene	inactive up to 60	−
Anthracene	inactive up to 60	−

jection of an aqueous solution of a dye) erythema appears immediately after irradiation and disappears again after some hours. In the second case, on the contrary,
erythema appears only after a latent period of several hours following irradiation,
lasts some days and is succeeded by a dark pigmentation. In table 8 are summarized
the skin-photosensitizing properties of a number of polycyclic hydrocarbons [4].

These different effects have been correlated to the hydrophylic properties of the
dyes and the lipophylic ones of the hydrocarbons; in the first case the site of action
may be extracellular, in the second it can be endocellular [10].

The possibility of interaction between polycyclic aromatic hydrocarbons and
nucleic acids has been known for a long time. Many studies have been made on the
formation of complexes [7, 13, 14]; the hypothesis of an intercalation of planar
molecules of hydrocarbons between the planes of the bases in DNA has recently
been confirmed [78] by means of flow-dichroism measurements; it was demonstrated that in the complex the plane of the molecule of benz(a)pyrene and some
other analogous compounds is parallel to the plane determined by the purine and
pyrimidine bases in the rigid structure of native DNA.

Tso and P. Lu in 1965 [107] found that after irradiation of an aqueous solution
of DNA containing ³H-benz(a)pyrene, DNA isolated from the solution was radioactive, indicating thus that a stable combination between the two substances took
place.

Successively Rice [91], by irradiating solutions containing benz(a)pyrene and
the purine and pyrimidine bases commonly present in nucleic acids, obtained some
photoadducts. Both pyrimidine and purine bases photoreacted; to the photoadduct
formed in the irradiation of benz(a)pyrene and cytosine he attributed the structure
XIII.

Antonello, Carlassare and Musajo, after studying the photooxidation products formed by the irradiation of benz(a)pyrene at 365 nm [2], reinvestigated also the photoreaction at 365 nm between benz(a)pyrene and pyrimidine or purine bases [4]. From the irradiation products of benz(a)pyrene and thymine they isolated and analyzed a substance which corresponded to those previously obtained by Rice [91]; however they demonstrated by means of spectroscopic evidence that the structure which must be attributed to this photoadduct was that represented by formulas XIV and XV.

Photoadducts between benz(a)pyrene and other pyrimidine and purine bases had a very similar spectrophotometric behaviour and therefore the above mentioned authors attributed an analogous structure also to these compounds.

Moreover Antonello, Carlassare and Musajo [5] from a sample of DNA irradiated in the presence of benz(a)pyrene have been able to isolate very small amounts of some photoadducts very similar to those obtained from the small molecules; among these the photoadduct between benz(a)pyrene and thymine has definitely been identified.

References

[1] C. Antonello, Sintesi di metilxantotossine. Gazz. Chim. Ital. *88* (1958) 415–429.

[2] C. Antonello and F. Carlassare, Prodotti di fotoossidazione del benzo(a)pirene in luce UV, Atti Ist. Veneto Sci. Lett. Arti *122* (1963) 9–19.

[3] C. Antonello, F. Carlassare, G. Rodighiero and L. Musajo, La fotoreazione tra flavin-adenin-dinucleotide e benz(a)pirene. Gazz. Chim. Ital. *94* (1964) 1093–1100.

[4] C. Antonello, F. Carlassare and L. Musajo, Fotoreazione a 3655 Å tra benzo(a)pirene e basi pirimidiniche e puriniche. Gazz. Chim. Ital. *98* (1968) 30–41.

[5] C. Antonello and F. Carlassare, Isolation of a benz(a)pyrene-thymine photoadduct from DNA hydrolized after irradiation at 365 nm in the presence of benz(a)pyrene. Z. Natur-forschg. *25b* (1970) 1269–1271.

[6] G.F. Azzone, G. Rodighiero, L. Musajo, F. Dall'Acqua, U. Fornasiero and G. Malesani, Ricerche enzimologiche sui flavin-fotocomposti. Gazz. Chim. Ital. *94* (1964) 1101–1107.

[7] J.K. Ball, J.A. McCarter and F.M. Smith, The interaction in vitro of polycyclic aromatic hydrocarbons with deoxyribonucleic acid. Biochim. Biophys. Acta *103* (1965) 275–285.

[8] J.S. Bellin and L.I. Grossman, Photodynamic degradation of nucleic acids. Photochem. Photobiol. *4* (1965) 45–53.

[9] J.S. Bellin and G.A. Yankus, Effects of photodynamic degradation on the viscosity of deoxyribonucleic acid. Biochim. Biophys. Acta *112* (1966) 363–371.

[10] A. Bergamasco, Studio comparativo sull'azione di alcuni fotosensibilizzatori. Arch. Ital. Dermatol. Sifil. Venereol. *16* (1940) 131–154.

[11] R. Beukers and A. Berends, Isolation and identification of the irradiation product of thymine. Biochim. Biophys. Acta *41* (1960) 550–551.

[12] F. Bordin, L. Musajo and R. Bevilacqua, Fluorescent and non-fluorescent C₄-cycload-ducts in the photoreaction at 365 nm between psoralen-³H and DNA. Z. Naturfschg. *24b* (1969) 691–693.

[13] E. Boyland and B. Green, The effect of formaldehyde and thermal denaturation on the solubilisation of polycyclic hydrocarbons by aqueous solutions of deoxyribonucleic acid. Biochem. J. *92* (1964) 4–7 C.

[14] E. Boyland, B. Green and S.L. Liu, Factors affecting the interaction of polycyclic hydrocarbons and deoxyribonucleic acid. Biochim. Biophys. Acta *87* (1964) 653–663.

[15] H.N. Buck, I.A. Magnus and A.D. Porter, The action spectrum of 8-methoxy-psoralen
for erythema on human skin. Preliminary studies with a monochromator. Brit. J. Dermatol. *72* (1960) 249–255.

[16] G. Caporale and C. Antonello, Sintesi di alcuni derivati furocumarinici. Farmaco Ed. Sci.
13 (1958) 363–367.

[17] G. Caporale, Derivati del bergaptene e dell'allobergaptene. Ann. Chim. (Roma) *50* (1960)
1135–1149.

[18] G. Caporale, G. Rodighiero, C. Giacomelli and C. Ballotta, La fotoreazione tra flavin-
mono-nucleotide e metilpsoraleni. Gazz. Chim. Ital. *95* (1965) 513–532.

[19] G. Caporale, L. Musajo, G. Rodighiero and F. Baccichetti, Skin-photosensitizing activity
of some methyl-psoralens. Experientia *23* (1967) 985–986.

[20] G. Caporale and A.M. Bareggi, Su alcuni metilpsoraleni. Gazz. Chim. Ital. *98* (1968)
444–457.

[21] P. Chandra and A. Wacker, Photodynamic effects on the template activity of nucleic
acids. Z. Naturforschg. *21b* (1968) 663–666.

[22] G. Ciamician and P. Silber, Chemische Lichtwirkungen. Chem. Ber. *35* (1902) 4128–4131.

[23] G. Colombo, A.G. Levis and V. Torlone, Photosensitization of mammalian cells and of
animal viruses by furocoumarins. Prog. Biochem. Pharmacol. *1* (1965) 392–399.

[24] G. Colombo, Photosensitization of sea-urchin sperm to long-wave ultraviolet light by
psoralen. Exp. Cell Res. *48* (1967) 167–169.

[25] E.S. Cook, M.I. Hart and R.A. Joly, The effect of 1,2,5,6-dibenzanthracene on the growth
and respiration of yeast. Science *87* (1938) 331.

[26] F. Dall'Acqua and G. Rodighiero, The dark-interaction between furocoumarins and nucleic acids. Rend. Accad. Naz. Lincei (Roma) *40* (1966) 411–422.

[27] F. Dall'Acqua and G. Rodighiero, Changes in the melting curve of DNA after the photoreaction with skin-photosensitizing furocoumarins. Rend. Accad. Naz. Lincei (Roma)
40 (1966) 595–600.

[28] F. Dall'Acqua, S. Marciani, F. Bordin and R. Bevilacqua, Studies on the photoreaction
(365 nm) between psoralen and thymine. Ric. Sci. (Roma) *38* (1968) 1094–1099.

[29] F. Dall'Acqua, M. Terbojevich and F. Benvenuto, Light-scattering and viscosimetric
studies on DNA after the photoreaction with some furocoumarins. Z. Naturforschg. *23b*
(1968) 943–945.

[30] F. Dall'Acqua, S. Marciani and G. Rodighiero, The action spectrum of xanthotoxin and
bergapten for the photoreaction with native DNA. Z. Naturforschg. *24b* (1969) 667–671.

[31] F. Dall'Acqua, S. Marciani and G. Rodighiero, Photoreactivity (3655 Å) of various skin-
photosensitizing furocoumarins with nucleic acids. Z. Naturforschg. *24b* (1969) 307–314.

[32] J. Eisinger and A.A. Lamola, Mechanism of thymine photodimerization. Mol. Photochem. *1* (1969) 209–223.

[33] A.M. El Mofty, Vitiligo and Psoralens (Pergamon Press, London, 1968) pp. 1–221.

[34] S.S. Epstein, The photodynamic activity of polycyclic hydrocarbons carcinogens. Acta
Union International contre le cancer *19* (1963) 599–601.

[35] S.S. Epstein, M. Small, H.L. Falk and N. Mantel, On the association between photodynamic and carcinogenic activities in polycyclic compounds. Cancer Res. *24* (1964)
855–865.

[36] S. Farid and C.H. Krauch, Photochemical and biological reactions of the furocoumarins.
Radiation Research 1966 (North-Holland, Amsterdam, 1967) pp. 870–885.

[37] T.B. Fitzpatrick et al., Symposium on psoralens and radiant energy, Kalamazoo (Michigan,
U.S.A.) 1958; proceedings published in J. Invest. Dermatol. *32* (1959) 132.

[38] E. Fornasari and G. Rodighiero, Ricerche polarografiche sulla fotoossidazione delle proteine del siero di sangue in presenza di furocumarine e di altre sostanze fotodinamiche. Farmaco Ed. Sci. *13* (1958) 379–384.

[39] E. Fornasari, G. Rodighiero and U. Fornasiero, Ricerche polarografiche sulla fotoossidazione di acidi grassi insaturi in presenza di furocumarine e di altre sostanze fotodinamiche. Farmaco. Ed. Sci. *14* (1959) 734–738.

[40] W.L. Fowlks, D.G. Griffith and E.L. Oginsky, Photosensitization of bacteria by furocoumarins and related compounds. Nature *181* (1958) 571–572.

[41] K. Hamaguchi and F.P. Geiduschek, The effect of electrolytes on the stability of the deoxyribonucleate helix. J. Am. Chem. Soc. *84* (1962) 1329–1338.

[42] S.A. Henry and D.P.H. Canton, Further observations on dermatitis due to celery in vegetable canning. Brit. J. Dermatol. Syphil. *50* (1938) 342–351.

[43] G. Kahn and B.P. Davies, Photoreactions of the skin photosensitizer chlorpromazine with nucleic acids, purines and pyrimidines. Book of Abstracts of the 5th International Congress on Photobiology, Hanover, New Hampshire, U.S.A. (1968) N. Ge-8.

[44] K.D. Kaufman, Synthetic furocoumarins. I. A new synthesis of methyl-substituted psoralens and isopsoralens. J. Org. Chem. *26* (1960) 117–121.

[45] K.D. Kaufman, F.G. Gaiser, T.D. Leth and L.R. Worden, Synthetic furocoumarins. II. Synthesis of several alkylated psoralens and of a dihydroisopsoralen. J. Org. Chem. *26* (1960) 2443–2446.

[46] H. Koffler and I.L. Markert, Effect of photodynamic action on the viscosity of desoxyribonucleic acid. Proc. Soc. Exp. Biol. Med. *76* (1951) 90–92.

[47] C.H. Krauch, S. Kraft and A. Wacker, Zum Wirkungsmechanismus photodynamischer Furocumarine. Biophysik *2* (1965) 301–302.

[48] C.H. Krauch, S. Farid and G.O. Schenck, Photo-C₄-cyclodimerisation von Cumarin. Chem. Ber. *99* (1966) 625–633.

[49] C.H. Krauch and S. Farid, Photo-cycloadditionen mit Furocumarinen und Furochromonen. Chem. Ber. *100* (1967) 1685–1695.

[50] C.H. Krauch, D.M. Krämer and A. Wacker, Zum Wirkungsmechanismus photodynamischer Furocumarine, Photoreaktion von Psoralen (4-¹⁴C) mit DNA, RNS, Homopolynucleotiden und Nucleosiden. Photochem. Photobiol. *6* (1967) 341–354.

[51] H. Kuske, Experimentelle Untersuchungen zur Photosensibilisierung der Haut durch pflanziche Wirkstoffe. Archiv. Dermatol. Syphil. *178* (1938) 112–123.

[52] M.M. Mathews, Comparative studies of lethal photosensitization of Sarcina lutea by 8-methoxy-psoralen and by toluidine blue. J. Bacteriol. *85* (1963) 322–331.

[53] A.M. Moore, Ultraviolet irradiation of pyrimidine derivatives. II. Note on the synthesis of the product of reversible photolysis of uracil. Can. J. Chem. *36* (1958) 281–283.

[54] J.C. Mottram and I. Doniach, Photodynamic action of carcinogenic agents. Nature *140* (1937) 933–934.

[55] L. Musajo, G. Rodighiero and G. Caporale, L'activité photodynamique des coumarines naturelles. Bull. Soc. Chim. Biol. *36* (1954) 1213–1224.

[56] L. Musajo, G. Caporale and G. Rodighiero, Isolamento del bergaptene dal sedano e dal prezzemolo. Gazz. Chim. Ital. *84* (1954) 870–873.

[57] L. Musajo, Interessanti proprietà delle furocumarine naturali. Farmaco Ed. Sci. *10* (1955) 539–558.

[58] L. Musajo, G. Rodighiero and L. Santamaria, Le sostanze fotodinamiche con particolare riguardo alle furocumarine. Atti. Soc. Ital. Patol. *5* (1957) 1–70;

[59] L. Musajo, G. Rodighiero, G. Caporale and C. Antonello, Ulteriori ricerche sui rapporti tra costituzione e proprietà fotodinamiche nel campo delle furocumarine. Farmaco Ed. Sci. *13* (1958) 355–362.

[60] L. Musajo and G. Rodighiero, A new photoreaction between some furocoumarins and flavin-mononucleotide. Nature *190* (1961) 1109−1110.

[61] L. Musajo and G. Rodighiero, The skin-photosensitizing furocoumarins. Experientia *18* (1962) 153−161.

[62] L. Musajo, Photoreactions between flavin-coenzymes and skin-photosensitizing agents. Pure Appl. Chem. *6* (1963) 369−384.

[63] L. Musajo, G. Rodighiero, G. Caporale, U. Fornasiero, G. Malesani, F. Dall'Acqua and C. Giacomelli, La fotoreazione tra bergaptene e flavin-mononucleotide. Gazz. Chim. Ital. *94* (1964) 1054−1072.

[64] L. Musajo, G. Rodighiero, G. Colombo, V. Torlone and F. Dall'Acqua, Photosensitizing furocoumarins: Interactions with DNA and photo-inactivation of DNA containing viruses. Experientia *21* (1965) 22−24.

[65] L. Musajo, G. Rodighiero and F. Dall'Acqua, Evidences of a photoreaction of the photo-sensitizing furocoumarins with DNA and with pyrimidine nucleosides and nucleotides. Experientia *21* (1965) 24−25.

[66] L. Musajo and G. Rodighiero, Sul meccanismo d'azione delle furocumarine fotosensibi-lizzatrici. Rend. Accad. Naz. Lincei (Roma) *38* (1965) 591−599.

[67] L. Musajo, G. Rodighiero, A. Breccia, F. Dall'Acqua and G. Malesani, The photoreaction between DNA and the skin-photosensitizing furocoumarins studied using labelled berg-apten. Experientia *22* (1966) 75.

[68] L. Musajo, G. Rodighiero, A. Breccia, F. Dall'Acqua and G. Malesani, Skin-photosensitizing furocoumarins: photochemical interaction between DNA and $-O^{14}CH_3$ bergapten (5-methoxypsoralen). Photochem. Photobiol. *5* (1966) 739−745.

[69] L. Musajo, F. Bordin, G. Caporale, S. Marciani and G. Rigatti, Photoreactions at 3655 Å between pyrimidine bases and skin-photosensitizing furocoumarins. Photochem. Photo-biol. *6* (1967) 711−719.

[70] L. Musajo, F. Bordin and R. Bevilacqua, Photoreactions at 3655 Å linking the 3,4-double bond of furocoumarins with pyrimidine bases. Photochem. Photobiol. *6* (1967) 927−931.

[71] L. Musajo and G. Rodighiero, The mechanism of action of the skin-photosensitizing furocoumarins. Acta Derm. Venereol. (Helsingfors) *47* (1967) 298−303.

[72] L. Musajo, Photochemical interaction between skin-photosensitizing furocoumarins and DNA. In: Radiation Research 1966 (North-Holland, Amsterdam, 1967) pp. 803−812.

[73] L. Musajo, G. Rodighiero, F. Dall'Acqua, F. Bordin, S. Marciani and R. Bevilacqua, Prodotti di fotocicloaddizione a basi pirimidiniche isolati da DNA idrolizzato dopo irradiazione a 3655 Å in presenza di psoralene. Rend. Accad. Naz. Lincei (Roma) *42* (1967) 457−468.

[74] L. Musajo, P. Visentini, F. Baccichetti and M.A. Razzi, Photoinactivation of Ehrlich ascites tumor cells in vitro obtained with skin-photosensitizing furocoumarins. Experientia (Basel) *23* (1967) 335−336.

[75] L. Musajo, F. Bordin, F. Baccichetti and R. Bevilacqua, Psoralen thymine C_4-cycloadduct formed in vitro in the photoinactivation with psoralen of Ehrlich ascites tumor cells. Rend. Accad. Naz. Lincei (Roma) *43* (1967) 442−447.

[76a] L. Musajo, M. Petek and F. Baccichetti, (1969) unpublished results.

[76b] L. Musajo, P.A. Visentini and F. Baccichetti, Preliminary report on the protection from Ehrlich ascites tumor with tumor cells photoinactivated in the presence of psoralen or 8-methyl-psoralen. Z. Naturforschg. *25b* (1970) 642−644.

[77] A. Mustafa, Furopyrans and furopyrones (Interscience, London, 1967) pp. 14−90.

[78] C. Nagata, M. Kodama, T. Yusaku and I. Akira, Interaction of polynuclear aromatic hydro-carbons, 4-nitroquinoline-1-oxide and various dyes with DNA. Biopolymers *4* (1966) 409−427.

[79] B. Nicoletti and G. Trippa, Sull'azione mutagena del psoralene irradiato con luce ultra-violetta in *Drosophila melanogaster*. Rend. Accad. Naz. Lincei (Roma) *43* (1967) 259−263.

[80] E.L. Oginsky, G.S. Green, D.G. Griffith and W.L. Fowlks, Lethal photosensitization of bacteria with 8-methoxy-psoralen to long wavelength ultraviolet radiation. J. Bacteriol. *78* (1959) 821–833.

[81] D.W. Owens, J.M. Glicksman, R.G. Freeman and R. Carnes, Biologic action spectra of 8-methoxy-psoralen determined by monochromatic light. J. Investig. Dermatol. *52* (1968) 435–440.

[82] M.A. Pathak and T.B. Fitzpatrick, Relationship of molecular configuration to the activity of furocoumarins which increase the cutaneous responses following long wave ultraviolet radiation. J. Investig. Dermatol. *32* (1959) 255–262.

[83] M.A. Pathak and T.B. Fitzpatrick, Bioassay of natural and synthetic furocoumarins (psoralens). J. Investig. Dermatol. *32* (1959) 509–518.

[84] M.A. Pathak, J.H. Fellman and K.D. Kaufman, The effect of structural alterations on the erythemal activity of furocoumarins: psoralens. J. Investig. Dermatol. *35* (1960) 165–183.

[85] M.A. Pathak and J.H. Fellman, Activating and fluorescent wavelengths of furocoumarins: psoralen. Nature *185* (1960) 382–383.

[86] M.A. Pathak and J.H. Fellman, Photosensitization by furocoumarins: psoralens. In: Progress in Photobiology, Proc. 3rd Inter. Congress Photobiol., Copenhagen 1960, eds. B.Ch. Christensen and B. Bauchmann (Elsevier, Amsterdam, 1961) pp. 552–554.

[87] M.A. Pathak, B. Allen, D.I.E. Ingram and J.H. Fellman, Photosensitization and the effect of ultraviolet radiation on the production of unpaired electrons in the presence of furocoumarins (psoralens). Biochem. Biophys. Acta *54* (1961) 506–515.

[88] M.A. Pathak, Mechanism of psoralen photosensitization and in vivo biological action spectrum of 8-methoxy-psoralen. J. Invest. Dermatol. *37* (1961) 397–407.

[89] M.A. Pathak, L.R. Worden and K.D. Kaufman, Effect of structural alterations on the photosensitizing potency of furocoumarins (psoralens) and related compounds. J. Investig. Dermatol. *48* (1967) 103–118.

[90] M.A. Pathak and D.M. Krämer, Photosensitization of skin in vivo by furocoumarins (psoralen). Biochim. Biophys. Acta *195* (1969) 197.

[91] J.M. Rice, Photodynamic addition of benz(a)pyrene to pyrimidine derivatives. J. Amer. Chem. Soc. *86* (1964) 1444–1446.

[92] G. Rodighiero and C. Bergamasco, Ricerche sulla fotoossidazione dell'α-terpinene ad ascaridolo in presenza di furocumarine ed altre sostanze fotodinamiche. Farmaco Ed. Sci. *13* (1968) 268–272.

[93] G. Rodighiero and C. Caporale, Furocumarine ed emolisi fotodinamica. Farmaco Ed. Sci. *13* (1958) 373–378.

[94] G. Rodighiero and V. Cappellina, Ricerche sulla fotodimerizzazione di alcune furocumarine. Gazz. Chim. Ital. *91* (1961) 103–114.

[95] G. Rodighiero, G. Caporale and T. Dolcher, Osservazioni sull'attivita citologica di alcune furocumarine e sul loro comportamento di fronte all'acido desossiribonucleinico. Rend. Accad. Naz. Lincei (Roma) *30* (1961) 84–89.

[96] G. Rodighiero, L. Musajo, F. Dall'Acqua and G. Caporale, La fotoreazione tra psoralene e flavin-mononucleotide. Gazz. Chim. Ital. *94* (1964) 1074–1083.

[97] G. Rodighiero, L. Musajo, U. Fornasiero, G. Caporale, G. Malesani and G. Chiarelotto, La fotoreazione tra xantotossina e flavin-mononucleotide. Gazz. Chim. Ital. *94* (1964) 1084–1092.

[98] G. Rodighiero, F. Dall'Acqua and G. Chiesa, Ricerche sul comportamento del bergaptene di fronte ad acidi ribonucleinici. Rend. Accad. Naz. Lincei (Roma) *42* (1967) 510–514.

[99] G. Rodighiero, F. Dall'Acqua and G. Chimenti, Fotodimerizzazione della cumarina e di alcune furocumarine in matrice di ghiaccio. Ann. Chim. (Roma) *58* (1968) 551–561.

[100] G. Rodighiero, L. Musajo, F. Dall'Acqua, S. Marciani, G. Caporale and L. Ciavatta, A comparison between the photoreactivity of some furocoumarins with native DNA and their skin-photosensitizing activity. Experientia *25* (1969) 479–481.

[101] L. Santamaria, A. Angelotti and L. Monaco, Misura dell'attivita fotodinamica di idro-carburi cancerogeni e di furocumarine mediante elettroforesi di siero fotoossidato. Atti Soc. It. Patol. *5*, parte 2^a (1957) 487–501.

[102] L. Santamaria, Photodynamic action and carcinogenicity. In: Recent contributions to cancer research in Italy, Vol. I, eds. P. Bucalossi and U. Veronesi (1960) pp. 167–288.

[103] G.O. Schenck, I. von Wilucki and C.H. Krauck, Photosensibilisierte Cyclodimerisation von Cumarin. Chem. Ber. *95* (1962) 1409–1412.

[104] K.C. Smith, Photochemical reactions of thymine, uracil, uridine, cytosine and bromou-racil in frozen solution and in dried films. Photochem. Photobiol. *2* (1963) 503–517.

[105] K.C. Smith, Biochemical effects of ultraviolet light on DNA. In: The biologic effects of ultraviolet radiation, ed. F. Urbach (Pergamon Press, London, 1969) pp. 47–55.

[106] E. Späth, Die naturlichen Cumarine, Chem. Ber. *70* (1937) 83–117.

[107] P.O.P. Ts'o and P. Lu, Interaction of nucleic acids, II. Chemical linkage of the carcinogen 3,4-benzpyrene to DNA induced by photoradiation. Proc. Nat. Acad. Sci. (Wash.) *51* (1964) 272–280.

[108] I. Von Wilucki, H. Matthäus and C.H. Krauch, Photosensibilisierte Cyclodimersation von Thymine Losung. Photochem. Photobiol. *6* (1967) 497–500.

[109] F. Wessely and K. Dinjaski, Uber die Lichteinwirkung auf Stoffe vom Thypus der Furocoumarine. Monatsh. *64* (1934) 131–142.

[110] F. Wessely and J. Kotlan, Uber die Photodimerisierung von Furocumarinen und das Sphondylin. Monatsh. *86* (1955) 430–436.

[111] L.R. Worden, K.D. Kaufman, J.A. Weis and T.K. Schaof, Synthetic furocoumarins. IX. A new synthetic route to psoralen. J. Org. Chem. *34* (1969) 2311–2313.

[112] E. Yeargers and L. Augenstein, Absorption and emission spectra of psoralen and 8-methoxypsoralen in powders and in solutions. J. Investig. Dermatol. *44* (1965) 181–187.

[113] E. Yeargers and L. Augenstein, Molecular orbital calculations on photosensitizing pso-ralen derivatives. Nature *212* (1966) 251–253.

CHAPTER 11

THE DYE-SENSITIZED DEGRADATION OF NUCLEOTIDES

A. KNOWLES*

Chester Beatty Research Institute,
Fulham Road, London, S.W.3, UK

1. Introduction

Treatment of cells with a sensitizing substance and visible or UV radiation of wavelengths greater than 320 nm, which can only be absorbed by the sensitizer, may result in death or modification of the biological activity of the cells. It is impossible to generalize about the site of this photosensitized attack, for there are reports of damage to most cellular organelles, but there is no doubt that the most important damage is to the nucleus. Many of the observed mutagenic and lethal effects caused by photosensitizers are due to conformational changes, strand breakage or base deletion in nucleic acids. Some effects may be the result of changes in nucleoproteins, but little is known of this type of damage, and this survey will be concerned exclusively with the modification of nucleic acids.

Two main classes of sensitizer can be recognized: firstly, those high-energy sensitizers which absorb between 300 and 400 nm. The action of these resembles that of direct UV irradiation since it is the pyrimidine bases of the nucleic acid which react. This class includes agents like acetone, which act be generating electronic excited states of the nucleotides, and those like the psoralens [19, 56] and possibly the polycyclic hydrocarbons [73] which form adducts with the pyrimidine bases. The action of these does not require oxygen. The second class are compounds, such as dyes, which absorb in the visible region of the spectrum, i.e. 400–700 nm. Their sensitizing effect does require oxygen and is true photodynamic action, as defined by Blum [9]. The larger polycyclic hydrocarbons absorb in this region, but it is not yet established whether their sensitizing effect is upon nucleic acid and if oxygen is involved in their action [27, 55].

Only the dye-sensitized reactions of the nucleic acids will be considered here. These have two remarkable features: (i) light which is of very low energy in photochemical terms can be effective. Thus the methylene blue molecule excited by absorption in its long wavelength band at 665 nm is a very effective photodynamic agent and can disrupt molecules, although its energy is only about 40 kcal/mole

*Present address: MRC Vision Unit, University of Sussex, England.

which is less than most bond energies. (ii) The reaction is highly specific; only the purine residues of the nucleic acid are involved and generally of these, only the guanine bases are damaged. The reaction is primarily oxidative, and the products are quite unlike those generated by high-energy sensitizers or by UV radiation.

Photodynamic effects are found with a wide range of dyes acting upon a variety of organisms. There is significance in the fact that it is dyes which are effective, for some of the best photodynamic agents are dyes which are used as histological stains, e.g. methylene blue and acridine orange, which suggests that ability to bind to a particular organelle within the cell is of importance to their action.

A few organisms which are not coloured have been found to be naturally sensitive to visible light, but in general, even cells which are strongly coloured by hemes or chlorophyll are not damaged by the absorption of light. The protection mechanism within coloured cells is of interest, for isolated hemes or chlorophylls have been shewn to act as photodynamic agents.

The significance of dye-binding and the reasons for the specific attack upon guanine residues in a nucleic acid is not fully understood, and this article is an attempt to survey our present knowledge.

2. Photodynamic action upon nucleic acids in situ

2.1. *Introduction*

Comprehensive reviews of the many organisms which have been studied have been made recently by Spikes [67] and Spikes and Livingstone [68]. Detailed investigation of photodynamic action upon a variety of viruses, bacteria and other small organisms have lead many workers to conclude that the most important site of attack is the nucleus, and the resulting genetics effects are reviewed by Zelle in this volume [87].

A convincing demonstration of this damage to nucleic acid has been made by Yamamoto [85] who inactivated bacteriophage by treatment with methylene blue and light showed that plaque-forming ability could be recovered by genetic recombination with a prophage. This shows that the inactivation was due to single strand breaks in the DNA which could be repaired by combination with the prophage. Similarly, Cramer and Uretz [18] have shown that the damage in bacteriophage inactivated by the photodynamic action of acridine orange is in the genetic material, for cross inactivation is possible. Their experiments show that the phage nucleic acid is still injected into the host cell, and so the damage is at specific sites in the DNA and does not involve breaks in both strands. The damage differs from that caused by UV irradiation, because in this case, reactivation is never so complete. Later experiments showed that the principal result of this treatment of phage is the formation of an alkali-labile bond in the nucleic acid backbone which lead to single strand breaks. This may result from complete removal of a base, or a modification of it which will make the N-glycosidic bond more labile [Friedfelder and Uretz, 22].

2.2. The importance of penetration by the dye

There is evidence which suggests that the dye must first penetrate the viral coat before it can show a sensitizing effect, and so the site of action must be inside the virus. Cramer and Uretz [17] have shown that the conditions of treatment of bacteriophage with acridine orange before irradiation can affect the degree of inactivation on irradiation. Incubation with dye in darkness increases the subsequent effect in the light, which suggests that the penetration of the dye takes an appreciable time. Fluorescence studies of the binding of the dye show that this increases regularly with increasing dye concentration in the medium up to 25 μg/ml. The photodynamic action also increases with dye concetration, but only up to 1 μg/ml, and then levels off. This means that dye binding at some site is necessary for sensitization to occur, but binding can also take place at other sites at which the dye is not active.

A difference in the degree of proflavin-sensitized inactivation of two strains of T4 bacteriophage was attributed by Ritchie [57] to a difference in the permeability of the head membrane to the dye. He also found that the inactivated virus will show multiple reactivation, which again implies single strand breaks which can undergo genetic recombination with other genomes. The optimum conditions for inactivation of three plant mosaic viruses by the action of toluidine blue, neutral red and acridine orange were investigated by Orlob [51]. Increasing the pH of the medium increases the action of the dyes and it is suggested that this is due to an enhanced penetration of the viral coat by the dye. A preparation of neutral red in methanol is more effective than one in water because, here again, penetration to the interior of the virus is facilitated. Bellin [3] has found that neutral red has little effect on *Euglena gracilis* at low pH, but that a sudden increase in photodynamic action is seen above pH 6, which corresponds to the pK at which the dye cation loses a proton to give a neutral molecule. She suggests that this is because the neutral molecule can penetrate the cell membrane more easily than the dye cation. Similar differences are seen between the neutral and ionised forms of methylene blue and rose bengal. Further evidence that damage to the nucleic acids is the principal result of photodynamic action is given by Orlob [51], for when the RNA is removed from a mosaic virus and then treated with dye and light, the loss of infectivity is greater than is found by the same treatment of the intact virus. In the case of the free RNA, the dye does not have to penetrate the viral coat, and there is less chance of the dye being bound at a site at which it has no sensizing effect. Sastry and Gordon [58] also showed that free tobacco mosaic virus (TMV) RNA is more susceptible to the photodynamic action of acridine orange than is the virus. They explain this in the basis of restricted penetration by the dye and suggest that the increased inactivation seen on increasing the pH of the medium is due to easier penetration of the viral coat.

A review of the photodynamic inactivation of animal viruses has been made by Wallis and Melnick [80] whose experiments show that only dye bound inside the virus particle is effective. A preparation of polio virus treated with neutral red,

proflavine or toluidine blue and passed down a cationic exchange column to remove
the excess dye, appears to be colourless. However, the small amount of bound dye
which remains causes a high degree of inactivation on exposure to light. They conclude that the active dye is not only tightly bound, but is bound internally, for if it
was located on the surface of the virus, it would cause the viral particles to be re-
tained on the column. Here again, an effect of pH was noted, the degree of inactivation falling at lower pH values. These workers conclude that this is due to dissociation of the dye—nucleic acid complex.

2.3. *The importance of dye binding*

Yamamoto [85] has shown that dyes which are not effective in the inactivation
of bacteriophage can also bind to active sites on the nucleic acid. Consequently,
when mixtures of effective and ineffective dyes are used, competition for binding
sires by an ineffective dye reduces the inactivation caused by an effective dye. The
importance of dye binding was demonstrated by Oster and McLaren [52], who
showed that only bound dye is effective in the acriflavine-sensitized inactivation of
TMV. Similarly, in the sensitized inactivation of tumour cells by a variety of dyes,
it is necessary for the dye to be bound, although Bellin, Mohos and Oster [6] show
that other factors also determine the inactivating power of a particular dye. For
example a relationship between the inactivating power of the dyes and their photo-
oxidation of a model compound was also demonstrated. Bellin and Oster [7] found
that similar factors determine the efficiency of various dyes in the sensitized inactivation of transforming principle from *Diplococcus pneumoniae*.

Bellin [2, 4] has shown that the properties of a dye change when it is absorbed
onto a macromolecule, and that the dye is more readily photoreduced when bound.
It is suggested that the lifetime of the excited dye molecule is increased on binding,
and this would also be expected to result in increased photo-oxidising power. However, in the photo-oxidation of histidine by methylene blue and rose bengal, the
quantum yield is found to fall on the addition of a polymer to the solution. The effect with rose bengal is even more difficult to explain, because in this case, the
quantum yield passes through a minimum and increases again as more dye is bound
to polymer. The inhibition on binding may be because the oxygen molecules required in the oxidation have difficulty in reaching the dye molecules if these are
held in convolutions of the polymer.

The significance of complex formation between dye and nucleic acid is discussed
by Sastry and Gordon [58] who find that two complexes can form between acridine
orange and RNA, depending upon the dye concentration in the medium. Only the
complex formed at lower concentrations is effective in the photo-inactivation of
the nucleic acid, and a very low concentration corresponding to one complexed dye
molecule to every 300 bases can cause loss of biological activity. If ions such as Cu^{2+},
Ni^{2+} or Co^{2+} are added, the photodynamic activity is lost, and it is suggested that
the metal ions add on to form a ternary complex which is inactive.

An interesting observation has been made by Ito et al. [30] who measured the action spectrum for the inactivation of TMV-RNA by acridine orange. They found that this did not correspond to the absorption spectrum of the dye, but to that of the acridine orange-RNA complex. This seems good evidence that only complexed dye is effective.

2.4. The participation of oxygen

Few workers have demonstrated that oxygen is consumed in the supposed photodynamic reaction of biological systems. The uptake of oxygen during the irradiation of phage DNA in the presence of of methylene blue was shown by Simon and Van Vunakis [63] to be related to the loss of activity of the nucleic acid. Yamamoto [85] showed an oxygen dependence in the sensitized inactivation of bacteriophage and Nakai and Saeki [49] have shown that the mutation of a strain of *Escherichia coli* can be induced by irradiation in the presence of several dyes in air, but there is no mutagenic effect in a nitrogen atmosphere.

2.5. The site of damage in the nucleic acid

Examination of the damaged nucleic acids from inactivated organisms have been made in only a few cases, and the majority of these studies were on viruses. Wacker et al. [79] carried out a base analysis of radioactivity-labelled DNA from bacteria which has been killed by treatment with thiopyronine and light. This showed that low-level illumination destroyed only the guanine residues, while higher light levels also caused some damage of the bases cytidine and thymine, but no damage to adenine or uracil could be detected.

Simon and Van Vunakis [63] found that the only damage to bacteriophage DNA inactivated by methylene blue and light was at the guanine residues. Up to 80% of the guanosine in TMV-RNA can be destroyed by the sensitizing action of proflavine and thiopyronine [Singer and Fraenkel-Conrat, 66], while less than 20% of the other bases are lost. When the RNA is given more mild treatment, so that only 20% of the guanine bases are destroyed, there is little loss of activity. This might be due to the test conditions, for Cramer and Uretz [18] have shown that genetic reactivation can overcome the effects of photodynamic action. Specific destruction of guanine in transforming DNA labelled with radioactive guanine and adenine was demonstrated by Sussenbach and Berends [69] after photosensitization by lumichrome.

There seems no doubt that photodynamic action on biological systems results in damage to the genetic material in many cases, resulting in death or mutation. Penetration of the dye to the nucleic acid and its attachment there seem to be necessary steps in the reaction, but these factors, coupled with the multitude of alternative binding sites in the cell, make quantitative studies very difficult. Quantitative investigations can be carried out more precisely with isolated nucleic acids, but except for those from viruses, subsequent tests for biological activity are difficult to make.

3. Investigations with isolated nucleic acids

Changes in physcial properties, such as absorption spectra or sedimentation pattern, and in chemical properties, such as base composition and product analysis, following photodynamic treatment have been made with a wide range of isolated nucleic acids. The participation of oxygen, the effects of changes in environment and the binding of the dye can be studied more easily outside the cell.

3.1. Changes in physcial properties

Breakage of the backbone of the DNA molecule by the photodynamic action of a series of dyes has been followed by the resulting changes in viscosity by Bellin and Yankus [8]. Methylene blue and rose bengal, which are two of the most powerful sensitizers in destroying biological activity, were also found to be effective in DNA chain scission. These workers propose that destruction of neighbouring guanine residues leads to weakening of bonds in the DNA backbone. Changes in the conformation of nucleic acids can be followed from changes in the 'melting temperature', the temperature at which a large change in the UV absorbance is seen. Bellin and Grossman [5] measured the photodynamic effects of a series of dyes upon the melting temperature of DNA from various sources. Methylene blue was again found to be one of the most active dyes in reducing the melting temperature and changing the shape of the melting curve. A plot of the reduction in melting temperature versus the guanine–cytosine (G–C) content of the nucleic acids shows a remarkable linearity, the degree of degradation of the nucleic acid increasing with G–C content. This is further evidence that physical changes result from attack on the guanine residues. The only exception to this linear relationship was with DNA from T4 bacteriophage, which had the lowest G–C content of the nucleic acids tested, but was relatively sensitive to photodynamic attack. This is probably because this molecule contains hydroxymethyl cytosine instead of cytosine, and its bond with guanine is more sensitive than the G–C bond.

Loss of biological activity does not necessarily involve chain scission. Simon, Grossman and Van Vunakis [65] used the synthetic polynucleotides poly-UG, UC, UA and UAC as messenger RNA's to direct the incorporation of amino acids into polypeptides and the binding of amino acyl-sRNA to ribosomes. They show that irradiation in the presence of methylene blue results in changes in absorption spectra only in those polynucleotides containing guanine. A level of treatment sufficient to cause loss of biological activity and of 260 nm absorbance does not result in any breakage of the chains. Further, photodynamic treatment of poly-UG does not affect its digestion by pancreatic ribonuclease, although the digestion by ribonuclease T_1 is inhibited. This is because the pancreatic enzyme works specifically on phosphodiester bonds adjacent to the pyrimidine bases, which are unaffected by the methylene blue sensitized reaction. Ribonuclease T_1 breaks bonds adjacent to guanine residues, and its action is inhibited because it can no longer recognise the damaged guanine residues.

Sastry and Gordon [58] have shown that TMV-RNA treated with acridine orange and light to give 95% inactivation still shows the normal sedimentation pattern of the nucleic acid, and so the loss of infectivity does not result from chain breakage. It appears that mild photodynamic treatment of a nucleic acid leads to specific damage to the guanine residues, while more drastic conditions of dye concentration and light intensity cause chain scission. In an intact organism, probably only the mild conditions are necessary for mutation or death. This probably explains why Belin and Grossman [5] find that neutral red does not give chain breakage, although it is as effective as methylene blue in the inactivation of transforming principle and tumour cells.

3.2. *The relative sensitivity of the bases*

Several groups have studied the sensitivities of isolated nucleotides and their results all confirm the conclusion that guanine is particularly sensitive. Simon and Van Vunakis [64] compared the reactivities of adenine, guanine and other purines to methylene blue sensitized oxidation. Guanine was the most sensitive base, but uric acid is considerably more reactive and xanthine slightly more so. A different set of purines were tested by Wacker et al. [79] in the methylene blue sensitized reaction. The compounds were of the general type:

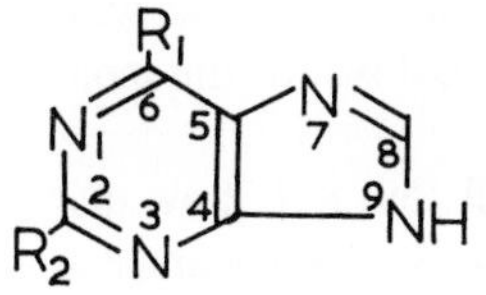

Guanine (R_1 = OH; R_2 = NH$_2$) was again found to most readily destroyed. Replacement of R_1 to form 2-amino purine (R_1 = H) or 2,6-diamino purine (R_1 = NH$_2$) leads to a small reduction in sensitivity. Changing R_2 has a greater effect, iso-guanine (R_1 = NH$_2$; R_2 = OH) and xanthine (R_1OH; R_2 = OH) showing half of the sensitivity of guanine. When there is no substituent at R_2, e.g. in adenine (R_1 = NH$_2$; R_2 = H) and hypoxanthine (R_1 = OH; R_2 = H), the compounds are no longer damaged by the photosensitized reaction.

The methylene blue degradation of a wider range of purine and pyrimidine bases and nucleotides was followed by changes in UV absorbance by Zenda et al. [88]. They found no pyrimidine to be sensitive and the reactive purines have substituents R_1 and R_2 so that the molecule can exist in the lactam form:

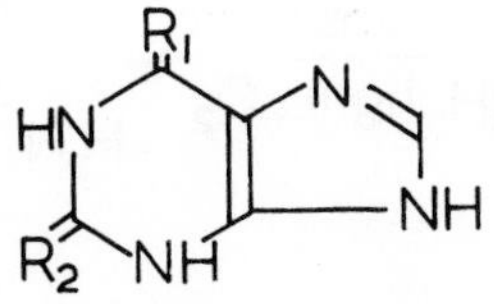

which agrees with the findings of Wacker et al. [79], that hydroxyl and amino substituents are the only ones which make the purine reactive. Substitution at 8-C or 9-C has little effect, the nucleotide beings as sensitive as the base. On the other hand, when 8-C of guanine is replaced by nitrogen to form 8-azaguanine, the sensitivity is completely lost.

Comparison of the results of different workers should include a scrunity of the pH conditions of the reaction. Several workers have shown that the sensitivities of nucleotides vary with pH. For example, Simon and Van Vunakis [64] found that the sensitivity of deoxyguanosine increased by a factor of 25 in going from pH = 7 to 10. The mid-point of this change lies at pH = 9.2, which corresponds to the de-protonation of 1-N in guanine, and so the anion is considerably more sensitive than the neutral molecule.

3.3. Products of the destruction of the bases

Little is known of the exact nature or distribution of the products of the sensitized degradation of the nucleotides, or how this related to chain breakage of the nucleic acid. This degradation can be quite drastic, resulting in complete fragmentation of the bases. Oxygen is consumed during the reaction, which must therefore be oxidative in character. Wacker et al. [79] ran paper chromatograms of the products from the irradiation of 2-^{14}C and 8-^{14}C guanine after treatment with methylene blue and light. A different series of peaks was obtained with the two compounds, which suggests a complete breakdown of the molecule, although no products were isolated or identified.

Sastry, Gordon and Waskell [59, 81] show that acridine orange and methylene blue cause breakage of both rings of the guanosine molecule under their 'drastic' conditions of high light intensity and dye concentration. These reaction conditions do not affect adenosine, cytosine or uridine. Four main products were identified: urea, ribosyl urea, guanidine and ribose, but no free guanine, which suggests that scission of the purine—sugar bond is not the first step in the reaction, as one might suppose.

The evolution of carbon dioxide on the photo-oxidation of transforming DNA by lumichrome was noted by Sussenbach and Berends [69]. Subsequent examination [70] of the products of this reaction upon guanines labelled with ^{14}C at 2-C, 6-C and 8-C show that the principal reaction involves cleavage of the six-membered ring to give guanidine, parabanic acid and carbon dioxide:

Table 1

I. R=H II. R=Me III.

IV. CO₂Na V.

VI.

VII. VIII.

Up to 80% of the activity of 8-^{14}C-guanine degraded in the reaction could be recovered as radioactive parabanic acid.

Matsuura and Saito [46] investigated the products from the rose bengal sensitized oxidation of xanthine (I), 8-methyl xanthine (II) and uric acid (III) in alkaline media (table 1). Xanthine gave allantoin (IV) as the principal product, with triuret (V) as the second product, while 8-methyl xanthine gave different products, sodium oxonate (VI) being detected in 25% yield. Uric acid gave a 20% yield of triuret, with sodium oxonate (VI) as a second product. The sensitized oxidation of N-substituted xanthines and uric acids in methanol was studied by the same workers [47], who found another type of product, substituted 4,5-dimethoxyuric acid (VII).

It is suggested that all of these products might arise from a zwitterionic intermediate of the type VIII. Formation of allantoin from xanthine would be a cyclic peroxide formed across the five-membered ring. The significance of such an intermediate will be discussed in a later section. The sensitized reaction by methylene blue in chloroform gives different products, though these probably come from the same type of intermediate.

Simon and Van Vunakis [64] report the production of 1,3 dimethyl allantoin from theophylline (1,3-dimethyl xanthine) on methylene blue sensitization. The major product from the sensitization of adenine by riboflavin was shown to be hypoxanthine by Uehara et al. [77]. Amman and Lynch [1] isolated seven products from the chlorophyll-sensitized degradation of uric acid in ethanol. Of these, allantoin, urea, parabanic acid and cyanuric acid were identified. Under their reaction conditions, xanthine, hypoxanthine and adenine were not degraded.

Preliminary experiments reported by Fujita and Yamazaki [22a] confirm that the purine ring system of deoxyguanosine monophosphate is the primary site of photosensitized attack by methylene blue. When the reaction is performed in D_2O, the NMR signal from the purine ring protons is seen to fall dramatically relative to that from the protons of the deoxyribose ring.

3.4. *The relative efficiencies of the sensitizing dyes*

It is apparent that many factors determine the efficiency of a dye in sensitizing biological material, such as the penetration of membranes and binding at an active site, beyond the power of the dye as a simple oxidizing agent. There is also the possibility that dyes can cause inactivation in the dark. Chessin [14] found that treatment of TMV-RNA with 6×10^{-4} M acridine orange causes inactivation without any illumination. This is probably due to a change in the conformation of the RNA, because the acridine orange is known to intercalate between the bases of nucleic acids. This unusual complex is probably not formed with other dyes and may explain why acridine orange shows a greater photo-inactivating effect on biological systems than its power as an oxidizing agent would suggest. Sastry and Gordon [58] showed that acridine orange is nearly as powerful as methylene blue in the inactivation of TMV-RNA, but is far less effective in the destruction of isolated guanosine.

One difficulty in experiments designed to compare the relative effects of dyes is that although the treatment of the biological material with different dyes may be adjusted so that a uniform amount of light is absorbed by each, there is no way to determine what proportion of the dye molecules are located at active sites, and this factor may vary greatly between dyes.

Oster, Bellin and co-workers [2–8] have made sustained efforts to establish the relative effects of a wide range of dyes upon the biological activity and physical properties of nucleic acids. Methylene blue emerges as one of the most effective dyes in all the systems studied, namely the inactivation of transforming principle tumour cells and *Euglena*, in the reduction of viscosity and melting temperature of isolated DNA and in the photo-oxidation of a model compound, p-toluene diamine. Other thiazine dyes, e.g. azure-A, -B, and -C, and toluidine blue are also generally effective. Acridine orange, proflavine and riboflavin will sensitize the model oxidation reaction and the biological inactivations, but do not cause chain scission to the same extent. Crystal violet and neutral red will inactivate the biological systems but do not lower the melting temperature of DNA. All of these dyes bind to DNA at pH = 7, and in general, the inactive dyes do not bind. Exceptions to this are the xanthene dyes: rose bengal, eosin and pyronine, which do not bind, but have appreciable effects upon both the biological and physical properties of nucleic acids.

Measurement of the uptake of oxygen by deoxyguanosine on irradiation with a similar range of dyes was made by Simon and Van Vunakis [64]. Methylene blue gave the highest rate of uptake, followed by thionine, toluidine blue, rose bengal and eosin. Acridine orange and proflavine gave negligible rates of oxygen consumption despite their effect on biological systems. Thiopyronine has been shown to be more effective than methylene blue by Wacker et al. [79], but this dye is not commonly used as a photodynamic agent.

The common acceptance of guanine as the sole site of photodynamic action has been questioned by Uehara et al. [74, 76], who found that riboflavin is more effective than methylene blue in the inactivation of nicotinamide adenine dinucleotide and adenine-containing coenzyme A. They show also that, under the conditions of their experiments, riboflavin is a better sensitizer than methylene blue in destroying the amino acid acceptor activity of sRNA and the infectivity of TMV-RNA. They suggest that this is because riboflavin, unlike methylene blue, will also degrade the adenine residues, and thus show a greater biological effect.

No simple test can be used to predict the effect of a dye on a biological system. It is necessary to consider all of the factors involved, for example: (i) is the biological effect due only to damage of nucleic acid? (ii) can the dye penetrate to the nucleic acid, and bind to it at an active site? (iii) is the dye good oxidizing agent under the conditions of pH, etc. found within the cell? The behaviour of a dye as a simple-oxidizing agent is probably the easiest property to deal with in a quantitative manner, and the remainder of this article will be devoted to this topic.

4. Possible mechanisms of sensitized photo-oxidations

4.1. *Introduction*

Several mechanisms have been proposed for sensitized photo-oxidation reactions, and these have been reviewed recently by Spikes and Livingston [68] and Grossweiner [26]. In all of the photodynamic systems that have been studied, there seems no doubt that the initial reaction is the absorption of a quantum of light by the sensitizing dye. This will excite from its ground state (D) to a singlet excited state. This may not be the lowest singlet excited state, but radiationless decay to the lowest singlet is so fast, that for our purposes, all of the excited molecules may be regarded as being initially in the lowest singlet state (^1D). The absorption process may be represented:

$$D + h\nu \rightarrow {}^1D.$$

The reactions which follow this result in the oxidation of the substrate, with the consumption of molecular oxygen, and the return of the excited dye molecule to its ground-state. The reaction schemes suggested for this process fall into four main types, which will be described in the following sub-sections.

4.2. *Singlet–singlet energy transfer*

The singlet excited molecule may lose its excitation energy by one of three spontaneous processes:

$^1D \rightarrow D$ radiationless decay;
$^1D \rightarrow D + h\nu$ fluorescence;
$^1D \rightarrow {}^3D$ inter-system crossing.

The molecule may decay to the ground-state by a radiationless process, the energy probably being converted to vibrational energy and being distributed amongst the solvent molecules. The excitation energy can be re-emitted as fluorescence or undergo 'inter-system crossing' to form the triplet excited state of the molecule (^3D). Development of techniques for the study of excited states has lead to recognition of the importance of inter-system crossing in many molecules, the proportion of singlet excited molecules crossing over to form the triplet often being very high. All of these processes are seen in dye molecules, and their relative importance in a particular dye will depend upon its structure and environment. For a dye in fluid solution, the lifetime of ^1D is of the order of 10^{-8} sec. A fourth process, transfer of excitation energy to a second molecule, can also take place:

$$^1D + SH \rightarrow D + {}^1SH.$$

This results in the return of the dye molecule to the ground-state and excitation of the second molecule to a singlet excited state. This may be a second dye molecule, or the substrate (SH) in the sensitized reaction. If the molecules are freely distributed in fluid solution, there is little chance of such a bimolecular reaction taking place before ^1D decays spontaneously by one of the unimolecular reactions. If the dye molecule is in the form of a complex with the substrate before excitation, the transfer of energy becomes more feasible. In many cases where a dye is complexed with the substrate molecule, the quantum yield of fluorescence of the dye is diminished. This could be due to energy transfer competing with the fluorescence process, but is more probably due to increased rates of radiationless decay or inter-system crossing.

The oxidation step would then be a reaction of excited substrate with oxygen:

$$^1SH + O_2 \rightarrow \dot{S}O + \dot{O}H \text{ etc.}$$

This might be expected to lead to the same products as direct excitation of ^1SH by UV radiation. In the case of the nucleotides, it has been shown that dye sensitized oxidation leads to completely different products to those from direct UV excitation, so this mechanism seems unlikely. It also seems unlikely on energetic grounds, because it is necessary for ^1SH to be of equal or lower energy than ^1D. These quantities are readily derived from the absorption spectra, and for methylene blue and guanine are about 43 and 113 kcal/mole respectively, and so singlet transfer from methylene blue to guanine appears to be impossible.

4.3. *Triplet energy transfer*

The triplet excited state is of much greater lifetime than the singlet-typically 10^{-4} sec for a dye molecule — and is far more likely to take part in reactions involving collision with a second molecule. It has been shown to be the reactive intermediate in the majority of photochemical reactions which have been studied in detail, and many instances are known of energy transfer to excite a second molecule into its triplet state:

$$^3D + SH \rightarrow D + {}^3SH.$$

This process can be of high efficiency in fluid solutions, every encounter leading to transfer in some cases. The resulting excited substrate molecule could then react with molecular oxygen in solution:

$$^3SH + O_2 \rightarrow \dot{S}O + \dot{O}H \text{ etc.}$$

Here again, the transfer will be facilitated by the molecules being complexed, and ^3D should be of higher energy than ^3SH. The energies of triplet states in fluid solution at room temperature are difficult to obtain, and in most cases the only data available is that derived from the phosphorescence of frozen solutions at $77°K$. The phosphorescence of nucleotides has been measured by Longworth et al. [44], and this places the triplet energies of the purine nucleotides at about 65 kcal/mole. The lowest triplet of most dyes will be less than 50 kcal/mole which means that the energy deficiency is less than with the singlet states, but is still an obstacle to energy transfer.

It has been demonstrated by Terenin et al. [71] that dye molecules can sensitize high-energy reactions by a process involving the absorption by the dye of two quanta nearly simultaneously. The first quantum excites the dye into the lowest triplet state, and it is this which absorbs the second quantum to generate a triplet state of higher energy ($^3D^*$)

$$D + h\nu \rightarrow {}^1D \rightarrow {}^3D; {}^3D + h\nu \rightarrow {}^3D^*.$$

Thus higher triplet state can have more than double the energy of ^3D. Terenin et al. [71] detected methyl radicals generated from methyl iodide and t-butanol by the scission of C–I and C–C bonds in a reaction sensitized by acridine yellow and fluorescein. These are strong bonds which can normally only be broken by UV irradiation. The lowest triplet level of fluorescein is at 48 kcal/mole which is too low to effect the reaction, but the absorption of a second quantum could give a higher triplet level which is more than 120 kcal/mole above the ground state. They propose that this energy can be transferred to excite methyl iodide or t-butanol, which then dissociates to give methyl radicals. This type of biphotonic mechanism can be tested by varying the light intensity. If the rate of the reaction is dependent upon the square of the light intensity, the near-simultaneous absorption of two quanta is involved.

Another instance of 'impossible' energy transfer has been investigated by Yang et al. [86]. The isomerisation of substituted ethylene takes place through the triplet state, and this can be excited by energy transfer from a series of carbonyl sensitizers with energies in the range 68 to 72 kcal/mole. The triplet state of the ethylenes are known to be of considerably higher energy than this, but quantum yields for the sensitized reaction can be as high as 0.6. It is suggested that this type of sensitizer can excite some lower lying ethylene triplet level which is not accessible by direct irradiation.

The excitation energy of a triplet can also be lost by encounter with oxygen molecules, and the concentration of oxygen in fluid solution at room temperature will generally reduce the lifetime of the triplet to less than 10^{-6} sec. This means that observation of the triplet must be carried out in the absence of air, or in frozen solution. The lifetime is also reduced by the presence of heavy atom, such as iodide, which may be added in the form of ethyl iodide in organic solvents or as iodide ions in aqueous solutions. Paramagnetic ions also have a quenching effect, and these have been used by some workers to test for triplet participation in sensitized reactions. Sussenbach and Berends [69] show that the sensitized inactivation of transforming DNA by lumichrome is inhibited by Co^{2+}, Mn^{2+}, Cu^{2+} and Ni^{2+}, and they conclude that this is due to quenching of the triplet state of the dye. This is probably the explanation for the observation by Sastry and Gordon [58] that similar ions will prevent the acridine orange inactivation of TMV-RNA.

4.4. *Hydrogen abstraction*

Most sensitizing dyes contain carbonyl groups, and a characteristic reaction of the triplet excited states of many aromatic ketones in their abstraction of hydrogen atoms or electrons from any reducing agent in the solution (see Calvert and Pitts [12]). This results in the formation of two radicals, the semi-reduced dye and the semi-oxidised substrate radicals:

$$^3D + SH \rightarrow \dot{D}H + \dot{S}.$$

The substrate radical may react with oxygen as the first step in the formation of the oxidised product:

$$\dot{S} + O_2 \rightarrow \dot{S}OO \text{ etc.}$$

while the dye radicals react to form the leuco-dye (DH_2):

$$\dot{D}H + \dot{D}H \rightarrow D + DH_2.$$

In the presence of air, this is immediately re-oxidised to give the dye molecule in its original condition:

$$DH_2 + \tfrac{1}{2} O_2 \rightarrow D + H_2O.$$

Reduction to the leuco dye will change the absorption spectrum, often resulting in a complete loss of colour. If the reaction is carried out in the absence of oxygen, the dye will eventually all be converted to the leuco form and thus become totally bleached. Bleaching of the dye when it is irradiated in the presence of a sunstrate in air-free solution is a reasonable indication that a hydrogen abstraction reaction has taken place. This information is not generally available from the descriptions of photodynamic reactions.

Knowles [36a] has compared the quantum yields of destruction of a series of purine and pyrimidine nucleotides sensitized by methylene blue in air, with the anaerobic photoreduction of the dye in the presence of the same nucleotides. Inosine, adenine, cytosine and uridine monophosphates, which are insensitive to the photodynamic action of methylene blue, do not act as reducing agents of the photoexcited dye. On the other hand, xanthosine monophosphate is degraded and is seen to reduce the dye under anaerobic conditions, and the quantum yields of the two processes are similar. This leads to the conclusion that methylene blue brings about photo-oxidation by a hydrogen abstraction mechanism in the case of the nucleotides. Hydrogen abstraction has also been proposed as a common mechanism in the rose bengal-sensitized oxidation of mono-, di- and tributylamine by Schaefer and Zimmerman [59a], on the basis of their very careful analysis of the reaction products. This is surprising in view of the conclusions of others that rose bengal is a very highly efficient generator of another oxidising species, singlet oxygen (see section 4.5).

4.5. *Oxygen sensitization*

It has been assumed so far in this account that the interaction of dye and substrate is the first step in the reaction sequence. However the efficient quenching of triplet dye molecules by oxygen in solution suggests that a dye-oxygen reaction may be important. It was proposed many years ago that excited dye molecules could form an active oxidizing species with molecular oxygen – the 'moloxide' (DO_2), which could then oxidize a substrate molecule:

$$^3D + O_2 \rightarrow DO_2^*$$
$$DO_2^* + SH \rightarrow SO_2 + \dot{D}H \text{ etc.}$$

Many sensitized oxidation reactions have been explained on the basis of this mechanism by Schenck and co-workers [24]. The exact nature of this complex has been the subject of much speculation, and some believe it to be a radical, e.g. $DO\dot{O}$, and others, an excited charge transfer complex, $D^+ \ldots O_2^-$ [38, 60].

Another proposal is that the oxygen molecule can be excited by the dye and this excited oxygen can migrate away from the dye before it reacts with the substrate [23]. Kearns and Khan [34] have demonstrated that it is feasible for triplet dye molecules of quite low energy to transfer to an oxygen molecule. The oxygen molecule is paramagnetic, and thus its ground state is triplet in character. The excited species formed by transfer must then be a singlet state, by the rules of quantum mechanics:

$$^3D + {}^3O_2 \rightarrow D + {}^1O_2$$
$$^1O_2 + SH \rightarrow SO_2 \text{ etc.}$$

Schnuringer et al. [61] have shown that this species can travel some distance from the exciting dye before reacting with the substrate molecule. The methylene blue photo-oxidation of rubrene and the eosin sensitized oxidation of diphenyl anthracene takes place even when the dye and substrate are separated by a barrier of several stearate monolayers, and they suggest that singlet oxygen is the intermediate which can diffuse through the barrier.

Singlet oxygen molecules can be generated by other means. Corey and Taylor [15] passed oxygen through a 6.7 MHz discharge, which generates a glow characteristic of singlet oxygen. The emerging gas was bubbled into solutions of substituted anthracenes and was found to give 9,10-endo-peroxides, identical with those from the UV irradiation of the anthracenes. Chemical reactions can also generate singlet oxygen, for example, the reaction of sodium hypochlorite and hydrogen peroxide [20], the thermal decomposition of 9,10-dephenyl anthracene peroxide [82] and of the adduct of dephenyl phosphote and ozone [48]. Foote et al. [20] have compared the products of dye-sensitized oxidation of a number of compounds with those from chemically-produced singlet oxygen oxidation, and have shown a remarkable degree of similarity, both in the nature and relative amounts of the products. This is clear evidence of a common intermediate in the two types of reaction.

The singlet oxygen reaction gives characteristic products which are either peroxides or appear to come from peroxide intermediates, for example:

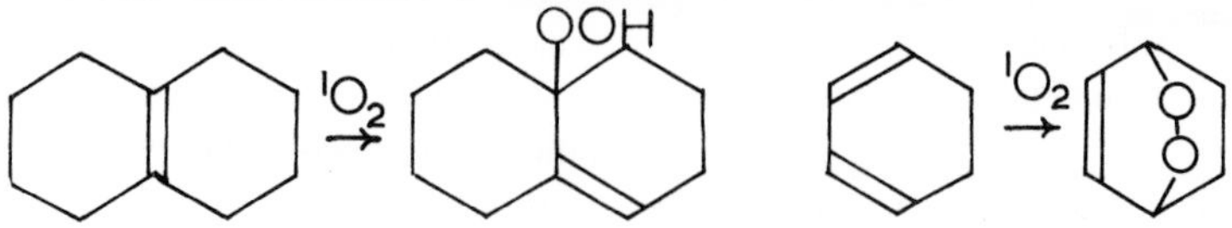

The products and the reaction intermediate described by Matsuura and Saito [47] in the rose bengal sensitized oxidation of purines, seem to be typical of a singlet oxygen reaction.

Wasserman [83] has isolated a wide variety of products from a series of substituted imidazoles on reaction with singlet oxygen, some of which are of the type which is expected from photosensitized oxidation of the purines, i.e. involving cleavage of the ring system:

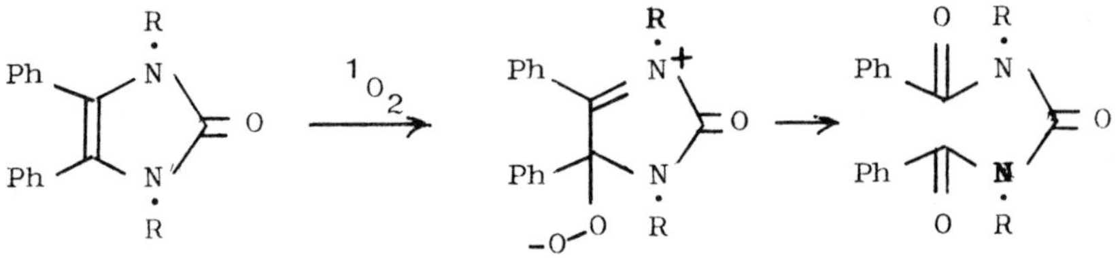

Kearns [31] found two types of product in the photo-oxidation of cholestenol and suggested that these were typical of two types of singlet oxygen, $^1\Delta_g$ and $^1\Sigma_g$ and it was supposed that the ratio of these products formed by reaction with a given sensitizer was dependent upon the triplet excitation energy of the latter. It has been demonstrated, however, that the lifetime of $^1\Sigma_g$ oxygen is very short since it readily forms the $^1\Delta_g$ state, so that sensitized reactions involving the former are unlikely [1a]. Gollnick et al. [23a] have studied a number of oxygenation reactions employing a wide range of sensitizers and conclude that a common oxidizing intermediate is involved in the majority of them. This appears to be $^1\Delta_g$ singlet oxygen, irrespective of the triplet energy of the sensitizer, and they show that it can be generated in large amounts by some sensitizers, e.g. the formation of singlet oxygen by rose bengal has a quantum yield of 0.8.

The oxidation of a molecule by singlet oxygen can be carried out without the use of a light-excited sensitizer if hydrogen peroxide and sodium hypochlorite are added to a solution of the molecule. While singlet oxygen is generated efficiently, these reactants are themselves oxidising agents and one might also expect other reactive species, such as peroxy and hydroxyl radicals to be present in the system. Thus the destruction of nucleotides reported by Hallett et al. [26a] need not necessarily be due to singlet oxygen oxidation. The lifetime of $^1\Delta_g$ oxygen is such that it may be generated in a radio-frequency discharge or chemically in a separate vessel, and then passed in a stream of gas into a solution of the reactant. This technique should remove all uncertainty about the nature of the oxidizing agent, for $^1\Sigma_g$ oxygen is not expected to be of sufficient lifetime to survive this transfer.

Foote et al. [21] have shown that solvent composition can have a major effect upon the products of sensitized reactions, and there may be even greater changes in going from solution to the conditions inside the cell.

5. The sensitized oxidation of nucleotides by flavins — a model photodynamic system

5.1. *Introduction*

It seems unlikely that any one mechanism can explain all of the observed properties of the sensitized oxidation of nucleic acids and their nucleotides by different dyes. Little quantitative data is available and so the remainder of this article will be devoted to a discussion of the results obtained in this laboratory using a single type of sensitizer and isolated nucleotides. Flavins are not considered to be such good sensitizers as methylene blue or acridine orange, but their wide occurrence in biological systems, and the possible relationship of their sensitizer function with their role as the prosthetic group of many oxidising enzymes, make them of particular interest.

Simon and Van Vunakis [64] tested a range of dyes, including riboflavin, by the uptake of oxygen on their irradiation in the presence of deoxyguanosine and found that the flavin caused less oxidation than methylene blue. Riboflavin was also tested by Bellin et al. [5], who show that it is active in their tests on the inactivation of tumours and transforming principle, and in the reduction of the melting temperature and viscosity of DNA. Bradley [10] found riboflavin to be quite effective in the inactivation of actinophage, and Uehara et al. [76] show that it is slightly more effective than methylene blue in the inactivation of nicotinamide adenine dinucleotide and coenzyme A. Reports of the action of riboflavin fall into the general pattern of photodynamic action; oxygen is required and it is principally the guanine bases which are destroyed.

Before looking at the sensitization reaction, it is first necessary to gain some insight into the photochemical behaviour of the dye. When an oxygen-free solution of riboflavin is irradiated with blue light, i.e. about 450 nm, the colour is bleached. The mechanism is apparently one of hydrogen abstraction by the excited molecule, either from the solvent, or from the ribose side-chain of the same or a different molecule (see the review by Penzer and Radda [53]). We find that lumiflavin, which has no side-chain, is bleached more slowly [35]. When a reducing agent such as methionine [11] or ethylenediamine tetra-acetic acid [54] is added to the solution, the rate of bleaching is greatly increased, the added molecule being a better source of hydrogen atoms than the solvent. The bleaching is inhibited by small amounts of triplet state quenchers and so the flavin triplet state appears to be the reactive species. When air is admitted to the solution, the colour is largely restored by reoxidation of the leuco dye [29].

The generation of an excited triplet state on the irradiation of flavins has been demonstrated by: (i) phosphorescence emission on decay of the triplet [43], (ii) typical triplet e.p.r. signals [62] and (iii) by the flash-photolysis technique [35]. The hydrogen abstraction reaction by the triplet can result in oxidation of substrate molecules added to the solution. Nathanson et al. [50] have shown that riboflavin can oxidise indoleacetic acid with a quantum yield of up to 0.71. This reaction only requires oxygen for the re-oxidation of the reduced flavin, and will show the same quantum yield in the absence of air until the flavin is all reduced.

Kearns et al. [32] tested riboflavin in their experiments on the oxidation of cholestenol and concluded that the products were those from a singlet oxygen intermediate, although the product distribution did not fit into the pattern of the other sensitizers. It is suggested that this is because a hydrogen abstraction reaction is taking place at the same time.

Transfer of triplet energy to excite the triplet state of the substrate molecule is suggested by Sussenbach and Berends [69] as the first step in the photo-oxidation of guanine by lumichrome, a dye which is related to the flavins. Yet another possible mechanism arises from a mechanism for biological oxidations proposed by Mager and Berends [45]. They suggest that the flavin semiquinone radical ($\dot{F}H$) will react with oxygen to give a hydroperoxide, which can oxidise a substrate molecule:

$$\dot{F}H \xrightarrow{O_2} FHOOH$$
$$FHOOH + SH \rightarrow S.OH + F + H_2O.$$

Penzer [53a] has measured the rate of oxygen uptake during the irradiation of flavin mononucleotide in the presence of a variety of substrates, and distinguishes between those compounds like thymine and cytosine with which the rate of uptake increases as the oxygen concentrations falls (Type I reaction), and those like adenosine and guanosine, where the rate falls with oxygen concentration. This 'Type II' reaction is typical of aromatic substrates. It is suggested that in the Type I reaction, the rate-determining step is hydrogen abstraction by the flavin triplet from the substrate, while the Type II reaction possibly involves oxygen sensitization.

Most of the mechanism ever proposed for sensitized photo-oxidation reactions have thus been employed to explain the oxidizing action of flavins. A factor which should also be considered in the case of the flavin sensitization of nucleotides is the formation of complexes between these species. Weber [84] showed that flavins

readily form complexes with purines in solution, and this may enhance the subsequent oxidation of the purine on exposure to light. Turnbull [75] believes that
such ground-state complexes are not important in the flavin-sensitized oxidation of
amino acids and suggest that complexes formed between the excited flavin molecule
and the substrate may be more important to the reaction.

5.2. Flash photolysis experiments

When a solution of the flavin is subjected to a brief, intense flash of light, a large
proportion of the molecules are excited into the singlet state and then cross over
to populate the triplet state. The triplet excited molecules have an absorption spectrum different to that of the ground-state, and so a considerable difference is seen
between the absorption spectra measured before and immediately after the flash.
Observations made at a series of increasing times after the photolysis flash enable
a complete picture to be built up of the absorption spectra of the transients and
their reactions. Flash excitation of the simplest flavin, lumiflavin, with visible light
generates at least two short-lived species, the spectra of which have been calculated
[35] and are shown in fig. 1. The transient showing an absorption maximum at
about 380 nm decays by a first-order process, with a half-life of 60 μsec. This species is formed in very high yield, it is quenched by potassium iodide and oxygen,
and it is believed to be the flavin triplet. The second transient has a rather broad
absorption spectrum and decays more slowly by a second-order process. It can still
be detected in the solution at 2 msec after the flash and is probably the flavin semi-

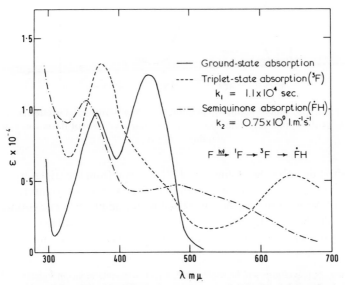

Fig. 1. Absorption spectrum of lumiflavin in water (——), the flavin triplet excited state (- - -)
and semiquinone radical (—·—) derived from flash-photolysis experiments [35].

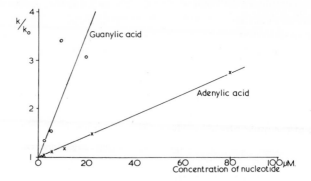

Fig. 2. The increase in the first-order rate constant of lumiflavin triplet decay with nucleotide concentration.

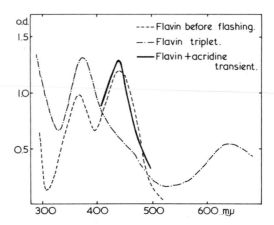

Fig. 3. Absorption spectrum of lumiflavin in water (- - -), the flavin triplet excited state (- . -) and the transient absorption spectrum observed on flashing a solution of flavin containing acridine (−) [35].

quinone radical described by Holmström [28], formed from the triplet by hydrogen abstraction. This mechanism has also been studied by Green and Tollin [25].

The rate of decay of the triplet is increased by the addition of nucleotides to the solution. This is shown in fig. 2 for GMP (guanylic acid) and AMP (adenylic acid)*. Since the lifetime of the triplet is decreased, but not its initial concentration, the

*The following abbreaviations will be used: AMP, 2'(3')-adenosine monophosphate; GMP, 2'(3')-guanosine monophosphate; CMP, 2'(3')-cytidine monophosphate; UMP, 2'(3')-uridine monophosphate; IMP, 5'-inosine monophosphate; XMP, 5'-xanthosine monophosphate.

nucleotide apparently reacts with the flavin triplet and not its precursor, the singlet state. Triplet quenching by heavy atoms and paramagnetic ions is well known, but this quenching by nucleotides, which may be a primary step in their sensitization, is best explained as a transfer of energy from flavin triplet to nucleotide. This depends upon some property of the nucleotide, for GMP is a better quencher than AMP, while the pyrimidines have negligible effect. This order of reactivity of the nucleotides with the flavin triplet resembles that of their relative ease of sensitized photo-oxidation.

Quenching of the flavin triplet is also found with acridine, another heterocyclic molecule [35]. Fig. 3 shows the flavin ground-state absorption spectrum. On flashing the solution, conversion to the triplet causes a major reduction in the height of the 440 nm absorption maximum due to conversion to the triplet. When acridine is added to the solution, the absorption spectrum after the flash is then found to lie above the ground-state peak at 440 nm. This is explained by the rapid quenching of the flavin triplet back to the ground-state and the simultaneous generation of the acridine triplet, which is known to absorb at 437 nm. In this experiment, as in all of the experiments described in this section, the photolysis flash is filtered so that only the long-wavelength absorption band of the flavin is excited and no radiation which could be absorbed by the acridine enters the solution. Thus a transfer of triplet energy has taken place to excite the acridine triplet. This process is feasible on energetic grounds, since the flavin triplet lies at 47 kcal/mole and the acridine triplet just below this at 46 kcal/mole.

This mechanism cannot be so readily applied in the case of the nucleotides. No sign of a nucleotide triplet has been in similar transfer experiments, although this is not surprising, as their triplet lifetimes are believed to be short. Their triplet energies are about 65 kcal/mole (44), which means an energy gap of about 18 kcal/mole, which is unlikely to be made up by thermal energy in the solution.

The lifetime of the flavin triplet is reduced by oxygen quenching when these reactions are carried out in air so that no transients can be detected.

Flash photolysis experiments give much information about the early stages of photochemical reactions, but to understand the complete mechanism, this data must be related to the results of long-term irradiation experiments.

5.3. Irradiation under anaerobic conditions

The bleaching of a flavin solution on long-term irradiation in the absence of air is a reaction of the triplet state of the dye, and any change in the rate of the reaction will be an indication of the behaviour of the triplet. When potassium iodide is added to the solution, the quantum yield of the bleaching reaction falls, as shown in fig. 4.This also shows the quenching effects caused by a series of nucleotides, and it is apparent that GMP and XMP are nearly as effective as potassium iodide, while the pyrimidines have little effect. This exactly parallels the triplet quenching seen in the flash experiments. As far as we can tell, the nucleotide is not damaged in these experiments. It seems that the sensitized photo-oxidation cannot be due to

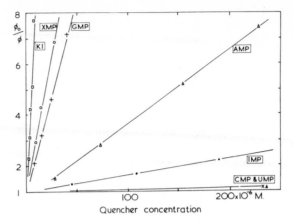

Fig. 4. The reduction in quantum yield of the anaerobic photoreduction of lumiflavin on the addition of nucleotides [36b]

hydrogen abstraction, for this would be expected to take place in the absence of air, and would increase the rate of bleaching of the flavin.

5.4. *Irradiation in air*

Under the conditions of these experiments, no bleaching is observed in the presence of air. This is probably due to triplet quenching by the oxygen, but might be due to rapid oxidation of the leuco dye. When purine nucleotides are added to the solution, their characteristic UV absorption spectra are lost on irradiation, due to the breaking of the ring system. The quantum yield for this reaction has been determined for various concentrations of the nucleotides [37], and fig. 5 shows that a linear relationship exists between the reciprocals of the nucleotide concentration and the quantum yield of its destruction.

The quantum yields for GMP are consistently above those for AMP and reach a maximum value of 0.01. The relationship is explained on the basis of the following reactions of the flavin triplet (^3F):

$$^3F \quad \rightarrow F \tag{1}$$
$$^3F + N \rightarrow F + N^* \tag{2}$$
$$N^* + O_2 \rightarrow Products \tag{3}$$
$$N^* \quad \rightarrow N \tag{4}$$
$$^3F + (H) \rightarrow \dot{F}H \tag{5}$$
$$^3F + O_2 \rightarrow F + O_2 \tag{6}$$
$$N^* + O_2 \rightarrow N + O_2 \tag{7}$$

The flavin triplet can return to the ground-state by the natural processes of phosphoresence and radiationless decay, which are included in reaction 1, and by the oxygen quenching (reaction 6). In can react with unspecified sources of hydrogen

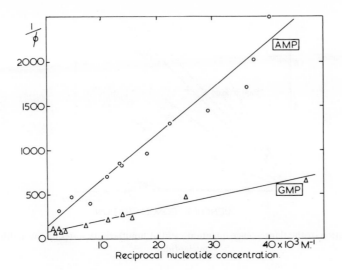

Fig. 5. The double-reciprocal plot of the variation in the quantum yield of flavin-sensitized photo-oxidation of nucleotides with nucleotides concentration [37].

in the solution to form the flavin semiquinone radical (FH) (reaction 5) and it is
assumed that a transfer of energy can take place to cause the electronic excitation
of the nucleotide (reaction 2). This excited nucleotide (N*) can decay to the ground-
state (reaction 4), react with oxygen in the solution as the first stage in its oxidation (reaction 3), or be quenched by oxygen (reaction 7). It was shown in the last
section that reaction 6 could lead to singlet excited oxygen or a moloxide adduct
which might oxidise the nucleotide, and differences in the oxidation rates of the
nucleotides would then depend upon their rate of reaction with this oxidizing species. This is possible, but we believe that the close correspondence of the overall
rate of photo-oxidation of the various nucleotides with their quenching effect upon
the flavin triplet, seen in both flash and long-term experiments, means that reaction
2 is important, and its rate constant (k_2) determines the relative rates of oxidation
of the different nucleotides.

The possibility that simultaneous absorption of two photons could excite a
flavin triplet state of higher energy has been excluded by experiments showing that
the quantum yield is independent of the light intensity [36].

An alternative explanation of the specificity of the reaction with GMP might be
that the flavin is preferentially bound to this nucleotide, and will thus be more
likely to react with it. Flavin molecules which form ground-state complexes no
longer fluoresce and so the reduction of fluorescence quantum yield can be used as
measure of complex formation. Fig. 6. shows the fall in fluorescence quantum
yield on complex formation by the nucleotides [36b]. This plot shows that GMP
forms stronger complexes with flavins than the other nucleotides, and it does resem-

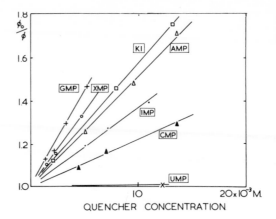

Fig. 6. Reduction of the fluorescence quantum yield of lumiflavin in water on the addition of nucleotides.

ble the inhibition of the photoreduction of the flavin shown in fig. 4. However it is important to note the concentration scales, for a GMP concentration of 5×10^{-6} M will halve the rate of flavin photoreduction, while a concentration of 11×10^{-3} M is necessary for 50% of the flavin molecules to be complexed. The photo-oxidation of the nucleotide has an appreciable rate at concentrations of the order of 10^{-6} M when only a very small amount of the molecules are complexed, and it is concluded that it is not necessary for a ground-state complex to form between flavin and nucleotide for the photo-oxidation to take place.

Fig. 7 illustrates these points by comparing the quantum yield of photo-oxidation with the quenching constants (k_Q) for the quenching of fluorescence and inhibition of photoreduction, from the relation:

$$\frac{\phi_0}{\phi} = 1 + k_Q[N].$$

In the case of the fluorescence quenching, if the complexed molecules do not fluoresce, k_Q is identical with the association constant of the complex

$$k_Q = K_A = \frac{[F \cdot N]}{[F] \cdot [N]}.$$

Comparing the height of the bars for different nucleotides, it is apparent that there is a good quantitative relationship between their inhibition of the photoreduction reaction — that is, interaction with the flavin triplet — and the ease of their sensitized photo-oxidation. It does not seem as if ease of complex formation determines reactivity for AMP, IMP and CMP complex well, but are not readily photo-oxidised.

For these reasons, the reaction scheme involving triplet—triplet transfer has been adopted. An expression for the quantum yield can be derived from the seven equations above in the following way:

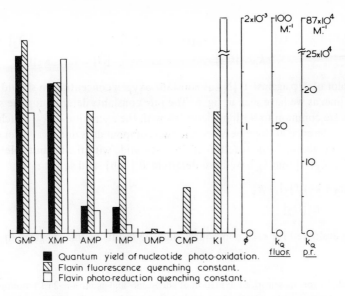

Fig. 7. Comparison of the quantum yields of flavin-sensitized oxidation of nucleotides with their quenching constants for the reduction of lumiflavin fluorescence and anaerobic photoreduction.

$$\phi = \frac{\text{Rate of information of product}}{\text{Rate of absorption of light}}.$$

If we assume that every quantum of light absorbed by the flavin results in the generation of a triplet, which is a fair assumption in the case of the flavins, then the rate of absorption of light will equal the rate of formation of triplets. Under steady-state conditions of constant light intensity, the rate of information of the triplet-excited flavin molecules is equal to their rate of decay by reactions 1, 2, 5 and 6, therefore:

$$\phi = \frac{k_3[N^*]\cdot[O_2]}{[^3F]\,(k_1 + k_2[N] + k_5[H] + k_6[O_2])}$$

similarly the rate of formation of N* will equal its rate of reaction, i.e.

$$k_2[^3F]\,[N] = [N^*]\,(k_3[O_2] + k_4 + k_7[O_2])$$

and these may be combined to give the relation:

$$\phi = \frac{k_2[N]\cdot k_3[O_2]}{(k_1 + k_2[N] + k_5[H] + k_6[O_2])\cdot(k_3[O_2] + k_4 + k_7[O_2])}. \qquad (1)$$

This may be rearranged:

$$\frac{1}{\phi} = \frac{\{(k_3 + k_7)[O_2] + k_4\}\cdot\{k_1 + k_5[H] + k_6[O_2]\}}{k_2 k_3 [O_2]} \cdot \frac{1}{[N]} + \frac{(k_3 + k_7)[O_2] + k_4}{k_3 [O_2]}.$$

Thus a plot of $1/\phi$ against $1/[N]$ at constant oxygen concentration should give a straight line, as we have seen in fig. 5. The rate constants determining the slopes of the lines are common to both nucleotides, with the exception of k_2, which suggests that the difference in slope between the two compounds is due to a difference in the rate of reaction 2, the reaction of the nucleotide with the flavin triplet. Values for $k_1, k_2, k_5[H]$ and k_6 have been determined [36b], and so the ratio:

$$\alpha = \frac{(k_3 + k_7)[O_2] + k_4}{k_3 [O_2]}$$

could be evaluated for each compound, and the results of this are given in table 2.

Table 2

Rate constants for the sensitized oxidation of adenosine- and guanosine-2'(3')-monophosphate by lumiflavin

		GMP	AMP
$k_1 (s^{-1})$		1.10×10^4	1.10×10^4
$k_2 (M^{-1}s^{-1})$		1.39×10^9	0.32×10^9
$k_5[H] (s^{-1})$		55	55
$k_6 (M^{-1}s^{-1})$		2.65×10^9	2.65×10^9
$k_4/k_3[O_2]$		25.7	9.8
$(k_3 + k_7)/k_3$		31.7	44.0
α	(i)	57.4	53.8
	(ii)	43.5	25.0

(i) based on experiments in which $[O_2]$ was varied, and
(ii) based on experiments in which [N] was varied.

On the other hand, if the oxygen concentration is varied at constant nucleotide concentration, the quantum yield of photo-oxidation of both nucleotides is found to rise sharply with oxygen concentration at low oxygen concentrations, passes through a maximum at about 5.5×10^{-5} M and then falls at higher concentrations. A similar oxygen dependence has been found by Bellin and Yankus [8a] for the oxidation of histidine by rose bengal and methylene blue, and by Hodgson et al. [27a] for the flavin-sensitized inactivation of lysozome, trypsin and α-chymotrypsin. Our results are shown for guanosine monophosphate in fig. 8. Now, eq. (1) is of the form:

$$\phi = \frac{[O_2]}{a + b[O_2] + c[O_2]^2}$$

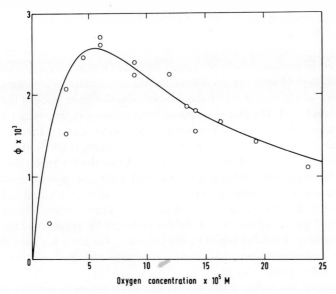

Fig. 8. The effect of oxygen concentration upon the quantum yield of photo-oxidation of guanosine-2'(3')-monophosphate by lumiflavin. The solid line is the calculated curve fitted to the experimental points. Taken from [37].

and the solid line in fig. 8 is a curve of this form fitted to the experimental points. The shape of this was largely determined by the values chose for the parameters a and c and these correspond to:

$$a = \frac{k_4(k_1 + k_2[N] + k_5[H])}{k_2 k_3[N]}.$$

$$c = \frac{k_6(k_3 + k_7)}{k_2 k_3[N]}.$$

Using the values of a obtained from the curve fitting with the values for k_1, k_2 and $k_5[H]$ obtained previously, it was possible to calculate values for $k_4/k_3[O_2]$ which are given in table 2. This represents the ratio of the rates of natural decay of the excited nucleotide and its reaction with oxygen. Similarly, the values for c, taken with values for k_2 and k_6 will give $(k_3 + k_7)/k_3$, which determines the rate of reaction of the excited nucleotide with oxygen in competition with its quenching by oxygen, The latter becomes significant at higher oxygen concentrations. The high values obtained for k_2 show that the reaction of the nucleotide with the flavin triplet is of high efficiency, and so the low overall quantum yields of sensitized oxidation arise from the spontaneous decay (reaction 4) and the oxygen quenching (reaction 7) of the excited nucleotide species. A value for α could also be calculated from these figures, and this is shown in table 2 to compare reasonably with the values from the other set of experiments.

If similar expressions are worked out for an oxygen-sensitization mechanism, in which the initial reaction is between triplet flavin and oxygen, the expressions relating quantum yield with oxygen and nucleotide concentrations are interchanged. Thus the plot of $1/\phi$ against $1/[N]$ should show a minimum in place of the straight line which was found. The plot of quantum yield against oxygen concentration should then be linear, with a positive slope at all concentrations, rather than the complex curve which was found. For this reason, a primary reaction of flavin triplet and oxygen was ruled out and it was concluded that the dye triplet reacts initially with the nucleotide. Bellin and Yankus [8a] reach the opposite conclusion from their study of the sensitized oxidation of histidine by rose bengal and methylene blue, which shows a similar oxygen dependence. Their reaction scheme includes an improbable quenching reaction of the dye moloxide by oxygen, and has no term for the quenching of the dye triplet by the substrate. The latter has been shown to be of major importance when considering the behaviour of the flavin triplet in the presence of nucleotides.

It was pointed out earlier in this section that flavin appear to sensitize by more than one mechanism. The interactions between flavins and nucleotides may be rather unusual and it would be foolish to suggest that all dye-sensitized oxidations of nucleic acids follow this mechanism.

6. Conclusions

Very little is known of the mechanism of the dye-sensitized oxidation of nucleic acids or the importance of naturally occurring sensitizers causing the malfunction of common biological systems under daylight conditions. There is no doubt that the reaction is quite specific for the guanine residues, but the genetic consequences of this kind of damage is only now becoming understood, and it is not yet known whether photodynamic treatment will of use in making controlled genetic modifications or in the determination of the structure of nucleic acids.

The factors controlling photodynamic action in biological systems are complex, and it is essential to investigate the photo-oxidation of more simple model systems before tackling problems such as the importance of dye penetration and binding, and the relative importance of conformational changes, base deletion and chain scission in nucleic acids. It seems most likely that more than one mechanism can operate and so any extrapolation from one dye to another, or from one type of organism to another, must be made with great caution.

References

[1] E.C.R. Ammann and V.H. Lynch, Biochem. Biophys. Acta *120* (1966) 181.

[1a] S.J. Arnold, M. Kubo and E.A. Ogryzlo, Advan. Chem. Ser. *77* (1968) 181.
[2] J.S. Bellin, Photochem. Photobiol. *4* (1965) 33.
[3] J.S. Bellin, in Radiation Research, ed. G. Silini (North-Holland, Amsterdam, 1967) pp. 896–904.
[4] J.S. Bellin, Photochem. Photobiol. *8* (1968) 383.
[5] J.S. Bellin and L.I. Grossman, Photochem. Photobiol. *4* (1965) 45.
[6] J.S. Bellin, S.C. Mohos and G. Oster, Cancer Res. *21* (1961) 1365.
[7] J.S. Bellin and G. Oster, Biochim. Biophys. Acta *42* (1960) 533.
[8] J.S. Bellin and C.A. Yankus, Biochim. Biophys. Acta *112* (1966) 363.
[8a] J.S. Bellin and C.A. Yankus, Arch. Biochem. Biophys. *123* (1968) 18.
[9] H.F. Blum, Photodynamic Action and Diseases Caused by Light (Reinhold, New York, 1941). Reprinted in 1964 by Hafner Publ., New York.
[10] S.G. Bradley, Proc. Soc. Exp. Biol. Med. *122* (1966) 877.
[11] P. Byrom and J.H. Turnbull, Photochem. Photobiol. *8* (1968) 243.
[12] J. Calvert and J.N. Pitts Jr., Photochemistry (Wiley, New York, 1966) p. 352.
[13] P. Chandra, this volume.
[14] M. Chessin, Science *132* (1960) 1840.
[15] E.J. Corey and W.C. Taylor, J. Amer. Chem. Soc. *86* (1964) 3881.
[16] W.A. Cramer and R.B. Uretz, Virology *28* (1966) 142.
[17] W.A. Cramer and R.B. Uretz, Virology *29* (1966) 462.
[18] W.A. Cramer and R.B. Uretz, Virology *29* (1966) 469.
[19] S. Farid and C.H. Krauch, in Radiation Research, ed. G. Silini (North-Holland, Amsterdam, 1967) pp. 869–886.
[20] C.S. Foote, Account Chem. Res. *1* (1968) 104.
[21] C.S. Foote, S. Wexler and W. Ando and R. Higgins, J. Amer. Chem. Soc. *90* (1968) 975.
[22] D. Freifelder and R.B. Uretz, Virology *30* (1966) 97.
[22a] H. Fujita and H. Yamazaki, Bull. Chem. Soc. Japan, *43* (1970) 1177.
[23] K. Gollnick, Advan. Chem. Ser. *77* (1968) 78.
[23a] K. Gollnick, T. Franken, G. Schade and G. Dörhöfer, Ann. N.Y. Acad. Sci. *171* (1970) 108.
[24] K. Gollnick and G.O. Schenck, Pure Appl. Chem. *9* (1964) 507.
[25] M. Green and G. Tollin, Photochem. Photobiol. *7* (1968) 129 and 145.
[26] L.I. Grossweiner, Photochem. Photobiol. *10* (1969) 183.
[26a] F.R. Hallett, B.P. Hallett and W. Snipes, Biophys. J. *10* (1970) 305.
[27] A.P. Harrison and V.E. Raabe, J. Bacteriol. *93* (1967) 618.
[27a] C.F. Hodgson, E.B. McVey and J,D. Spikes, Experienta *25* (1969) 1022.
[28] B. Holmström, Photochem. Photobiol. *3* (1964) 97.
[29] B. Holmström and G. Oster, J. Amer. Chem. Soc. *83* (1961) 1867.
[30] T. Ito, T. Shiroya, S. Kubo, K. Tomura and J. Amagasa, Curr. Mod. Biol. *1* (1967) 192.
[31] D.R. Kearns, R.A. Hollins, A.U. Kahn, R.W. Chambers and P. Radlick, J. Amer. Chem. Soc. *89* (1967) 5455.
[32] D.R. Kearns, R.A. Hollins, A.U. Kahn and P. Radlick, J. Amer. Chem. Soc. *89* (1967) 5456.
[33] D.R. Kearns, A.U. Khan, C.K. Duncan and A.H. Maki, J. Amer. Chem. Soc. *91* (1969) 1039.
[34] D.R. Kearns and A.U. Khan, Photochem. Photobiol. *10* (1969) 193.
[35] A. Knowles and E.M.F. Roe, Photochem. Photobiol. *7* (1968) 421.
[36] A. Knowles, Stud. Biophys. *3* (1967) 97.
[36a] A. Knowles, Photochem. Photobiol. *13* (1971) 473.
[36b] A. Knowles, Photochem. Photobiol. *13* (1971) 225.
[37] A. Knowles and G.M. Mautner, Photochem. Photobiol. (1971) in press.

[38] M. Koizumi and Y. Usui, Tetrahedron Letters (1968) 6011.
[39] C.H. Krauch, D.M. Kramer and A. Wacker, Photochem. Photobiol. *6* (1967) 341.
[40] H.E. Kubitschek, Science *155* (1967) 1545.
[41] Y. Kubota and M. Muira, Bull. Chem. Soc. Japan *40* (1967) 2989.
[42] S. Kumar and A.T. Natarajan, Mutation Res. *2* (1965) 11.
[43] J.M. Lhoste, A. Haug and P. Hemmerich, Biochemistry (U.S.A.) *5* (1966) 3290.
[44] J.W. Longworth, R.D. Rahn and R.G. Shulman, J. Chem. Phys. *45* (1966) 2930.
[45] H.I.X. Mager and W. Berends, Biochem. Biophys. Acta *118* (1966) 440.
[46] T. Matsuura and I. Saito, Tetrahedron *24* (1968) 6609.
[47] T. Matsuura and I. Saito, Tetrahedron Letters (1968) 3273
[48] R.W. Murray and M.L. Kaplan, J. Amer. Chem. Soc. *91* (1969) 5358.
[49] S. Nakai and T. Saeki, Genet. Res. *5* (1964) 158.
[50] B. Nathanson, M. Brody, S. Brody and S.B. Broyde, Photochem. Photobiol. *6* (1967) 177.
[51] G.B. Orlob, Virology *31* (1967) 402.
[52] G. Oster and A.D. McLaren, J. Gen. Physiol. *33* (1949) 215.
[53] G.R. Penzer and G.K. Radda, Quart. Rev. *21* (1967) 43.
[53a] G.R. Penzer, Biochem. J. *116* (1970) 733.
[54] G.K. Radda, Biochem. Biophys. Acta *112* (1966) 448.
[55] G. Reske and J. Stauff, Z. Naturforsch. *20B* (1965) 15.
[56] G. Rodighiero, this volume.
[57] D.A. Ritchie, Biochem. Biophys. Res. Commun. *20* (1965) 720.
[58] K.S. Sastry and M.P. Gordon, Biochim. Biophys. Acta *129* (1966) 32.
[59] K.S. Sastry and M.P. Gordon, Biochim. Biophys. Acta *129* (1966) 42.
[59a] F.C. Schaefer and W.D. Zimmerman, J. Org. Chem. *35* (1970) 2165.
[60] G.O. Schenck, this volume.
[61] B. Schnuriger, J. Bourdon and J. Bedu, Photochem. Photobiol. *8* (1968) 361.
[62] T. Shiga and L.H. Piette, Photochem. Photobiol. *4* (1965) 769.
[63] M.I. Simon and H. Van Vunakis, J. Mol. Biol. *4* (1962) 488.
[64] M.I. Simon and H. Van Vunakis, Arch. Biochem. Biophys. *105* (1964) 197.
[65] M.I. Simon. L. Grossman and H. Van Vunakis, J. Mol. Biol. *12* (1965) 50.
[66] B. Singer and H. Fraenkel-Conrat, Biochemistry *5* (1966) 2446.
[67] J.D. Spikes, in Photophysiology III, Current Topics, ed. A.C. Giese (Academic Press, New York, 1968) pp. 36–64.
[68] J.D. Spikes and R. Livingston, in Advances in Radiation Biology, Vol. 3, eds. L.G. Augenstein, R. Mason and M. Zelle (Academic Press, New York, 1969)
[69] J.S. Sussenbach and W. Berends, Biochim. Biophys. Acta *76* (1963) 154.
[70] J.S. Sussenbach and W. Berends, Biochim. Biophys. Acta *95* (1965) 184.
[71] A.N. Terenin, V. Rylkov and V. Kholmogorov, Photochem. Photobiol. *5* (1966) 543.
[72] L. Thiry, Virology *28* (1966) 543.
[73] P.O.P. T'so and P. Lu, Proc. Nat. Acad. Sci. US *51* (1964) 272.
[74] A. Tsugita, Y. Okada and K. Uehara, Biochim. Biophys. Acta *103* (1965) 360.
[75] J.H. Turnbull, Experienta *24* (1968) 409.
[76] K. Uehara, T. Mizoguchi and Y. Okada, J. Biochem. (Tokyo) *55* (1964) 685.
[77] K. Uehara, T. Mizoguchi and S. Hosomi, J. Biochem. (Tokyo) *59* (1966) 550.
[78] A. Wacker and P. Chandra, Z. Naturforsch. *21B* (1966) 663.
[79] A. Wacker, H. Dellweg, L. Trager, A. Kornhauser, E. Lodeman, G. Turck, H. Selzer, P. Chandra and H. Ishimoto, Photochem. Photobiol. *3* (1964) 369.
[80] C. Wallis and J.L. Melnick, Photochem. Photobiol. *4* (1965) 159.
[81] L.A. Waskell, K.S. Sastry and M.P. Gordon, Biochim. Biophys. Acta *129* (1966) 49.
[82] H.H. Wasserman and J.R, Scheffer, J. Amer. Chem. Soc. *89* (1967) 3073.
[83] H.H. Wasserman, Ann. N.Y. Acad. Sci. *171* (1970) 108.

[84]   G. Weber, Biochem. J. *47* (1950) 114.
[85]   N. Yamamoto, this volume.
[86]   N.C. Yang, J.I. Cohen and A. Shani, J. Amer. Chem. Soc. *90* (1968) 3264.
[87]   M. Zelle, this volume.
[88]   K. Zenda, M. Saneyoshi and G. Chihara, Chem. Pharm. Bull (Tokyo) *13* (1965) 1108.

CHAPTER 12

ENERGY TRANSFER IN DYE–NUCLEIC ACID COMPLEXES

M. DELMELLE and J. DUCHESNE

Department of Atomic and Molecular Physics,
University of Liège,
Sart-Tilman par Liège 1, Belgium

1. Introduction

By definition, photodynamic action concerns the effects of visible light. It refers to the damage incurred in the presence of oxygen by a biological medium, usually called the substrate, into which a sensitizing agent has been introduced.

These phenomena have received a great deal of attention during the past years and a considerable number of photodynamic reactions have been studied [26, 28, 29]. The most delicate experimental techniques have been utilized in an attempt to elucidate the mechanisms involved. Although substantial results have been obtained in analyzing the damage incurred by biomolecules studied in isolation and by living systems, this mass of data has unfortunately not yet been coordinated in a truly satisfactory manner [29]. It seems clear that the application of other techniques, such as electron paramagnetic resonance or nuclear magnetic resonance spectroscopy, should enable us to make some progress in studying problems of photodynamic action.

In order to illustrate the capabilities of electron paramagnetic resonance (EPR) spectroscopy, we would like to discuss the results of a number of experiments carried out in our laboratory with the aid of this technique. Our goal was to find out whether, in a photodynamic system, free radicals can be produced in the substrate, whether the structure of the spectrum associated with these radicals reveals a preferential locus for damage in the substrate, and whether the spectra vary with the sensitizer used. Finally, we wished to consider the possibility that the photodynamic effects detected in vivo may be linked to radical mechanisms. These problems may actually be considered from two perspectives: molecular and biological. These two orientations explain the progression in the course of our work. Initially we were interested in studying the sensitization of nucleosides and nucleotides. The next step was to explore the nucleic acids themselves, particularly deoxyribonucleic acid. Finally, we tackled the problem of a more complex system composed of viruses.

Photosensitization of the various substrates was achieved with the use of dyes of

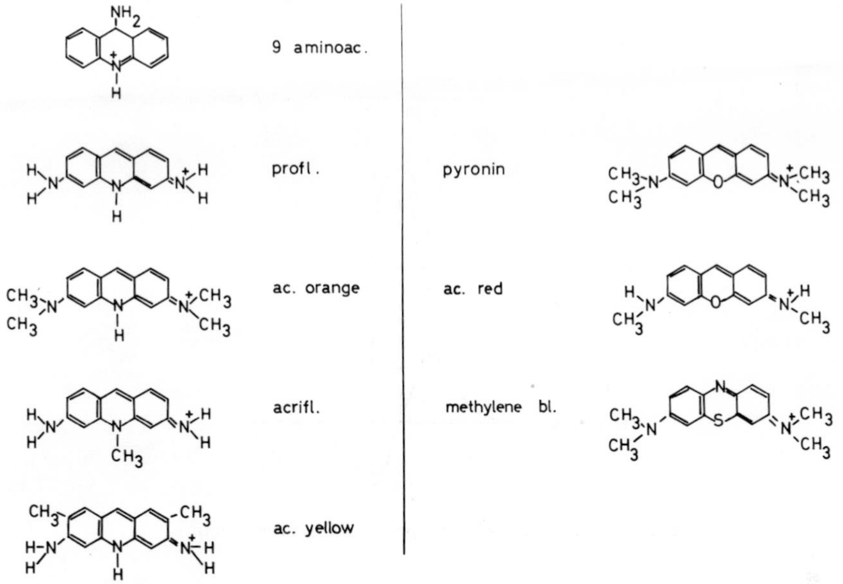

Fig. 1. Chemical structures of the dyes used. 9-aminoac. = 9 aminoacridine; profl. = proflavin; ac. orange = acridine orange; acrifl. = acriflavine; ac. yellow = acridine yellow; ac. red = acridine red; methylene bl. = methylene blue.

the acridine, xanthene and thiazine families. The chemical structures of the dyes used are shown in fig. 1. It will be noted that heteroatoms have been substituted in each family: nitrogen in the case of the acridine dyes, oxygen for the xanthene derivatives and nitrogen and sulfur for methylene blue.

But before we begin to discuss our results, we would like to describe briefly the principle of EPR spectroscopy.

2. Electron paramagnetic resonance

We know that a free radical is characterized by the presence of an unpaired electron. In the absence of a magnetic field, the magnetic moments of such electrons, have no preferential orientation. However, if a sample of the substance is exposed to a magnetic field, the magnetic moments associated with the system become oriented parallel to the direction of the field. According to the theory of the Zeeman effect, there is a lifting of the degeneracy which affected the energy levels in the absence of the field. This phenomenon is illustrated for a free electron in fig. 2. When the magnetic moments are aligned parallel to the direction of field H, the energy of interaction equals $-\frac{1}{2} g\beta H$. In the antiparallel position, this energy equals $+\frac{1}{2} g\beta H$. In these expressions, β represents the Bohr magneton and g the spectro-

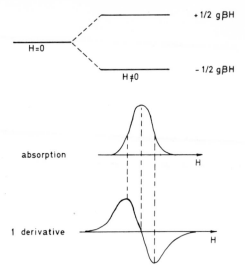

Fig. 2. Energy levels and EPR spectrum for a free electron.

scopic splitting factor which takes into account the anomaly of spin in the gyro-magnetic effect.

EPR spectroscopy makes it possible to observe the transitions between these sub-levels [1, 8, 20]. A weak hyperfrequency field is applied to the sample, the function of which is to provide the system with the quanta necessary for effecting the transi-. tions. Associated with these transitions are energies equal to:

$$h\nu = g\beta H.$$

Thus, for each transition there is a corresponding absorption of energy, as shown in fig. 2.

In actual fact, the method of detection utilized the first derivative of the absorption curve. This is depicted schematically in the lower part of the figure.

3. Results

3.1. *Study of dyes alone*

Before we could approach a system that may be termed complete, i.e. a solution containing both a sensitizer and a substrate, it was necessary first to investigate the effects of visible light on a solution containing only the photosensitizer.

Irradiation was performed with a 500 W mercury vapor lamp filtered so as to transmit only wavelengths of more than 3600 Å. The conditions of irradiation were identical for all observations.

The solutions to be studied were introduced into quartz tubes with an internal diameter of about 3 mm; the samples were brought to a temperature of $77°K$ and

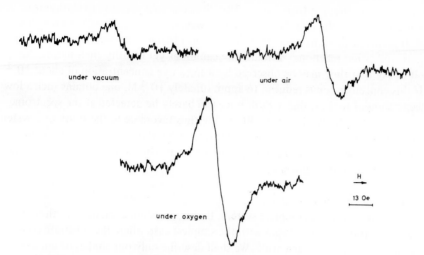

under vacuum

under air

under oxygen

H

13 Oe

Fig. 3. Comparison of the effect of air, oxygen and vacuum on the intensity of paramagnetic line observed on a solution of proflavin in water, irradiated in the visible range.

irradiated. After irradiation, they were transferred to the spectrometer and observations were made at about 100°K.

We found that when an aqueous solution of proflavin, for example, is kept at a low temperature and exposed to visible light, free radicals are generated in the system. Fig. 3. shows the spectrum obtained after one hour of illumination. We are dealing here with a singlet signal whose width at the points of inflexion is 13 Oe; g, measured at the center of the spectrum, has a value of 2.003.

In view of the fact that we intended later to study the influence of oxygen on the substrate, it was important to ascertain what would happen under the same conditions with the dye alone. We compared three samples sealed respectively under a vacuum of 10^{-3} mm Hg, under air at atmospheric pressure, and under an oxygen atmosphere of 500 mm Hg. The oxygen-sealed sample was found to give a spectrum approximately 5 times as intense, clearly indicating that oxygen is involved in the formation of the radicals.

Let us see how these results may be interpreted on the basis of previous data gathered by other workers. Utilizing flash photolysis, Rosenberg showed that, at low temperature, oxygen interacts with proflavin when the dye is excited in its first triplet state [25]. This interaction produces an inactivation of the dye, with transfer of the excitation energy toward the oxygen molecules. This can·be represented by the reaction:

$$PF(T_1) + O_2 = PF(S_0) + O_2^*.$$

However, following this reaction, the excited oxygen can itself react with excited dye molecules in the first triplet state. This gives rise to a parasitic oxidation reaction which destroys the dye. The reaction, however, is characterized by a very low yield. It seems justified to assert that the radicals observed are associated with these parasitic oxidation reactions. Although we cannot as yet identify their nature, suffice it to say that they may be observed by using a dye concentration of about 10^{-2} M. If this concentration is reduced to approximately 10^{-4} M, one obtains such a low concentration of radicals that a weak trace can barely be detected at the spectrometer's limits of sensitivity. These conditions are thus favorable to the study of a system which contains both a sensitizer and a substrate. Nevertheless, the fact that radicals are no longer observed with the dyes alone in no way affects the possibility that they may be detected in the substrate.

3.2. Nucleosides and nucleotides

Let us now consider a complete system, that is, a system containing both sensitizer and substrate. We may begin with the simplest case where the substrate consists of a nucleoside or a nucleotide. We shall describe only our studies of nucleotides; identical results were obtained with the nucleosides. This, incidentally, shows that the important part of the molecule is the organic portion; the phosphate group, at least qualitatively speaking, plays no role whatsoever.

When an aqueous solution containing a nucleotide and proflavin is irradiated, at low temperature, in the visible range, one obtains an intense paramagnetic spectrum characteristic of the nucleotide. Of course, in the absence of a sensitizer, no free radicals are observed, obviously because the light used for irradiation fails to be absorbed by the substrate.

Fig. 4a shows the spectrum obtained with the system proflavin–thymidine monophosphate. This signal gives rise to an absolutely characteristic hyperfine structure. The separation between each components is 20 Oe, the same as the width of the central peak. For this experiment, a nucleotide concentration of about 10^{-2} M was used; the concentration of proflavin was near 10^{-4} M, as mentioned earlier. The

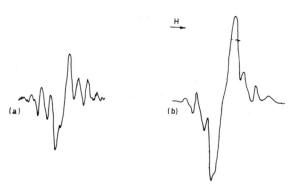

Fig. 4. (a) Spectrum obtained with the system proflavin–thymidine monophosphate. (b) Spectrum obtained with thymidine monophosphate irradiated with X rays. [14].

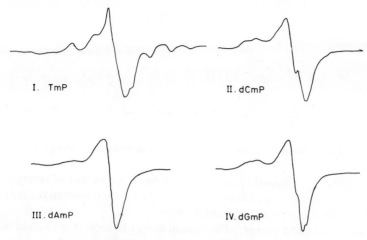

Fig. 5. Photosensitization of nucleotides by proflavin.

striking point here is that this spectrum is comparable to that obtained when the nucleotide is irradiated, either in solid form with X rays (fig. 4b), or in aqueous solution in the ultraviolet absorption band [14, 18]. Since X ray studies have made it possible to show that the free radical is localized in the conjugated part of the molecule, i.e. in the base, and since, as we have just seen, the spectra are identical, this automatically resolves the problem of where the radicals generated by photosensitization are localized.

The results obtained with the other nucleotides are illustrated in fig. 5. We have just described the TmP spectrum in detail. In the case of dCmP, there is an inflexion in the central part. The width of the spectrum measured at the points of inflexion is 22 Oe. This value is in good agreement with the one obtained with direct irradiation of the nucleotide either with X rays or ultraviolet light [14]. The analogy holds true for the two other nucleotides. All three forms of irradiation yield signals of the same width, about 16 Oe with dAmP and 14 Oe with dGmP. These spectra were recorded after one hour of illumination. As irradiation progresses, the intensity of the signal approaches saturation. We shall return to this phenomenon later. Insofar as the stability of the radicals is concerned, we have found that the intensity of the signals drops when the samples are heated. At around −60°C, the spectrum disappears altogether.

We shall now show that these observations satisfy the criteria established for photodynamic action. The problem as posed consists of understanding how visible light is able to induce, via the intermediate of a sensitizer, radical damage within the nucleotides, since the energy of the light used is actually too low to produce electronic excitation of the substrate directly. This is shown schematically in fig. 6. On the left, we have shown the first excited levels of the proflavin. The first excited singlet is 2.8 eV above the ground state, the first triplet being situated at 2.1 eV. On

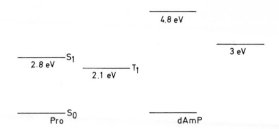

Fig. 6. Energy levels of proflavin (left) and deoxyadenosine monophosphate (right).

the right side of the figure the levels of dAmP are shown by way of example. The first excited singlet is located at an energy of 4.8 eV with respect to the ground state, while the first triplet is at about 3 eV.

In order for a direct transfer of excitation energy to occur, the levels of the donor must be situated at energies at least as great as the corresponding levels of the acceptor. One can immediately see that the S_1 level of the sensitizer is located at a lower energy than the S_1 level of the substrate. The same is true of the corresponding triplet states. Finally, a transfer of energy between the proflavin S_1 level and the substrate T_1 level seems highly improbable. First of all, the overlap between the two levels is slight; moreover, no studies to date have been able to demonstrate an energy transfer of the form

$$D_2^* + A_S = D_S + A_T^*.$$

Intramolecular deactivation via fluorescence or internal conversion is indeed more likely than an intermolecular energy transfer which implies an intersystem singlet– triplet transition. Thus, the observation just described cannot be interpreted with reference to a model involving direct transfer of excitation from the sensitizer to the substrate.

One way to resolve this problem would be to postulate the intervention of a biphotonic process. During biphotonic sensitization, the sensitizer is excited to the first singlet state, after which there is an intersystem crossing to the corresponding triplet state. As this state is relatively long–lived at a low temperature, the sensitizer molecule can absorb a second photon, thus making it possible for a photochemical reaction ot occur. Several biphotonic reactions have been observed with the aid of EPR spectroscopy. The first experiments were performed by Samaller [27] and by Terenin and co-workers [12] . These early investigators demonstrated phenomena associated with the sensitization of alcohols by aromatic amines. Later, Terenin et al. [30] detected a photosensitized dissociation of methyl iodide, a reaction which these authors likewise interpreted on the basis of a biphotonic mechanism. However, all these biphotonic reactions in no way imply that oxygen is involved. On the contrary, the majority of studies were made with samples sealed under vac-

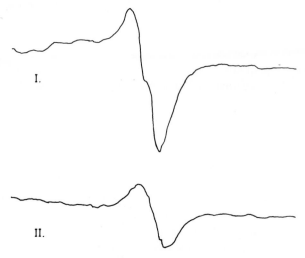

Fig. 7. Effect of oxygen on the photosensitization of deoxycytidine morophosphate by proflavin. (I) sample sealed under air, (II) sample sealed under vacuum (10^{-3} mm Hg).

uum so as to reduce to a minimum deactivation of the triplet states by oxygen.
During sensitization of the nucleotides, however, oxygen plays a predominant role,
as fig. 7 makes clear. Here, a sample of dCmP sealed under air at atmospheric pressure is compared with a sample sealed under a vacuum of approximately 10^{-3} mm
Hg. We see that the signal corresponding to the vacuum-sealed sample is considerably
less intense. This finding would appear to rule out the possibility that the mechanism involved is a biphotonic effect.

These phenomena do, however, appear largely consistent with the criteria defined
for photodynamic effects since oxygen plays an essential role in the formation of the
the radicals.

Basically, three different models of interaction between the sensitizer and oxygen have been proposed. These are shown in fig. 8. The first model involves a trans-

$$\text{I} \quad D_T + O_2 \longrightarrow D_{S_o} + O_2^* $$
$$O_2^* + X \longrightarrow XO_2$$

$$\text{II} \quad D_T + O_2 \longrightarrow D\cdots O_2$$
$$D\cdots O_2 + X \longrightarrow XO_2 + D$$

$$\text{III} \quad D_T + O_2 \longrightarrow D_{OX}^+ + O_2^-$$
$$D_{OX}^+ + X \longrightarrow D + X_{OX}^+$$

Fig. 8. Mechanism proposed for the photodynamic action D = dye; X = substrate.

fer of excitation energy toward the oxygen molecule [10]. The second, implies a complex intermediate between the dye and the oxygen, probably a peroxide of the sensitizer [19]. Finally, the third schema postulates the involvement of a semire- duced form of the oxygen and a semioxidized form of the photosensitizer [9].

Let us begin by pointing out that studies with flash photolysis have failed to re- veal semireduced and semioxidized radicals in the case of acridine dyes [13]. As far as the second model is concerned, the intervention of a peroxide seems quite unlikely in the light of our observations. This model implies a diffusion of the peroxide and it is evident that such a diffusion would have to be very slight at the temperature at which our observations are carried out. Hence, the first model, which postulates the involvement of excited oxygen, fits best with our findings. As we mentioned earlier, these molecules of excited oxygen are produced during deactivation of the dye molecules by oxygen. Several workers have suggested that this excited state of the oxygen corresponds to the $^1\Delta_g$ state, which is situated 1 eV above the ground state [2, 4, 6]. It should also be noted that recent theoretical research carried out by Kawaoka [11] has shown that the deactivation of triplet states by molecular oxy- gen involves primarily a transfer of energy producing an electronic excitation of O_2 in the $^1\Delta_g$ state. It seems possible to assume that the molecules of excited oxygen are responsible for the radical damage induced in the nucleotides.

However, no direct proof concerning the participation of excited oxygen in these processes is available in the present time. The formation of free radicals may also proceed directly from a dehydrogenation reaction between the excited dye and the substrate. According to a suggestion of Professor Schenk, the phenomenon could be also interpreted in the following way:

$$D^* + SH \rightleftharpoons DH^{\cdot} + S^{\cdot}, \tag{1}$$
$$DH^{\cdot} + O_2 \rightarrow D + HO_2^{\cdot}, \tag{2}$$
$$S^{\cdot} + O_2 \rightarrow SO_2^{\cdot}. \tag{2'}$$

In presence of oxygen, the free radicals produced in reaction 1, react with oxygen molecules giving rise to secondary reactions 2 and 2'. On the other hand, in absence of oxygen, the deactivations of the radicals $DH^{\cdot}$ and $S^{\cdot}$ are so rapid that the missing of free radicals observations in anaerobic conditions could be explained.

In fact, although the participation of oxygen in the reaction of sensitisation seems quite clear, the problem of understanding how O_2 plays a part in it, is still a matter of discussion.

Let us now examine the case of a more complex substrate consisting of desoxy- ribonucleic acid.

3.3. *Desoxyribonucleic acid*

Illumination in the visible range of a solution containing a mixture of DNA and proflavin generates a considerable quantity of free radicals. Evidently, in the absence of dye, no free radical is detected. Fig. 9 shows the spectrum obtained in the case of

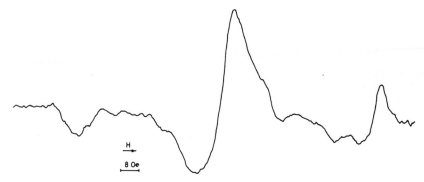

Fig. 9. Photosensitization of DNA by proflavin.

the sensitization of the DNA. One may note the presence of a peak asymetric with respect to the baseline and having a width of 20 Oe at the points of inflexion. On either side of this peak there appear two lateral components separated from one another by approximately 130 Oe. Unfortunately, it is impossible on the basis of this spectrum to specifiy the distribution of radicals in the different nucleotides. The spectrum probably results from the superposition of the signals corresponding to each of them, but given the absence of structure in the spectrum, we are unable to isolate the 4 components.

The influence of oxygen here, just as in the nucleotide experiments, is the determining factor. If irradiation is performed on a vacuum-sealed sample, the intensity of the signal is reduced by approximately 60% in comparison to a sample sealed at atmospheric pressure.

However, in the case of DNA photosensitization, we are confronted with a special problem which has not occured in the systems studied up to this point. The problem concerns the dimensions and structure of the DNA molecule, which control the capabilities of interaction between the sensitizer and substrate. We know, in fact, that Peacocke and Skerrett [21] have demonstrated two modes by which proflavin can interact with DNA. If r corresponds to the number of dye molecules bound to each phosphate group of the DNA, and c to the free dye concentration, each inflexion in the curve r versus c corresponds to the saturation of one form of interaction. Lerman [15, 16, 17] has suggested that the type of interaction that predominates at low dye concentrations corresponds to an intercalation of the dye molecules between the pairs of bases of the DNA. In contrast, at high concentrations, the predominant interaction consists of the fixation of dye cations on the phosphate groups of the DNA, outside the double helix.

We thought it would be interesting to attempt to determine which of these two forms of interactions is most conducive to photosensitization. To this end, we studied different systems in which the ratio between dye and substrate concentrations was modified; at each such ratio we then determined whether intercalation or aggregation predominated. Our results are shown in fig. 10. Curve I corresponds to a

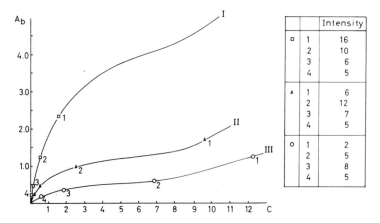

		Intensity
□	1	16
	2	10
	3	6
	4	5
▲	1	6
	2	12
	3	7
	4	5
○	1	2
	2	5
	3	8
	4	5

Fig. 10. Influence of intercalation and aggregation on the photosensitization. Evolution of the intensity of the EPR signal in function of the ratio between dye and substrate concentration.

DNA concentration of 1.5×10^{-3} M. Curves II and III correspond to concentrations of 5×10^{-4} and 2.5×10^{-4} M respectively. Free dye concentration is shown as the abscissa; the ordinate represents the concentration of bound dye. The various systems studied are represented by the various points on the three curves. Shown in the table are the intensities of the EPR spectra in the different cases. At a DNA concentration of 1.5×10^{-3}, the intercalation phenomenon was found to predominate. The signal's intensity varied directly with the amount of proflavin. When aggregation was operative, a different picture resulted. At a DNA concentration of 5×10^{-4} M, the intensity of the signal is maximum for which is equivalent to the saturation of intercalation. If the proflavin concentration is increased, aggregation occurs and the intensity of the spectrum diminishes. This conclusion is reinforced by the results obtained with the third DNA concentration. The spectra of the systems characterized by aggregation present lower intensities than this system which corresponds to the maximum of intercalation. We may infer from these observations that intercalation of the dye between the pairs of bases of DNA seems to be a condition favourable to photosensitization. In contrast, sensitization appears to be inhibited by aggregation, which occurs outside the helix. It is conceivable that these molecules form a protective screen which partially inhibits excitation of the intercalated molecules.

Now that we have determined which type of interaction between proflavin and DNA appears most favourable to sensitization, we may consider to what extent stabilization of the double helix is involved in this phenomenon. When DNA is denatured by heat, there is a marked decrease in the intensity of the paramagnetic signal. The spectrum is 50% less intense than with native DNA. Insofar as ionic strength is concerned, a more intense signal is obtained when the ionic strength of the solution is elevated (near 0.5). At a value of approximately 10^{-2} the signal is about twice as weak. These two experiments suggest that photosensitization takes place more readily as the structure of the macromolecule is more rigid.

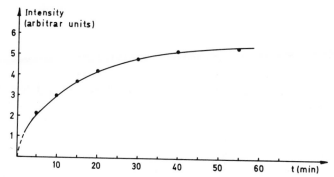

Fig. 11. Evolution of the intensity of the EPR signal obtained with the system DNA—proflavin, in function of the duration of irradiation.

Now that we have discussed the most important results obtained with proflavin, we may compare the activity of this sensitizer with the other dyes mentioned earlier. By way of example, we shall discuss here the spectra obtained with proflavin, acriflavine and pyronine. The shapes of the spectra are identical in all three cases. However, the intensities vary from one dye to the next. Proflavin yields the most intense signal, while acriflavine gives a spectrum about 20% less intense and finally, pyronin is characterized by a very low activity (80% less intense than with proflavin).

This comparison was made after a period of irradiation sufficiently long so that the plots of the signal intensities increasing over time were able to reach saturation. Fig. 11 shows the form of these curves. Similar curves have been obtained with nucleotides, as we pointed out earlier. Depending on the nature of the sensitizer, the curves are shifted to a greater or lessser extent and the saturation plateau thus reveals specific properties of each dye studied.

As far as methylene blue is concerned, the signal obtained cannot be considered characteristic of the substrate. The intensity of the spectrum is not modified by the presence of a nucleotide or DNA. This strongly suggests that the mode of action of this particular dye is different from that of proflavin.

However, we were interested in knowing whether the changes in the concentrations of radicals would parallel the differences in photodynamic activity detected in biological experiments. In an attempt to answer this question, we undertook a series of studies with the bacteriophage T4 in which we compared the magnitude of the biological effects with the concentrations of free radicals produced by photodynamic treatment.

3.4. *Bacteriophage T4*

In the presentation of the results, we shall discuss first, the biological measurements and then the physicochemical measurements. The biological studies were carried out by Dr. C.M. Calberg. Let us begin with the findings concerning proflavin.

3.4.1. *Biological measurements.* The inactivation of viruses by photodynamic action has been observed in a number of systems [29]. Recently, it has been shown that mutations may also be induced in the T4 phage by photodynamic treatment with proflavin [22, 23, 24], thiopyronin and psoralene [5]. Nevertheless, it is difficult to compare the data in the literature concerning the effects produced by different photosensitizers since the experimental conditions — irradiation, pH, ionic strength — have not been uniform. Hence, these studies were designed to determine the relative photodynamic activities of the various sensitizing dyes. For these experiments, irradiation was effected with a mercury vapor lamp of low power (42 W); the light was filtered so as to eliminate the ultraviolet region of the spectrum. Intensities of the lamp's emission lines were measured with a photomultiplier. As we shall discuss later on, these measurements were used to compare the various levels of interaction with respect to an identical dose of absorbed energy.

Let us first discuss the inactivation experiments [3]. As Helprin and Hiatt have shown [7], in order for the T-even phages to be sensitized, the viruses must be incubated with dye in the dark for several hours before being irradiated. Fig. 12 shows the change in the logarithm of the surviving phages per cubic centimeter as a function of the duration of irradiation. The line A corresponds to a sample incubated for 2.5 hr at $37°C$ in the dark with proflavin 1.5×10^{-5} M, diluted 100 times and irradiated. It may be observed that after about 4.5 min of irradiation, the number of active phages — originally 10^8 per ml — dropped to 10^6. The two other slopes reveal the effect of dye diffusion and show that a reduced contact between the phage DNA and the dye was accompanied by a decrease in inactivation. If the phage dye complex is diluted 100 times and further incubated in the dark to allow a dissociation of the complex, the photosensitivity of the phage decreases as shown in the slope B. The line C corresponds to a sample withdrawn after 3 min irradiation during experiment A, diluted 100 times, incubated overnight in the dark at $42°C$ and irradiated. The good parallelism between slopes B and C shows that the unbinding of the dye is not affected by a prior irradiation of the dye phage complex.

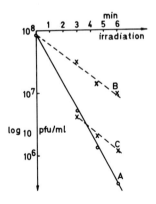

Fig. 12. Kinetics of T4 inactivation [3].

Dye	Phage inactivation				Free radicals induction
	Molar concentration	% inactivation during preincubation	Irradiation time for one average lethal hit (min)		Intensity of the E.P.R. signal
			Observed	Corrected	
Pyronine	5.10^{-5}	—	24	13,7	< 2
Acridine red	5.10^{-5}	—	13	4,70	< 2
9-Aminoacridine	5.10^{-5}	—	3,90	2,33	6,5
Acridine yellow	5.10^{-5}	—	0,80	0,73	5,5
Acriflavine	5.10^{-5}	67	0,37	0,34	8,5
	10^{-5}	—	0,75	0,70	
Acridine orange	5.10^{-5}	73	0,30	0,16	7,8
	10^{-5}	25	0,67	0,37	
Proflavine	5.10^{-5}	83	0,28	0,37	10,2
	10^{-5}	30	0,60	0,60	

After an incubation of about 20 hr, the level of inactivation reached a maximum. A similar pretreatment period was consequently used for the remainder of the experiments. It should be mentioned that inactivation ceased at the end of the period of irradiation. On the other hand, there is no inactivation of the control during the incubation without dye at 37°C nor during an irradiation as long as 15 min. Twenty hours preincubation with a high concentration of proflavin (5 × 10^{-5} M) leads to some inactivation in the dark. In each case, the dark reaction is found to be proportional to the light sensitivity exhibited by the phage–dye complex and infectivity is counted as initial before incubation.

Let us now compare the relative activities of the different sensitizers. The results are illustrated in table 1. This table indicates, for each sensitizer, the irradiation on time necessary to decrease the phage infectivity to 37% of its initial value. To be able to compare these inactivation rates, we have introduced a correction reducing to the same amount the energy absorbed per min by each phage–dye mixture. This correction includes two factors, the emission spectrum of the lamp and the absorption spectrum of the dye, and is calculated as follows:

(a) for every 10 mμ, in the absorption region of the dye, both the molar extinction coefficient (ϵ_i) and the light intensity of the lamp (I_{o_i}) are measured;

(b) the sum $\Sigma I_i = \Sigma I_{o_i} \exp(-\epsilon_i c l)$ and the difference $\Delta I = \Sigma I_i - \Sigma I_{o_i}$ are determined;

(c) the ratio of ΔI versus ΔI of proflavin is calculated and used to correct the time in the different inactivation rates.

The results are given in the column. Pyronin and acridine red appear to have very little activity. Among the acridines, two are slowly efficient, 9-aminoacridine and acridine yellow, three others having a great inactivating power. The results obtained with methylene blue are not reported in the table because the photoreaction pro-

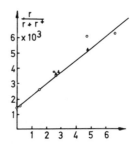

Fig. 13. Increase in r mutant frequency during inactivation with proflavin. Lethal hits are shown as the abscissa [3].

duced by this dye appears to be different. In fact, it is found that its inactivating power is greater than that of proflavin but the phage infectivity decreases for hours after the end of irradiation. In addition to inactivation, photodynamic treatment gives rise to a second phenomenon: a mutation effect [3]. The frequency of r plaques increases among the survivors during irradiation of T4 sensitized to light by proflavin. In fig. 13, the r frequencies are plotted as a function of the degree of inactivation. The points fit a straight line whose slope is calculated by the method of least squares. It is found that the increase of r mutant ratio is of 0.73×10^{-3} per lethal hit. The mutagenic activity of different dyes was then measured under conditions of visible light inactivation leading to around 5 lethal hits [3]. It appears that the mutagenic action of the dyes is comparable to that of proflavin. The proportion of mutants among the survivors rises with the level of inactivation. Thus, the photomutagenic activity is directly proportional to the rate of inactivation. The energy absorbed by the bound dye and transferred to DNA must lead to a lesion which either destroys the phage infectivity or gives rise to a mutation. Let us now proceed to the results obtained by EPR spectroscopy.

3.4.2. *EPR spectroscopy measurements*. When a phage suspension containing proflavin is irradiated in the visible range at low temperature, a paramagnetic spectrum is obtained as shown in fig. 14. In this case, a hyperfine structure appears in the form of the lateral components. This structure is strikingly reminiscent of the one observed with TmP, which suggests that the effect in question is operative at the level of the DNA. The intensity of the signal does indeed depend on the interaction between the dye and the nucleic acid molecule. This is illustrated in the fig. 15 where spectrum A corresponds to a sample which was incubated for several hours before being brought to a low temperature and irradiated. For spectrum B, the sample was irradiated without preliminary incubation. A decrease in the intensity of the signal is clearly visible in the case of the non-incubated system, where the dye was able to penetrate only slightly inside the phage. In spectrum C, the molecules of DNA were freed of their protein coat. Under these conditions incubation is no longer necessary and a signal is obtained equal in intensity to that of spectrum A.

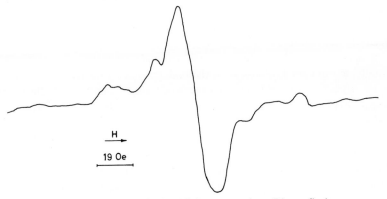

Fig. 14. Photosensitization of the system phage T4–proflavin.

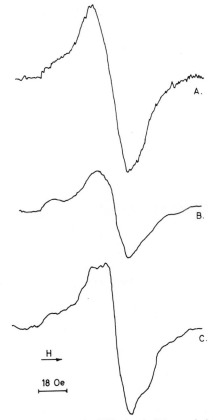

Fig. 15. Influence of the incubation on the EPR signal: (A) sample incubated overnight before irradiation, (B) sample irradiated without incubation, (C) after freezing and thawing of the system, the DNA molecules are freed, so that incubation is no longer necessary.

The process by which free radicals are formed in bacteriophages is indeed a photodynamic effect. If the intensities of the spectra obtained with three samples sealed respectively under nitrogen, air and oxygen are compared, unequivocably, the most intense signal corresponds to the sample sealed under oxygen.

Parallel to our earlier discussion of the biological experiments, we shall now compare the efficacity of the different dyes in producing free radicals. Proflavin yields the most intense signal, while that of acriflavine is about 20% weaker. The signal obtained with 9-aminoacridine is approximately 50% as strong as with proflavin. Finally, with derivatives of the xanthene family, pyronin and red acridine, the quantities of radicals produced are especially low.

Table 1 makes it possible to compare biological effects and free radical formation. Generally speaking, a substantial inactivation effect appears to be associated with a high yield of radicals. The variations observed within each of the subgroups cannot be considered significant. Nevertheless, on the basis of the results presented, it is tempting to conclude that radical processes are probably involved in the photodynamic phenomena of inactivation and photomutation.

In conclusion, the parallelism observed for bacteriophages between EPR and inactivation measurements is suggestive of a cause and effect relationship between the two phenomena. However, no decisive proof of this has yet been adduced.

References

[1] R.S. Alger, Electron paramagnetic resonance: techniques and applications (Interscience, Wiley, New York, 1968).

[2] J. Bourdon and B. Schnuriger, Mécanisme d'oxydation photosensibilisée dans un milieu rigide macromoléculaire. Photochem. Photobiol. 5 (1966) 507–514.

[3] C.M. Calberg-Bacq, M. Delmelle and J. Duchesne, Inactivation and mutagenesis due to the photodynamic action of acridines and related dyes on extracellular bacteriophage T_4B. Mutation Res. 6 (1968) 15–24.

[4] E.J. Corey and W.C. Taylor, A study of the peroxidation of organic compounds by externally generated singlet oxygen molecules. J. Amer. Chem. Soc. 86 (1964) 3881–3882.

[5] J.W. Drake and J. McGuire, Properties of r mutants of bacteriophage T4 photodynamically induced in the presence of thiopyronin and psoralen. J. Virology 1 (1967) 260–267.

[6] C.S. Foote and S. Wexler, Olefin oxidations with excited singlet molecular oxygen. J. Amer. Chem. Soc. 86 (1964) 3879–3880.

[7] J.J. Helprin and C.W. Hiatt, Photosensitization of T2 coliphage with toluidine blue. J. Bacteriol. 77 (1959) 502–505.

[8] D. Ingram, Free radicals as studied by electron spin resonance (Butterworths, London, 1958).

[9] V. Kasche, Radical intermediates in the fluorescein and eosin photosensitized autoxidation of L-tyrosine. Photochem. Photobiol. 6 (1967) 643–650.

[10] H. Kautsky, H. de Bruin, R. Neuwirth and W. Baumeister, Photosensibilisierte Oxydation als Wirkung eines aktiven, metastabilen Zustandes des Sauerstoff. Moleküls. Ber. 66 (1933) 1588–1600.

[11] K. Kawaoka, A.U. Khan and D.R. Kearns, Role of singlet excited states of molecular oxygen in the quenching of organic triplet states. J. Chem. Phys. 46 (1967) 1842–1853.

[12] V.E. Kholmogorov, E.V. Baranov and A.N. Terenin, Investigation of the sensitization of the photodehydration of alcohols at 77°K by the EPR method. Doklady Akademii Nauk SSSR *149* (1963) 142—145.

[13] K. Kikuchi and M. Koizumi, Photoreduction of proflavine in the aqueous solution. Bull. Chem. Soc. Japan *40* (1967) 736—742.

[14] M. Lacroix and A. van de Vorst, Radicaux libres induits par l'action du rayonnement ultra-violet sur certains constituants de l'acide desoxyribonucléique. Photochem. Photobiol. *7* (1968) 477—483.

[15] L.S. Lerman, Structural considerations in the interaction of DNA and acridines. J. Mol. Biol. *3* (1961) 18—30.

[16] L.S. Lerman, Amino group reactivity in DNA—amino-acridine complexes. J. Mol. Biol. *10* (1964) 367—380.

[17] L.S. Lerman, Acridine mutagens and DNA structure. J. Cell. Comp. Physiol. Suppl. *1* (1964) 1—18.

[18] A. Müller, Spektrographische Untersuchungen mittels paramagnetischer Elektronenreso-nanz über die Wirkung ionisierender Strahlen auf elementare biologische Objekte. Akade-mie der Wissenschaften und der Literatur - Mathematisch-Naturwissenschaftlichen Klasse, no. 5 (1964).

[19] G. Oster, J.S. Bellin, R.W. Kimball and M.E. Schrader, Dye-sensitized photooxidation. J. Amer. Chem. Soc. *81* (1959) 5095—5099.

[20] G. Pake, Paramagnetic resonance (Benjamin, New York, 1962).

[21] A.R. Peacocke and J.N.H. Skerrett, The interaction of aminoacridines with nucleic acids. Trans. Farad. Soc. *52* (1956) 261—279.

[22] D.A. Ritchie, Mutagenesis with light and proflavine in phage T4. Genet. Res. (Camb) *5* (1964) 168—169.

[23] D.A. Ritchie, Mutagenesis with light and proflavine in phage T4, II. Properties of the mu-tants. Genet. Res. (Camb) *6* (1965) 474—478.

[24] D.A. Ritchie, The photodynamic action of proflavine on phage T4. Biochem. Biophys. Res. Commun. *20* (1965) 720—726.

[25] J.L. Rosenberg and D.J. Shombert, The phosphorescence of adsorbed acriflavine. J. Amer. Chem. Soc. *82* (1959) 3253—3257.

[26] L. Santamaria and G. Prino, The photodynamic substances and their mechanism of action. Res. Progress. Org. Biol. Med. Chem. *1* (1964) 259—336.

[27] B. Smaller, Role of triplet state in photoreactions. Nature *195* (1962) 593—594.

[28] J.D. Spikes and C.A. Ghiron, Photodynamic effects in biological systems. Physical Pro-cesses in Radiation Biology (Academic Press, New York, 1964) pp. 309—338.

[29] J.D. Spikes and R. Straight, Sensitized photochemical processes in biological systems. Ann. Rev. Phys. Chem. *18* (1967) 409—436.

[30] A. Terenin, V. Rylkov and V. Kholmogorov, Biphotonic sensitization of bond splitting in organic molecules at 77°K: an EPR study. Photochem. Photobiol. *5* (1966) 543—553.

CHAPTER 13

PHOTODYNAMIC ACTION:
A VALUABLE TOOL IN MOLECULAR BIOLOGY

P. CHANDRA

Institüt für Therapeutische Biochemie
der Universität, Frankfurt (Main), Germany

1. Introduction

The maintenance of life is dependent on the structure and physical state of certain macromolecules. These macromolecules are essentially polysaccharides, proteins and nucleic acids. However, the communication, or the transfer of biological information takes place primarily through nucleic acids. This information transfer is based on the principle of base-pairing discovered by Watson and Crick [61, 62, 63] for DNA. Just as guanine pairs with cytosine in the two strands of the DNA double helix, so can guanine in a DNA strand pair with a cytosine in a ribonucleic acid strand (and vice versa). Similarly, the adenine of DNA can pair with the uracil in RNA, the thymine in DNA with the adenine in RNA. The basic hydrogen-bonding rules that apply to these DNA–RNA hybrid molecules are those that apply to the two strands of DNA itself. Two RNA strands also can form a duplex structure by similar rules.

These base-pairing rules provide a mechanism that is extremely important in specifying the amino acid sequence of a polypeptide chain (fig. 1). In protein biosyn-

ATP = adenosine triphosphate, DNA = deoxyribonucleic acid, RNA = ribonucleic acid, poly A = polymer of adenylic acid, poly C = polymer of cytidylic acid, poly U = polymer of uridylic acid, poly AG = copolymer of A and guanylic acid, poly UA = copolymer of U and A, poly UC = copolymer of U and C, poly UG = copolymer of U and G, mRNA = messenger RNA, tRNA = transfer of soluble RNA, rRNA = ribosomal RNA, S_{30} fraction = supernatant obtained after 30,000 × g centrifugation, S_{100} fraction = supernatant obtained after 100,000 × g centrifugation.

Amino-acid-RNA ligases [EC: 6.1.1], the enzyme catalyzes ATP-PPi exchange reaction, and the formation of aminoacyl-RNA.

RNA polymerase [Nucleoside triphosphate : RNA nucleotidyltransferase, EC: 2.7.7.6]. The enzyme requires ATP, CTP, GTP and UTP for activity and DNA as template. RNA synthesis is initiated by the incorporation of a ribonucleoside monophosphate to form the 5' terminus of the chain.

Polynucleotide phosphorylase [EC: 2.7.7.8], the enzyme catalyzes the formation of homo- and co-polymers of A, U, C and G. It is specific for ribonucleoside diphosphates and functions without any primer.

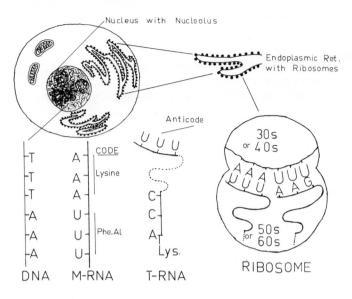

Fig. 1. Schematic representation of information transfer processes.

thesis, DNA serves as a template for a messenger RNA (mRNA) complement made in accordance with the pairing rules. The result is a DNA–RNA hybrid stage with a specific, inherited sequence of bases. The mRNA can then serve as a template upon which transfer RNA (tRNA) molecules are oriented, in a sequence that also is determined mainly by the Watson–Crick pairing rules. A slight modification of this rule has been suggested by Crick [8, 9] for the recognition of 3rd base at the anticodon site. As a result of this step the individual amino acids can be ordered in a specific sequence, for each tRNA molecule carries a specific amino acid. Finally, the amino acids are linked by peptide bonds to form a polypeptide chain. The amino acid sequence of this chain, then, has been dictated by the original DNA code, transcribed through mRNA, and translated by tRNA through the mechanism provided by the base-pairing rules.

An essential feature of base-pairing between purines and pyrimidines is their structural specificity. The base-pairing rules hold true for intact molecules. Thus any structural modification of a specific base in a nucleic acid may either block the information flow, or passed on false information. It is known that nucleic acids and proteins are not sensitive to visible light alone. However, in the presence of dyes and other photosensitizers they undergo structural modifications.

The studies described in the following pages concern the photodynamic action of several dyes*, furocoumarins and acetone. These compounds are known to have a

---

*The term 'dye' has been used arbitrarily to denote coloured compounds that absorb light in the photodynamic range and that have been studied as photosensitizers.

lethal action on bacteria in the presence of light. However, the mechanisms by which they act are highly specific, and are different for each one. For example, dyes destroy specifically guanine [3, 4, 48, 58], furocoumarins form adducts with thymine and uracil [12, 39, 55] and acetone catalyzes dimer formation of pyrimidines [5, 34, 55, 65]. These reactions can therefore be used to study some problems of molecular biology. Our studies will be described in three sections: photodynamic action on cells, on subcellular components, and on macromolecules.

2. Experimentation

2.1. *Light sources*

Most of the experiments were carried out using a Philips mercury vapour lamp HPR, 125 W. The absolute energy distribution given by the company is as follows:

Wave length (nm)	313	365	391–408	436	546	570
Absolute energy (W)	0.1	2.1	1.0	1.6	2.2	2.0

This type of light will from now on be described as visible light.

Acetone- and some furocouramine-mediated reactions were carried out using monochromatic radiation. The selectivity of wavelength in these experiments was necessary to achieve a dose-dependent response. In most cases a grating monochromator designed by Bausch & Lomb, Rochester was used. This apparatus consists of 3 component parts: (a) the monochromatic unit, (b) the cover-grating assembly, and (c) the illuminating unit. The illuminating unit consists of a super-pressure mercury lamp (Osram, HBO 200 W), fused quartz condensor and a quartz collective lens. This instrument is provided with a grating 1200 grooves per mm. The grating mount is rotatable to place different wavelengths on the exit slit, and is linear with the wavelength in its rotation.

2.2. *Measurement of light intensity*

To determine the quantum requirement or efficiency of a photodynamic action a measurement of the incident light intensity is necessary. This was done either by chemical actinometry, or using a compensated thermopile.

Chemical actinometry was employed to measure the light intensity of monochromatic radiation. This was achieved by irradiating a solution of malachite green leucocyanide in ethyl alcohol. Leucocyanide was prepared and purified by the method of Calvert and Rechen [2]. Irradiation transforms the leucocyanide into an intensely colored dye with an absorption maximum at 620 nm. The amount of dye formed was followed spectrophotometrically. The intensity of visible light was measured by ferrioxalate actinometer. According to Parker [44] the potassium ferrioxalate actinometer is applicable over a wide range of wavelengths and intensities. The details of this method are described by Hatchard and Parker [23].

Compensated thermopile (CA 1 Ser. No. 650509, Kipp and Zonen, Delft, Holland) was also employed to measure the light intensity of monochromatic radiation. The calibration of this instrument was carried out by the company, and given as the electromotive force (EMF) produced by the pile for a certain amount of incident radiation. The voltage read on the attached galvanometer (type AL 4- Microva) is related to the EMF of the thermopile by the simple equation:

$$V_g = \frac{R_g}{R_s + R_g} \text{(EMF)}$$

$$\text{or(EMF)} = \frac{R_s + R_g}{R_g} V_g$$

where V_g is the voltage read on the galvanometer, R_g is the galvanometer resistance at the relevant range and R_s is the resistance of the thermopile.

2.3. Methods of irradiation

Bacterial suspensions were irradiated under continuous stirring. This was achieved by putting the petridish containing bacterial suspension on a magnetic rotor (fig. 2). The overheating of cells was avoided by putting two electric fans on either side of the lamp. The irradiations were carried out in a cold room. Under these conditions

Fig. 2. Arrangement for the irradiation of cells.

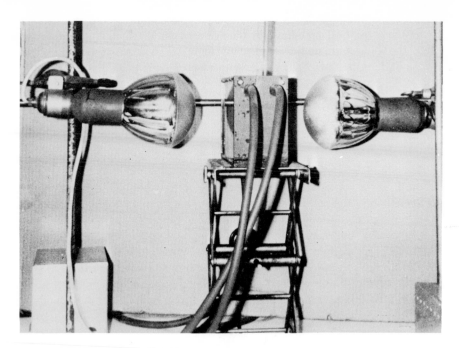

Fig. 3. Irradiation of sub-cellular fractions and macromolecules.

the temperature of the cell suspension remained below 25°C. Irradiation of sub-cellular fractions and macromolecules was done in a water bath having a water-flow
system. Test tubes containing fractions or solutions were fixed in the water bath,
and irradiated with two HPR 125 W lamps placed one on either side of the bath.
The lamps were placed at an equidistance to avoid errors in dose measurements.
The temperature of the reaction mixture was maintained constant by regulating the
water flow. An arrangement of the apparatus is shown in fig. 3.

2.4. *Removal of the excess of sensitizers*

The excess of sensitizer remaining after irradiation of the system should be removed. This is necessary to avoid non-specific effects observed during enzymatic
reactions in vitro. The method of removal of the sensitizer is dependent on the type
of system used. We have used various methods to remove sensitizers from cells, sub-
cellular fractions and macromolecules.

The excess of sensitizer from whole cells can be removed by washing them repeatedly with isotonic buffers. Our studies with labelled sensitizers have shown that
cells washed three times were completely free from the removable sensitizer. Studies
with isolated components need special methods since their sedimentation is very
difficult to achieve. Soluble enzyme fractions were dialyzed in cold buffers for 3–6
hr after their treatment. The dialyzing medium was changed 2–3 times. In some

cases DEAE-cellulose chromatography was used. Ribosomes on dialysis lost some of their biological activity. To remove excess of sensitizers from ribosomes we used discontinuous sucrose-gradient centrifugation. Under these conditions the sensitizers remain in the medium, whereas ribosomes settle down at the bottom. Macromolecules were freed from excess sensitizers by column chromatographic procedures. This was achieved by gel-filtration on sephadex columns.

3. Factors influencing photodynamic action

Some of the factors which influence a photodynamic response are: (a) action spectrum of the sensitizer, (b) concentration of the sensitizer, (c) intensity of radiation dose, (d) oxygen, and pH of the sensitizer and (e) the material or the system being sensitized. Hiatt [27] has published an excellent review in which he examines the factors which influence the photodynamic inactivation of viruses.

The problem of action spectrum has been dealt with by Dr. A. Kleczkowski in this book. It is essentially the wavelength at which a sensitizer exerts its maximum effect. This wavelength mostly corresponds to the absorption maxima of the sensitizer. Weil and Maher [64] found that the action spectra of photooxidation with methylene blue have maxima at 670 nm corresponding to a maximum in the absorption spectrum of methylene blue. Hinemets, Vinegar and Taylor [24] found two peaks in the absorption spectrum of methylene blue absorbed to *E. coli* B, the usual 670 nm maximum of methylene blue and a 610 nm peak due to the dye—bacteria complex. Irradiation with light of the wavelength of either peak was equally effective in killing the bacteria. This principle has recently been employed by many scientists in an effort to identify the nature of the chromophore in PR-enzyme.

The effect of varying the concentration of the sensitizer has been studied in too many cases to mention individually. In general one observes an intensification of the action as the sensitizer concentration increases until a maximum is reached. A further increase in the concentration could reduce the effectiveness. Besides such concentrations may produce non-specific effects, e.g. toxicity in dark and sensitizer—sensitizer interaction by light. That the photodynamic response is proportional to the dose of light is a well documented fact and does not need any explanation here.

Besides the factors mentioned above there are conditions dependent on the nature of the system under investigation. For example, permeability of a sensitizer may be very important for the photodynamic response at cellular level. Since most of the studies on photodynamic action record the survival fractions or bacterial lethality the problem of permeability to sensitizers deserves attention. A certain selectivity with respect to the type of microorganism has been observed for various sensitizers. Is the cell wall of the microorganism responsible for this selectivity, or are there internal components of the cell which react selectively?

4. Photodynamic action on cells

Many types of photodynamic phenomena have been studied at the cellular level, for example, loss of colony-forming ability, delays and abnormalities in cell-division, induction of mutations, decrease in motility, alteration of membrane permeability and active transport, loss of capacity to support bacteriophage growth and so on. These studies have been carried out mostly with single-celled organisms such as bacteria (see Harrison [22]), yeast [19, 36], algae [53], and protozoans [11, 12, 45]. In addition to these, photodynamic effects have been studied on mammalian cells in culture. A variety of effects have been reported, including cessation of growth [29], destruction of cellular enzymes [29], interruption of mitosis [35], and killing [31]. Most of these studies record a particular effect or damage produced in the presence of a sensitizer and light. Little is known, however, about the basic mechanism involved in these photodynamic processes. An attempt is therefore, made to correlate the photodynamic damage at cellular level to its individual components.

E. coli B cells were grown in a medium described previously [6] and harvested at an early log phase. The cells were resuspended in the same medium, 10 μg/ml of thiopyronin or psoralen added and the whole suspension was transferred into a petri dish. The petri dish was placed on a magnetic rotor and stirred well with a magnetic stirrer. This suspension was then irradiated with a Philips mercury vapour lamp HPR 125 W. The total radiation dose in these experiments was 6.6 $\times$ 10^6 erg /mm^2. The treated and untreated cells were washed three times to remove the excess of the sensitizers. The cells were broken with glass beads (0.11−0.12 mm ϕ) in a Braun homogenizer and protein-synthesizing fractions (ribosomes, tRNA and pH 5-fraction) isolated by the usual procedures. These fractions were tested for their biological activity by the method of Nirenberg and Matthaei [42].

As follows from table 1 the polymerization of phenylalanine under the influence

Table 1

In vivo effect of thiopyronin and visible light on the biological activity of ribosomes, tRNA and pH 5-enzymens

System from untreated bacteria			System from treated bacteria			Incorporation (%)
S^X_{100}	tRNA	Ribosomes	S^X_{100}	tRNA	Ribosomes	
+	+	+				100[xx]
	+	+	+			78
+	+				+	57
+	−	+		+		31
			+	+	+	0.4

[x] Designates the supernatant fraction obtained after 100,000 $\times$ g centrifugation.
[xx] 100% value designates the incorporation of 2507 cpm of phenylalanine in one mg if ribosomal protein.

of thiopyronin and visible light is completely inhibited. These experiments contained ribosomes, tRNA and pH 5-enzymes from thiopyronin-treated bacteria. To localise the site of maximum inactivation an exchange was made between the individual fractions of the treated and non-treated bacteria. A maximum inhibition of the phenylalanine incorporation was obtained when tRNA from the treated bacteria was used in the reaction mxiture. The invivo inactivation of tRNA by dyes was found to be the same for most of the amino acids tested. We, therefore, conclude that the inactivation is not specifically due to guanine destruction, but also the secondary structure is altered. The inhibition of the biological activity of ribosomes by thiopyronin may be due to modification of guanine of non-helical regions of rRNA, and photooxidation of several amino acids (see Spikes and Livingston [52]) in proteins of ribosomes. These aspects will be discussed later. That some interaction between proteins and dyes does take place in vivo is shown in the enzyme fraction of treated bacteria.

Table 2

In vivo effect of psoralen and visible light on the biological activity of ribosomes, tRNA and pH 5-enzymes

System from untreated bacteria			System from treated bacteria			Incorporation (%)
S^X_{100}	tRNA	Ribosomes	S^X_{100}	tRNA	Ribosomes	
+	+	+				100[XX]
	+	+	+			99
+	+				+	78
+		+		+		41
			+	+	+	8

[X] Designates the supernatant fraction obtained after 100,000 $\times$ g centrifugation.
[XX] 100% value designates the incorporation of 2448 cmp of phenylalanine per mg of ribosomal protein.

While the dyes react photodynamically with proteins and nucleic acids, furocoumarins react only with nucleic acids. The in-vivo effect of psoralen and light on the activity of ribosomes, tRNA and enzymes is shown in table 2. A complete system containing all fractions from treated bacteria shows 92% inhibition of phenylalanine incorporation. By cross-reacting the individual fractions from treated and un-treated bacteria we observed a maximum inhibition for tRNA, followed by ribosomes, but no inhibition of enzyme activity. The biological inactivation of tRNA by thiopyronin (see table 1) and psoralen are very similar, whereas ribosomes at the same radiation dose are more sensitive to thiopyronin.

Garvin, Julian and Rogers [18] have recently studied the action of various dyes on the photodynamic inactivation of ribosomes in vitro. They conclude that a protein or proteins are the primary substrates being altered by the photooxidation, and it is therefore unlikely that guanine residues of the RNA are involved in the inactivation process. Little is known about the internal anatomy of ribosomes. According

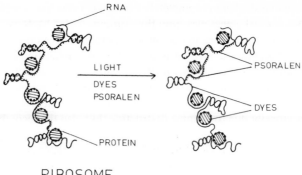

RIBOSOME

Fig. 4. Hypothetical scheme for the photodynamic inactivation of ribosomes by dyes and psoralen.

to Cotter, McPhie and Gratzer [7] the ribosomal proteins are supposed to be in the interior and associated with non-helical regions of rRNA (fig. 4). The conformation of the RNA in the ribosomes is similar to that in the free state, and contains about 60% of paired bases in short helical segments. On the other hand, the dye stacking data of Furano et al. [16] indicate that the RNA in ribosomes has much less helical content than the isolated ribosomal RNA. They have proposed a model of the ribo-some in which the RNA is held in a single stranded configuration by some kind of interaction between the nucleotide bases and the ribosomal protein. They also ob-served that the amount of dye bound to ribosomes is independent of the protein content. This was checked using various ribosomal preparations.

The comparative studies on the inactivation of ribosomal fractions from thiopy-ronin and psoralen treated *E. coli* B indicate that rRNA as well as proteins are in-volved in their inactivation process. We have not been able to detect any photody-namic damage to proteins caused by psoralen. This has been studied by us using labelled psoralen. *E. coli* B was grown with psoralen-[4-^{14}C], centrifuged and frac-tionated. During the treatment with psoralen the cells were irradiated with visible light (6.6×10^6 erg/mm^2). Under these conditions we observed an absolute uptake of 15% of the total radioactivity by whole cells. On fractionation, 13.6% of the total activity was recovered in nucleic acids. This explains that psoralen in vivo reacts only with nucleic acids, and protein components are not damaged. On the other hand, thiopyronin damages nucleic acids (guanine) as well as proteins (histidine, tyrosine and so on). This could account for the higher inactivation of ribosomes from thio-pyronin-treated bacteria.

Pyrimidines on irradiation with UV light undergo some typical photochemical reactions. Uracil [49, 60] yields a dimer and a water-addition product, i.e. 5,6-di-hydro-6-hydroxyuracil, whereas thymine [1, 56] gives its dimer as the main product. A similar phenomenon has recently been observed under photosensitized conditions [10, 34, 65], acetone or acetophenone being the most commonly used sensitizers. We have tried to isolate the photoproducts from bacteria after their irradiation at 315 nm in the presence of acetone.

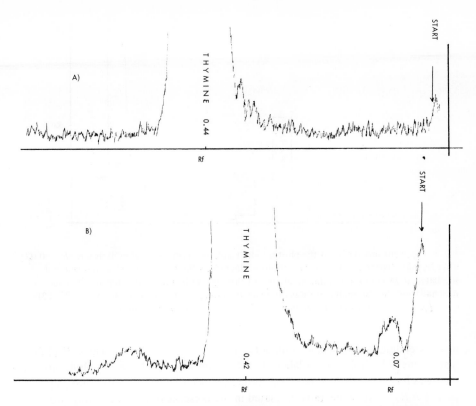

Fig. 5. Paper chromatogram of a perchloric acid hydrolysate of irradiated bacteria containing thymine-2-[14]C. *E. coli* 15T⁻ was grown in citrate medium containing 2 μg/ml 2-^{14}C-labelled thymine. Bacteria were washed 2× with 0.9% NaCl and irradiated in the same solution containing acetone (10%). Irradiation was made at 315 mμ using a grating monochromator. After irradiation (1.13 × 10⁵ erg/mm²) the bacteria were centrifuged, washed 3× with ethanol:ether (1:1, v/v) and 1× with ether alone, dried, suspended in 0.03 ml HClO₄ (70%) and heated in a boiling water bath for 60 min. The suspension was neutralized with 10 N KOH and applied on 2043 b filter paper (Schleicher & Schüll). The chromatograms were run in an ascending manner using n-butano:water (87:13, v/v) system. Ordinate: radioactivity (cpm), (A) without acetone, (B) with acetone.

    E. coli 15T⁻ was grown in citrate medium containing 2 μg/ml of 2-^{14}C-labelled thymine. Bacteria were washed twice with 0.9% NaCl and irradiated in the same solution containing 10% acetone. Irradiation was made at 315 nm using a grating monochromator. After irradiation (1.13 × 10⁵ erg/mm²) the bacteria were centrifuged, washed three times with ethanol:ether [1:1, v/v] and once with ether alone. The dried pellet was suspended in 0.03 ml perchloric acid (70%) and heated in a boiling water bath for 60 min. The suspension was neutralized with 10 N KOH and applied on 2043 filter paper (Schleicher and Schüll). The chromatograms were run

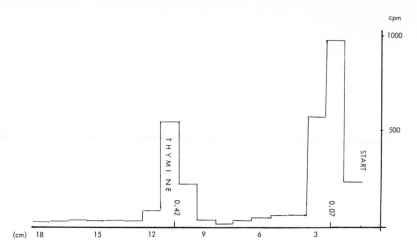

Fig. 6. Paper chromatogram of the photoproduct isolated from bacterial hydrolysate (R_f = 0.07) after its reirradiation at 240 nm. The photoproduct (R_f = 0.07) was eluated with water and irradiated at 240 mμ for 60 min. An aliquot of this was applied on a 2043 b paper (Schleicher & Schüll) and the chromatogram was run in an ascending manner in n-butano : water (87 : 13) system. Ordinate: radioactivity (cpm). Abscissa: paper strips in cm.

in an ascending manner using n-butanol : water (87 : 13, v/v) system. The radiochromatograms of hydrolysates of labelled bacteria are shown in fig. 5. No photoproduct was formed in the absence of acetone (fig. 5a), whereas a photoproduct appeared at R_f = 0.07 in bacteria irradiated in the presence of acetone. This photoproduct was eluted with water and reirradiated at 240 nm for 60 min (fig. 6). Rechromatography of the eluate shows the presence of thymine, indicating that the photoproduct is a dimer of thymine. The importance of this dimer in photoreactivation studies has been described by Dr. Adolf Wacker in this book.

5. Photodynamic action on sub-cellular components

Since the sub-cellular components vary greatly in composition and organisational structure, it is reasonable to think that the photodynamic action at this level would be selective, i.e., that with a particular sensitizer and set of reaction conditions, some components would be more sensitive to photodynamic action than others. Although very little work has been done at this level, some selective effects have been observed. Of particular interest are the studies on mitochondria. Santamaria [47] has shown that photodynamic treatment of isolated mitochondria from rat liver using polycyclic hydrocarbons as sensitizers causes extensive swelling of mitochondria. These studies have been substantiated by Spikes and Haga [51] who reported the effects of photodynamic treatment with methylene blue, eosin Y and thiopyronin as sensitizers on the oxygen-consumption, oxidative phosphorylation, succinic dehydrogenase activity and cytochrome oxidase activity of isolated rat liver mitochondria.

The studies reported in preceding pages concern mainly the photodynamic action on isolated ribosomes and transfer RNA.

Table 3
Nucleotide composition of the illuminated RNA in the presence of dyes

Nucleotide	Yeast RNA[+] (TP) (mole% nucleotide)		tRNA[++] (MB) (mole% of total phosphate)	
	Control	Irradiated	Control	Irradiated
GMP	27.9	11.0	25.1	10.7
AMP	26.1	−[†]	19.1	19.2
CMP	19.5	−	26.5	25.5
UMP	26.4	−	17.4	17.3

[+] Yeast RNA was irradiated at 1.2×10^7 erg/mm^2 with thiopyronin (10 μg/ml). The samples were hydrolyzed with alkali and nucleotides separated by electrophoresis.
[†] The values obtained did not show any significant change. These results are from W. Opree (M.D. Thesis, Frankfurt 1967).
[++] Data taken from Kuwano et al. [33]. MB = methylene blue.

5.1. Transfer RNA

Table 3 shows the photodynamic action of thiopyronin and methylene blue on RNA. A comparison of base composition shows selective destruction of guanine in irradiated samples. This offers an excellent possibility of studying the effect of guanine destruction in tRNA molecule on its functional activities.

Transfer RNA carries out at least two separate functions: it recognizes a particular amino acid-RNA-synthetase so that it can accept an appropriate activated amino acid and it codes with the messenger RNA in such a way as to ensure that the amino acid that it carries is correctly placed in sequence in the growing polypeptide chain. The site or loop responsible for the latter reaction is called anticodon loop, and carries a base-triplet complimentary to the code in mRNA. Thus the amino acids containing cytosine in their codes should possess guanine in their anticodons. One would therefore expect that such site should preferably be inactivated under the photodynamic action of dyes.

Table 4 shows the effect of dyes and light on the acceptor functions of tRNA for various amino acids. A highest inhibition is observed for proline followed by leucine, lysine and phenylalanine. The table contains values of thiopyronin-inactivation [3] and methylene blue inactivation of tRNA [33]. Surprisingly in both experiments the amount of inhibition is very similar. This indicates that anticodon influences the recognition of a specific aminoacyl-synthetase. One might also think that the bases at the recognition site have a pattern similar to the ones at the anti-

Table 4

Inhibition of the amino acid acceptor activity of tRNA after illumination with dyes

| Amino acid | % Inhibition of the acceptor activity on illumination with: | |
	Thiopyronin[+]	Methylene blue[++]
Phenylalanine	8	–
Lysine	10	–
Leucine	36	45
Proline	88	80

[+] tRNA (a stripped mixture) was irradiated with thiopyronin (10 μg/ml) with a day light lamp
(6.6 × 10^6 erg/mm^2), data taken from [3].
[++] Data taken from Kuwano et al. [33].

codon site. Recent experiments on valine and phenylalanine tRNA molecules have
shown that anticodon loop is not necessary for the recognition of aminoacyl-synthetases [37, 66]. The work of these authors and some preliminary reports indicate
that the bases adjacent to the CCA-end of tRNA are primarily reponsible for the
specific recognition of aminoacyl-synthetases. The base sequences for leucine and
proline tRNA molecules are not yet known. Whether the inactivation observed in
our experiments and those of Kuwano et al. [33] is due to guanine destruction at
this region remains open.

Table 5 shows the photodynamic inactivation of tRNA by thiopyronin. These
studies were carried out in the presence of ribosomes, synthetic polynucleotides
and transferases. The synthesis of the polypeptide in this reaction depends on the
complimentary between synthetic polynucleotides used as messengers and the anticodon sequence of tRNA. As expected one gets the highest inactivation for proline
followed by leucine, lysine and phenylalanine respectively. The inhibition of proline tRNA in tables 4 and 5 are approximately the same.

In the last chapter it was described that bacterial cells irradiated with labelled
psoralen incorporate this compound selectively into nucleic acids. This is due to
the formation of an adduct between psoralen and thymine (DNA), or psoralen and
uracil (RNA). A schematic diagram of the modification of RNA by psoralen and
light is shown in fig. 7.

Table 5

Thiopyronin-sensitized inactivation of tRNA in polypeptide synthesis on ribosomes

Amino acid	mRNA used	% Inhibition
Phenylalanine	Poly U	6
Lysine	Poly A	4
Leucine	Poly UC[3.6 : 1]	22
Proline	Poly C	80

Data taken from Chandra and Wacker [4].

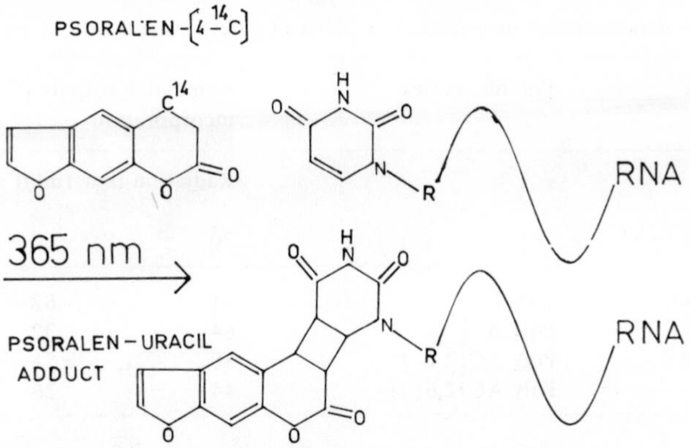

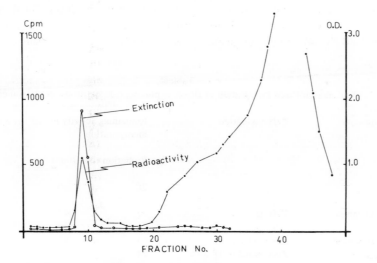

Fig. 7. Schematic representation of uracil modification in RNA by psoralen and light (365 nm).

Dr. Adolf Wacker has already shown the binding capacities of various synthetic polynucleotides, RNA and DNA for psoralen in the presence of light. Fig. 8 shows the binding of labelled psoralen to E. coli tRNA in the presence of light (365 nm). This was studied by irradiating a sample of tRNA with labelled psoralen at 365 nm, and separating the complex on sephadex G_{25} columns. The bound radioactivity is eluted with tRNA.

Fig. 8. Binding of psoralen to tRNA in the presence of light. The complex was separated on sephadex G-25 columns on a fraction collector; 64-drops fractions were collected and the radio-activity was measured in Packard liquid scintillation spectrometer (Model, 3375) using dioxane scintillator. Extinctions were measured at 260 nm.

Table 6
Psoralen-sensitized inactivation of tRNA in polypeptide synthesis on ribosomes

Amino acid	Polynucleotide	Remaining activity (%) of the amino acid incorporation	
		Radiation time (min)	
		20	60
Phenylalanine	Poly U	71	62
Lysine	Poly A	64	32
Glutamic acid	Poly AG(2.6:1)	88	51
Arginine	Poly AG(2.6:1)	44	26

Transfer RNA treated in the same manner but with cold psoralen was used for its activity determination. This was done by treating tRNA with 10 μg/ml psoralen and irradiating at various doses. These samples were used in the Nirenberg system [42] containing mRNA and ribosomes. Incorporation of phenylalanine, lysine, glutamic acid and arginine was measured in the presence of poly U, poly A, poly AG(2.5:1) and poly AG(2.5:1) respectively.

The percent remaining activity of amino acid incorporation is shown in table 6. At a lower dose the glumatic acid is most resistant, arginine most sensitive and no significant difference exists between phenylalanine and lysine. However, at a higher dose the lysine incorporation is only half as much as that of phenylalanine. Surprisingly, arginine incorporation is most sensitive at both the doses. The differential inhibition of phenylalanine and lysine may be due to destruction of anticodon in lysine, but the inhibitions for glumatic acid and arginine can not be only due to anti-

Table 7
Acetone-sensitized inactivation of tRNA in polypeptide synthesis on ribosomes

Amino acid	Polynucleotide	Remaining activity (%) of the amino acid incorporation	
		Radiation time (min)	
		20	60
Phenylalanine	Poly U	78	47
Lysine	Poly A	71	39
Glutamic acid	Poly AG(2.6:1)	61	43
Arginine	Poly AG(2.6:1)	58	42

tRNA (a stripped mixture) was irradiated with 20% acetone at 315 nm. Control samples were treated with acetone but without illumination. After irradiation acetone was removed under vacuum.

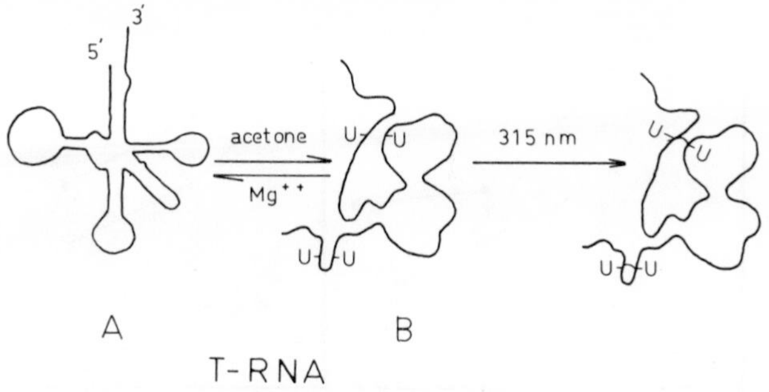

Fig. 9. Schematic representation of aceton-sensitized inactivation of tRNA.

code. Since the polynucleotides used in these systems are random copolymers, i.e., the occurrence of A and G does not follow any specific sequence, one would expect that not one but several tRNA molecules are acting in such systems. It may be interesting to examine whether different tRNA molecules for the same amino acid have different sensitivities towards the photodynamic action. This could be of a great importance in understanding evolution.

It has been previously described that the uracil moieties in RNA can be modified by the action of acetone and irradiation at 315 nm. Thus the inactivation of tRNA by psoralen and acetone is similar in that they both modify uracil. The former reaction forms an adduct with uracil and the latter catalyzes the formation of cyclobutane dimers of uracil. It was, therefore, interesting to study the inactivation of tRNA by acetone in polypeptide synthesis.

The percent remaining activity of amino acid incorporation is shown in table 7. Experimental conditions were exactly the same as in table 6. These results do not show any significant difference between the inhibitions observed for various amino acids. Thus, besides the modification of the anticode site there must be a damage occurring in all the tRNA types. A plausible explanation of these results can be given as follows: tRNA in 20% acetone loses its helical structure to form an open structure as shown in fig. 9. If acetone is evaporated and tRNA taken in Mg^{++}-containing solutions the helical structure is regenerated (controls); irradiation of the acetone solutions of tRNA could form inter-loop dimers or transdimers so that the regeneration of the helical structure is no more possible.

5.2. Ribosomes

Ribosomes are nucleoprotein particles composed of proteins and ribosomal RNA (see fig. 4). They play a vital part in the process of protein synthesis during which they become attached to a strand of messenger RNA to form complexes known as polysomes. However, the manner in which the protein part and the ribosomal RNA influence the functional activity of ribosomes is not yet known. Photosensitized reactions can be employed to study such problems.

Studies reported in the last chapter indicate that the protein part as well as nu-

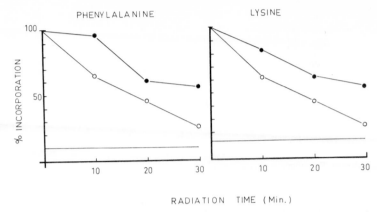

RADIATION   TIME (Min.)

Fig. 10. Acetone-sensitized inactivation of ribosomes in polypeptide synthesis. ●——● Acetone
without light. ○——○ Acetone with light.

cleic acids (rRNA) are essential for the functional activity of ribosomes. This was
shown by the photodynamic inactivation of whole cells by dyes and psoralen. The
advantage of using these two reactions is that one acts specifically on RNA bases
(psoralen), whereas the other acts on proteins as well as nucleic acids. This in vivo
behavior of photosensitizers is demonstratable on isolated ribosomes [3] . However,
acetone-sensitized effect on isolated ribosomes has not been studied as yet.

Fig. 10 shows the photodynamic action of acetone on ribosomes. Ribosomes
isolated from *E. coli* B were irradiated at various doses in the presence of 10% acetone. They were lyophilized and resuspended in Mg^{++}-containing buffer. These ribosomes were used to polymerize phenylalanine and lysine in the presence of poly U
and poly A respectively. Their inactivation is not specific to a particular amino acid,
i.e., the rate of inactivation of ribosomes for phenylalanine and lysine is about the
same. Surprisingly, acetone alone damages the functional activity of ribosomes.
This indicates that acetone at this concentration disturbes the helicity of rRNA required for the integrity of the internal structure of ribosomes.

6. Photodynamic action on macromolecules

Soon after the discovery of photodynamic action attention was focussed on proteins as the possible site of the action. Jodlbauer and Von Tappeiner [30] reported
the eosin-sensitized inactivation of several enzymes. Since 1960 this field has become very important, particularly due to our knowledge of the active centers in
enzymes. Photodynamic studies on enzymes are an excellent tool for locating the
active centers where histidine and aromatic amino acids play a vital role. These
amino acids are fortunately very sensitive to dye-catalyzed photodynamic reactions.
The work on photodynamic action on proteins and polypeptides has recently been
reviewed by Spikes and Straight [54] , Spikes and Livingston [52] and Spikes [50] .

The studies described in the preceeding pages concern mainly DNA and the synthetic polynucleotides.

6.1. *DNA*

Besides a selective destruction of guanine many physico-chemical changes take place on treatment of DNA with sensitizers in the presence of light. Depolymerization of DNA on extensive photodynamic treatment with acridine orange has been reported by Freifelder and Mahler [15]. These authors also observed a decrease in the melting temperature of salmon sperm DNA on photodynamic treatment with acridine orange. Interestingly, the decrease in melting temperature of DNA from several sources under photosensitized conditions was found to be proportional to the guanine content of DNA. Simon and Van Vunakis [48] have observed a decrease in the heat stability of T4-DNA; heat denatured DNA was more easily photo-oxidized than native DNA.

As a carrier of genetic information the DNA molecule serves two major functions: (a) to make exact copies of itself in the process of duplication or replication, and (b) to pass on the information coded in it to mRNA in the process of transcription so that the mRNA in its turn may translate the information in nucleic acid letters into proteins. The formation of mRNA on a DNA template is catalyzed by an enzyme transcriptase or DNA-dependent RNA polymerase (E.C. 2.7.7.6). This has now been highly purified from several bacterial species. The effectiveness of DNA as a template in these reactions varies with the structural and physical changes in DNA. Therefore, this reactions is excellent for detecting structural and physcial changes in DNA caused by the photodynamic action.

The photodynamic action of pyronin, methylene blue and thiopyronin on the template efficiency of DNA in RNA-polymerase reaction is shown in table 8. The experiments were performed in such a way that for each dye a control was made where the treated samples were kept in dark. The inhibition obtained by these

Table 8

Photodynamic action of pyronin, methylene blue and thiopyronin on the template activity of DNA in RNA-polymerase reaction

DNA treated with	Visible light 6.6 $\times 10^6$ erg/mm^2	^{3}H-AMP incorporation (cpm)	Inhibition (%)
Control	−	1788	0
without DNA	−	249	86
Pyronin	−	1683	6
	+	1411	21
Methylene	−	1696	5
blue	+	1261	29
Thiopyronin	−	1552	13
	+	762	57

dyes in the presence of light was considerably higher than the controls where no
light was given. Although the concentration of all these dyes used was the same,
the range of inhibition differs remarkably from one dye to the other. Thus, for
example, pyronin at a concentration of 10 μg/ml shows an inhibition of 21% but
thiopyronin at the same concentration is three times more powerful at inhibiting
the incorporation of labelled AMP. These results are in good agreement with the
photodynamic destruction of guanine and the photodynamic inactivation of bacteria by these compounds [59]. The inhibition of DNA-template activity by pyronin contradicts earlier results [58] which showed that no guanine is destroyed
by the photodynamic action of pyronin. However, one may consider that the
amount of guanine destroyed was so low that it could not be chemically detected.
It is known that the priming ability of DNA in RNA-polymerase reactions provides
a system which is extremely sensitive to very small changes in DNA.

Table 9 shows the photodynamic inactivation of the template activity of DNA
by xanthotoxol, xanthotoxin and psoralen. These furocoumarins are structurally
very similar and differ only at 8-C substitution. If the molecular structure of psoralen is altered, for example, by substitution of the hydrogen at 8-C by a methoxy
group one gets xanthotoxin; a hydroxy group at the same carbon atom gives xanthotoxol. Biological experiments on skin [39, 40] and bacteria [13, 55, 59] have
shown that these substitutions alter the effectiveness of these sensitizers in photodynamic reactions. A similar alteration in their photodynamic effectivity is observed
on the template activity of DNA. The maximum inhibition is shown by psoralen
followed by xanthotoxin and xanthotoxol respectively. These differences are explained by their affinity to bind to DNA [32]. Experiments in this laboratory have
shown that the furocoumarin-mediated lethality of bacteria is higher at lower temperatures. Bacterial studies with psoralen, xanthotoxin and bergapten showed a
highest inactivation at $-30°$C. Studies between $+20°$C and $-30°$C exhibit almost a

Table 9

Photodynamic action of furocoumarins on the template activity of DNA in RNA-polymerase
reaction

DNA treated with	Visible light 6.6 $\times 10^6$ erg/mm^2	^{3}H-AMP incorporation (cpm)	Inhibition (%)
Control	−	1788	0
Control	+	1711	4
without DNA	−	249	86
Xanthotoxol	−	1335	25
	+	1230	31
Xanthotoxin	−	1332	25
	+	860	48
Psoralen	−	1265	19
	+	605	66

Table 10
Photodynamic activity of thiopyronin and psoralen at 2°C

DNA treated with	Visible light 6.6 × 10⁶ erg/mm²	³H-AMP incorporation (cpm)	Inhibition (%)
Control	−	1647	0
	+	1566	6
Thiopyronin	−	1411	14
	+	1177	29
Psoralen	−	1374	16
	+	746	55

linear increase in the lethal action of furocoumarins as the temperatures are lowered. Oginsky et al. [43] made similar observations using xanthotoxin between 0°C–45°C, i.e. +45°C and lowering till 0°C. On the other hand, the photodynamic inactivation of bacteria with dyes is a temperature-dependent reaction and at low temperatures no inactivation is achieved [58]. We have tested this temperature effect on the photodynamic inactivation of DNA by thiopyronin and psoralen. The sample were irradiated in an ice bath (2°C).

Results reported in tables 8 and 9 show that at room temperature or at 15°C (water bath) thiopyronin and psoralen are photodynamically almost equally effective. However, at low temperatures (table 10) psoralen is approximately twice as effective as thiopyronin. These results explain the temperature effect on the photodynamic sensitivity of bacteria towards dyes and furocoumarins. Rodighiero (personal communication) has recently studied the furocoumarin binding to DNA at various temperatures, and had similar results. However, in his experiments the state of DNA, i.e., native or denatured plays a vital role.

Table 11
Acetone-sensitized and UV inactivation of DNA in RNA-polymerase reaction

Source of inactivation	³H-AMP incorporation (mμ moles/10 min)	Incorporation (%)
None	4.40	100
UV-light		
2.4 × 10⁵ erg/mm²	2.25	53
4.8 × 10⁵ erg/mm²	1.22	29
9.6 × 10⁵ erg/mm²	0.72	17
Acetone + light [315 mμ]		
4 × 10⁵ erg/mm²	2.30	55
8 × 10⁵ erg/mm²	1.53	36.5
16 × 10⁵ erg/mm²	0.97	23

The preferential dimerization of adjacent thymine or uracil molecules in nucleic acids on irradiation with UV light is now well known. These reactions have been employed for studying various biological processes at molecular level [4, 20, 57]. A similar kind of dimerization occurs under photosensitized conditions (see IV). We have, therefore, compared these two reactions on the template activity of DNA in RNA-polymerase reaction.

The results in table 11 show a gradual inhibition of ^{3}H-AMP incorporation with increasing energy doses. However, the inactivation by UV light is higher than that by acetone and light (315 nm). This indicates that the UV inactivation of DNA is not entirely due to dimer formation but also other photochemical reactions taking place simultaneously. Such reactions may involve the formation of hetero-dimers, deamination of cytosine and so on. To test whether the inhibition of AMP incorporation after DNA irradiation with acetone is due to specific destruction of thymine in DNA, we studied the incorporation of all the labelled triphosphates. These experiments were repeated using each time a different labelled nucleotide and the three other triphosphates. We have found that the incorporation of all triphosphates is impared to the same extent, i.e., AMP incorporation is not preferentially inhibited. The process of transcription is very complicated, and is not yet well understood. The specific site of unwinding of DNA and the mechanism by which polymerase reads the sequence during transcription have still to be worked out.

6.2. Polynucleotides

The first clear indication of the mechanism by which RNA might be synthesized in vitro was obtained in 1955 by Ochoa and his associates [21], who isolated from the microorganism *Azotobacter vinelandii*, an enzyme which catalyses the formation of high molecular weight polyribonucleotides from nucleoside 5'-diphosphates with the release of inorganic phosphate. If the substrate employed is adenosine diphosphate, the polymer formed is a polyribonucleotide containing adenine as the only base. It is usually referred to as poly A. If the substrate is an equimolar mixture of ADP and UDP, the product is random copolymer of A and U, called poly AU. Synthesis of homo- and hetero-polymers with this reaction have helped us a great deal in the discovery of genetic code, and advanced our knowledge of the mechanism of information transfer in biological systems, for example, by hybridisation studies. The use of polyribonucleotides as messenger RNA in the cell-free protein synthesis is now well established. The homopolymers of adenylic acid, cytidylic acid and uridylic acid are known to serve as messengers for the polymerization of lysine, proline and phenylalanine respectively. Similarly, copolymers of A and G, and U and C stimulate the incorporation of lysine and leucine respectively. We, therefore, studied the photodynamic effect of thiopyronin on the messenger activity of these polymers (table 12).

As follows from these results the template activities of poly U, poly A and poly UC remain unchanged, whereas the template activity of poly AG decreases to 56%. That this inactivation is entirely, or specifically due to guanine destruction is exhib-

Table 12
Photodynamic effect of thiopyronin on polynucleotides in cell-free portein synthesis

System	Amino acid incorporation (%)				
	Phenylalanine Poly U	Lysine		Leucine Poly UC[+]	Proline Poly C
		Poly A	Poly AG[+]	Poly UC[+]	
Non-treated					
Complete	100	100	100	100	100
-mRNA	4	15	15	3	3
Treated fractions					
mRNA	110	98	56	96	89
tRNA	94	96	96	78	20
Ribosomes	34	65			

Conditions: fractions were treated with 10 μg/ml thiopyronin and irradiated with visible light
(6.6 $\times$ 10^6 erg/mm^2).
[+] Poly AG (5.5:1) and poly UC (4:1).

ited by tRNA studies. Since the messengers used for leucine and proline are poly
UC and poly C, one would expect that the complimentary codes in tRNA must contain guanine. This is the reason that treated tRNA shows a maximum inhibition
for proline followed by leucine.

In the above experiments the biological inactivation of polynucleotides was due
to guanine destruction by thiopyronin. Similar studies were made using acetone as

Table 13
Acetone-sensitized inactivation of polynucleotides in protein synthesis

Polynucleotide	Irradiated at 315 mμ (3 $\times$ 10^5 erg/mm^2) % extinction		Amino acid incorporation cpm/mg rib. protein		
	−acetone	+acetone	Lysine	Proline	Phe. Ala
Poly A	100		1248		
		98	1368		
Poly C	100			389	
		97		401	
Poly U	100				5535
		43			794

Polynucleotides were irradiated at a concentration of 1 mg/ml containing 23% acetone.
The reaction mixture contained 80 μg/ml poly U, or 120 μg/ml poly A or poly C.

Table 14
Acetone-sensitized inactivation of polyribonucleotide-primed RNA synthesis by RNA-polymerase

	and light	mµ moles/20 min		
		³H-AMP	³H-GMP	³H-UMP
Poly A	−	0.058	0.16	4.40
	+			4.25
Poly C	−	0.148	4.75	0.07
	+		4.55	
Poly U	−	2.28	0.14	0.16
	+	1.31		

the sensitizer. The solutions of polymers of adenylic acid, uridylic acid and cytidylic acid were irradiated at 315 nm in the presence of 23% acetone. Irradiation
of poly U at a dose 3×10^5 erg/mm^2 showed about 55% decrease in its extinction.
The polymers of adenylic acid and cytidylic acid under these conditions of irradiation remained nearly unaffected. The experimental blank which contained no acetone but irradiated at the same dose showed no decrease in extinction (table 13).
The polymerization of phenylalanine directed by photodynamically treated poly U
is reduced to 12%. The messenger activities of poly A and poly C on irradiation
with acetone are not affected in these experiments.

The use of polyribonucleotides as templates for the polymerase-dependent synthesis of RNA has been reported by Fox et al. [14]. They found that the RNA-
polymerase preparations from the microorganism *Micrococcus lysodeikticus* catalyze the ribonucleotide polymerization utilizing synthetic polymer of complimentary base as primer. We, therefore, studied the effect of acetone and light on the
priming ability of poly A, for UTP, poly C for GTP, and poly U for ATP (table 14).
These reactions catalyzed by homopolymers were studied in the presence of only
a complimentary ribonucleotide.

All homopolymers were irradiated at 315 nm (3×10^5 erg/mm^2) in the presence
of 23% acetone. Under these conditions of treatment poly U loses more than 50%
of its activity for the polymerization of adenylic acid. The priming activities of
poly A for UTP and poly C for GTP are not significantly affected.

6.3. Oligonucleotides

Heteropolymers of various bases synthesized by polynucleotide phosphorylase
have a random distribution of bases. As a result their use in genetic studies is limited.
The sequence assignments of code-triplets is based on chemically synthesized triplets, or triplets obtained by cleaving such polymers selectively, using specific enzymes. These triplets do not catalyze the synthetis of polypeptides; however, the
codon—anticodon specificity remains unaltered, and can be assayed by a ribosomal-
binding technique. This technique was developed by Nirenberg and Leder [41].

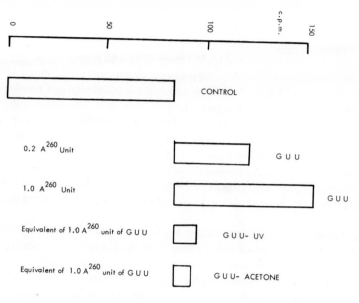

Fig. 11. Acetone-sensitized and UV inactivation of GUU.

Influence of UV light on uracil residues in the polynucleotide (poly U) is now well documented [20, 57]. This effect is due to dimerization of adjacent uracil residues and the hydration of uracil. Acetone-sensitized dimerization of uracil has been reported in the fore-going pages. However, the photodynamic alteration of uracil with acetone does not involve the formation of a water-addition product. It was, therefore, of interest to study the modification of guanylyl-uridylyl-uridine (GUU) under these conditions. The activity of GUU was assayed on the binding of valyl-tRNA to ribosomes.

GUU was irradiated at 257 nm using a low-pressure mercury lamp, or at 315 nm with a grating monochromator, in the presence of acetone (23%). Spectrophotometric measurements showed that UV light at a dose 2.4×10^5 erg/mm^2 causes a loss of 34% of extinction of GUU, whereas acetone and 315 nm-light (4.0×10^5 erg/mm^2) inhibit 25% only. Equivalent amounts of GUU irradiated at 257 nm or at 315 nm in the presence of acetone lose their capacity to stimulate the binding of valyl-tRNA to ribosomes. Surprisingly, GUU activity is more sensitive to the photodynamic action of acetone than to UV light. This may be explained by the fact that acetone-sensitized samples contain more dimer. The greater fall of extinction on UV irradiation is due to formation of dimers and a water-addition product. How far does the water-addition product contribute to such studies needs to be examined.

7. Conclusion

Photodynamic action has been used in several different ways as tool in viral research. This was first introduced by Herzberg [25] who reported that illumination in the presence of methylene blue suppresses vaccinia virus eruptions on the skin of rabbit. Hiatt [26] has studied the photodynamic sensitivity of a number of animal viruses towards toluidine blue and methylene blue. His studies indicate that photodynamic action could be useful for taxonomic purposes, i.e., susceptible viruses (adenoviruses, B-viruses, ECHO-10, rabbies, simian viruses, vaccinia and so on) and resistent viruses (Coxsackie, ECHO-1, poliovirus etc.). Furthermore, Hiatt and Moore [28] suggested a photodynamic tagging technique whereby photosensitive poliovirus produced by growth in the presence of proflavine can be used to distinguish between virus persisting from an inoculum and the resulting pyrogeny virus.

More recently, photodynamic action has been employed to study active centers in enzymes. Enzyme inactivation resulting from the photodynamic destruction of amino acid residues can involve two mechanisms, i.e., the destruction of amino acids located at the active center, or the destruction of amino acids elsewhere which are essential for the optimum confirmation of the enzyme. The use of photodynamic action for the three-dimensional mapping of the protein molecules has been proposed by Ray and Koshland [46]. Galiazzo and Jori [17] have recently proposed the irradiation of proteins containing the sensitizer covalently linked in the known positions of the molecule. These authors have employed this technique for elucidating the three-dimensional structures of ribonuclease A and some sulfhydryl-containing enzymes.

With the increasing availability of specific reactions catalyzed by various sensitizers, the photodynamic action can be used as a selective tool for studying various aspects of molecular biology. Studies described here concern the photodynamic action of dyes, furocoumarins and acetone. These sensitizers are highly selective in their mode of action. Furocoumarins and acetone react specifically with nucleic acid bases; whereas dyes react with amino acids as well. Making use of this selectivity, studies were made on the functional activity of informational macromolecules and subcellular components.

Acknowledgements

The author is grateful to Professor Adolf Wacker for his constant encouragement and the laboratory facilities. A part of the acetone work was done in collaboration with Dr. C.H. Krauch, Ludwigshafen, Germany.

References

[1] R. Beukers, J. IJlstra and W. Berends, Rec. Trav. Chim. 77 (1958) 729.

[2] J.G. Calvert and M.J.L. Rechen, J. Amer. Chem. Soc., *73* (1952) 2101.

[3] P. Chandra and A. Wacker, Z. Naturforschg. *21b* (1966) 663.

[4] P. Chandra and A. Wacker, Biophysik *3* (1966b) 214.

[5] P. Chandra, P. Mildner, H. Feller and A. Wacker, Europ. Photobiol. Symp. Hvar, Yugoslavia (1967) pp. 145.

[6] P. Chandra, A. Wacker, R. Süssmuth and F. Lingens, Z. Naturforschg. *22b* (1967) 512.

[7] R.I. Cotter, P. McPhie and W.B. Gratzer, Nature (Lond.) *216* (1967) 864.

[8] F.H.C. Crick, J. Mol. Biol. *19* (1966) 548.

[9] F.H.C. Crick, Proc. Roy. Soc. (B) *167* (1967) 331.

[10] D. Elad, C. Krüger and G.M.J. Schmidt, Photochem. Photobiol. *6* (1967) 495.

[11] S.S. Epstein, M. Burroughs and M. Small, Cancer Res. *23* (1963) 35.

[12] S.S. Epstein, M. Small, H.L. Falk and N. Mantel, Cancer Res. *24* (1964) 855.

[13] W.L. Fowlks, D.G. Griffith and E.L. Oginsky, Nature (Lond.) *181* (1958) 571.

[14] C.F. Fox, W.S. Robinson, R. Haselkorn and S. Weiss, J. Biol. Chem. *239* (1964) 186.

[15] D. Freifelder and H.R. Mahler, Biochem. Biophys. Acta *53* (1961) 199.

[16] A.V. Furano, D.F. Bradley and L.G. Childers, Biochemistry *5* (1966) 3044.

[17] G. Galiazzo and G. Jori, NATO- Symp., Sassari 1969, Abstracts (1969) pp. 40.

[18] R.T. Garvin, G.R. Julian and S.J. Rogers, Science *164* (1969) 583.

[19] E. Geissler, Naturwissenschaften *46* (1959) 376.

[20] L. Grossman, Proc. Nat. Acad. Sci. *48* (1962) 1609.

[21] M. Grunberg-Manago, P. Ortiz and S. Ochoa, Biochem. Biophys. Acta *20* (1956) 269.

[22] A.P. Harrison, Jr., Ann. Rev. Microbiol. *21* (1967) 143.

[23] C.G. Harchard and C.A. Parker, Proc. Roy. Soc. (A) *235* (1956) 518.

[24] F. Heinemets, R. Vineger and W. Taylor, J. Gen. Physiol. *36* (1952) 207.

[25] K. Herzberg, Klin. Wochschr. *10* (1931) 1626.

[26] C.W. Hiatt, Trans. N.Y. Acad. Sci. *23* (1960) 66.

[27] C.W. Hiatt, Bacteriol. Rev. *28* (1964) 150.

[28] C.W. Hiatt and D.E. Moore, Federation Proc. *21* (1962) 460.

[29] R.B. Hill, Jr., K.G. Bensch and D.W. King, Exp. Cell Res. *21* (1960) 106.

[30] A. Jodlbauer and H. Von Tappeiner, Muench. Med. Wochschr. *26* (1904) 1139.

[31] S.W. Klein and S.H. Goodgol, Science *130* (1959) 629.

[32] C.H. Krauch, D.M. Krämer and A. Wacker, Photochem. Photobiol. *6* (1967) 341.

[33] M. Kuwano, Y. Hayashi, M. Hayashi and K. Miura, J. Mol. Biol. *32* (1968) 659.

[34] A.A. Lamola and T. Yamane, Proc. Nat. Acad. Sci. *58* (1967) 443.

[35] M.R. Lewis, Anat. Rec. *91* (1945) 199.

[36] E.R. Lochmann, W. Stein and C. Umlauf, Z. Naturforschg. *20b* (1965) 778.

[37] A.D. Mirzabekov, I. Kazarinova, D. Lastity and A.A. Bayev, FEBS-Letters *3* (1969) 268.

[38] L. Masujo and G. Rodighiero, Experientia *18* (1962) 153.

[39] L. Musajo and G. Rodighiero, Progr. Biochem. Pharmacol. Vol. I (Karger, Basel, 1965) p. 366.

[40] L. Musajo, G. Rodighiero and G. Caporale, Bull. Soc. Chim. Biol. *36* (1954) 1213.

[41] M.W. Nirenberg and P. Leder, Science *145* (1964) 1399.

[42] M.W. Nirenberg and J.H. Matthaei, Proc. Nat. Acad. Sci. *47* (1961) 1588.

[43] L. Oginsky, G.S. Green, D.G. Griffith and W. Fowlks, J. Bacteriol. *78* (1959) 821.

[44] C.A. Parker, Proc. Roy. Soc. (A) *220* (1953) 104.

[45] O. Raab, Z. Biol. *39* (1900) 524.

[46] J.W. Ray and D.E. Koshland, J. Biol. Chem. *237* (1961) 2493.

[47] L. Santamaria, Recent Contributions to Cancer Res. in Italy (1960) pp. 167–287.

[48] M.I. Simon and H. Van Vunakis, J. Mol. Biol. *4* (1962) 488.

[49] R.L. Sinsheimer and R. Hastings, Science *110* (1949) 525.

[50] J.D. Spikes, NATO-Symp., Sassari 1969 (1969) Abstract pp. 91.

[51] J.D. Spikes and J.Y. Haga, NATO-Symp., Sassari 1969 (1969) Abstracts pp. 95.

[52] J.D. Spikes and R. Livingston, Advan. Radiat. Biol. *3* (1969) 29.

[53] J.D. Spikes and D.C. Hall, Nat. Res. Council, Misc. Publ. No. *1145* (1963) 733.

[54] J.D. Spikes and R. Straight, Ann. Rev. Phys. Chem. *18* (1967) 409.

[55] A. Wacker and P. Chandra, Stud. Biophys. *3* (1967) 239.

[56] A. Wacker, H.W. Dellweg and D. Weinblum, Naturwissenschaften *47* (1960) 477.

[57] A. Wacker, D. Jacherts and B. Jacherts, Angew. Chem. *74* (1962) 653.

[58] A. Wacker, G. Türck and A. Gerstenberger, Naturwissenschaften *50* (1963) 377.

[59] A. Wacker et al., Photochem. Photobiol. *3* (1964) 369.

[60] S.Y. Wang, M. Apicella and B.R. Stone, J. Amer. Chem. Soc. *78* (1956) 4180.

[61] J.D. Watson and F.H.C. Crick, Nature (Lond.) *171* (1953) 737.

[62] J.D. Watson and F.H.C. Crick, Nature (Lond.) *171* (1953) 964.

[63] J.D. Watson and F.H.C. Crick, Cold Spring Harb. Symp. Quant. Biol. *18* (1954) 123.

[64] L. Weil and J. Maher, Arch. Biochim. Biophys. *29* (1950) 241.

[65] I. Wilucki, H. Matthaus and C.H. Krauch, Photochem. Photobiol. *6* (1967) 497.

[66] M. Yoshida, Y. Kaziro and T. Ukita, Biochem. Biophys. Acta *166* (1968) 646.

CHAPTER 14

RHEOLOGICAL STUDY OF PHOTO-OXIDIZED CONTRACTILE PROTEINS

Roberto SANTAMARIA, Giuseppe GRANATO CORIGLIANO
and Rita SANTAMARIA

Institute of Human Physiology, University of Naples,
Naples, Italy

1. Introduction

Recently during comparative rheological analyses in studies on interactions of contractile protein materials with drugs we observed definite differences between actin and myosin systems obtained from skeletal and cardiac muscles [3]. These results do not support the idea that the physiological differences in muscles are only related to the differences in structures such as sarcoplasmic reticulum and T-system [11]. Actually our rheological analyses stress the importance of the molecular parameter in the physiological response, as might be deducted from the finding that the cysteine content in skeletal and cardiac meromyosins is not identical [1]. In an effort to verify the importance of the molecular parameter in the physiological re-sponse of the above actin and myosin systems we carried out a study on their sensi-tivity to photo-oxidation using haematoporphyrin as photosensitizer and rheological methods as tool of investigation. The results, reported in this paper, demonstrate a different sensitivity of the cardiac protein system. Additional support is given to the hypothesis that the physiology of skeletal and cardiac muscles is not only dependent on differences in organization and distribution of the sarcoplasmic reticulum and its relationship to the T-system within the muscle fibers but also on structures of the contractile protein materials.

2. Materials and methods

2.1. *Preparation of the proteins*
Myosin was extracted from bovine cardiac muscle according to Katz et al. [6]. Actin was extracted from the same source by the butanol-acetone method after T'sao and Bailey [12].

2.2. *Photo-oxidation and capillary viscometry*

Haematoporphyrin dihydrocloride (Hp) water solution at pH 7.0, was added to the protein solutions at a final concentration of 1×10^{-5} M. The samples were kept in test tubes immersed in a transparent water bath at about 4°C, and were illuminated with a 1000 W tungsten bulb (Osram). The light energy measured at the level of the samples was about 4.25×10^5 erg/cm^2/sec. The constant availability of oxygen in the solutions was ensured by gentle stirring. The exposure time ranged from 0 to 300 min. Soon after illumination the samples were mixed with equal volumes of the native partner protein solutions to give an actomyosin complex.

Viscosity measurements were carried out by Cannon-Fenske capillary viscometers n. 75 at 20°C in order to compute the ATP-sensitivity. The dissociation of actomyosin complex was tested by the addition of ATP at a final concentration 8×10^{-4} M.

The ATP-sensitivity was obtained according to the expression [13]

$$ATP_{sens} = \frac{\log \eta_{r(complex)} - \log \eta_{r(complex + ATP)}}{\log \eta_{r(complex + ATP)}} \times 100$$

2.3. *Photo-oxidation and extensive rheological analysis*

Newtonian, non-Newtonian, and other related rheological characterizations were tested using a Ferranti-Shirley cone-plate apparatus. This ensures variation of velocity gradient, automatic recording of shear stress, and constant temperature during the analysis.

The myosin concentration was 12 mg/ml in 0.55 M KCl buffered at pH 7.0 by 0.025 M phosphate. The actin concentration was 4 mg/ml in 0.1 M KCl + 0.001 M MgCl$_2$ at pH 7.0. The haematoporphyrin concentration in the protein solutions was 1×10^{-5} M. The illumination was carried out as above. The light exposure was 300 min to study a marked photo-oxidized product, since the Ferranti-Shirley apparatus is sensitive to rather viscous systems. Using this apparatus, the plots of velocity gradient versus shear stress permitted Newtonian or non-Newtonian characters to be ascertained. The non-Newtonian character was expressed by a non-linear trend. When the system was non-Newtonian, it was possible to check for plasticity, pseudo-plasticity, tixotropy, and rheopexy. Quantitative evaluations of tixotropy were made possible by the calculation of the area included between the *up curve* and the *down curve* on the plots. Quantitative evaluations of plastic viscosity (*U*) were obtained by extrapolation of the '*f*' value from the initial part of the *up curve* (see McKennel [7]).

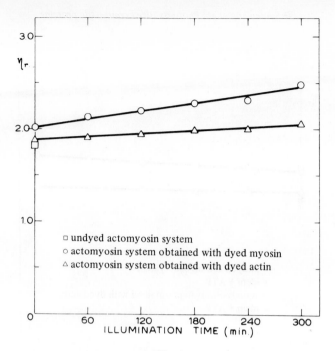

Fig. 1.

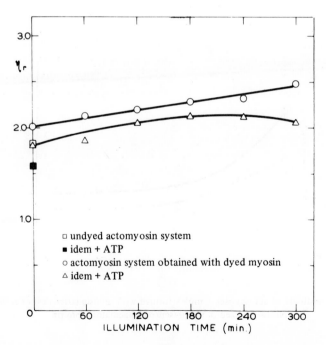

Fig. 2.

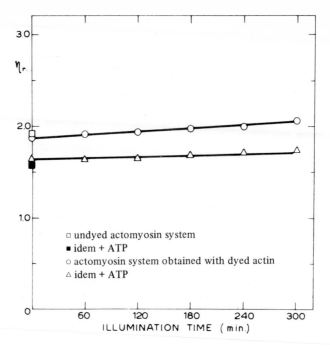

Fig. 3.

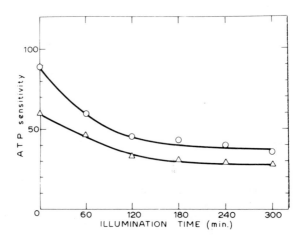

Fig. 4. ATP-sensitivity of actomyosin complex formed with myosin (triangles) or actin (circles) photo-oxidized in the presence of haematoporphyrin.

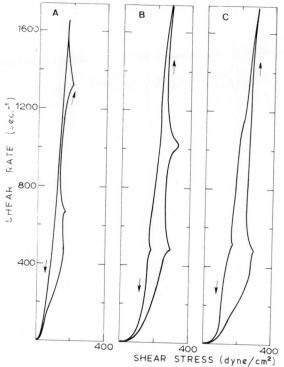

Fig. 5. Rheological behaviour of actomysin system obtained by mixing: (a) native myosin and actin; (b) pretreated myosin in the dark with native actin; (c) pretreated actin in the dark with native myosin.

3. Results and discussion

Fig. 1 shows the trends of relative viscosity of actomyosin systems formed alternatively with dyed myosin or dyed actin. It appears that the variation in relative viscosity of the actomyosin system is higher with photo-oxidized myosin than with photo-oxidized actin. Therefore, myosin is more sensitive than actin to the photodynamic effect as far as the protein complex formation is concerned. In this connection it should be pointed out that in a previous investigation the relative viscosity of skeletal protein systems was found to be lowered under identical photodynamic treatment, whereas in the present study the relative viscosity of cardiac protein systems is increased [2].

Figs. 2 and 3 report the decrease in relative viscosity produced by the addition of ATP to actomyosin formed with photo-oxidized myosin plus actin or myosin plus photo-oxidized actin. Apparently, the ATP addition is less effective on the protein complex formed with photo-oxidized myosin.

The ATP-sensitivity, evaluated according to the formula above reported (see

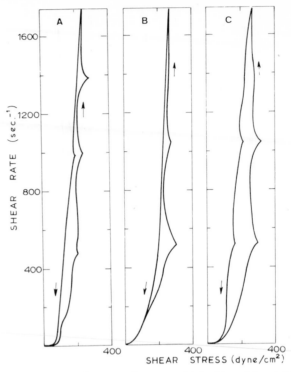

Fig. 6. Rheological behaviour of actomyosin system obtained by mixing: (a) native actin with myosin treated with Hp in the dark, for 30 min; (b) native actin with myosin treated with Hp in the dark, for 300 min; (c) photo-oxidized actin for 300 min, with native myosin.

Weber and Portzehl), gives the plot of fig. 4. Here, is also evident the minor effectiveness of ATP on the protein complex formed with photo-oxidized myosin.

Figs. 5 to 7 report the recordings obtained by the Ferranti-Shirley apparatus relative to the actomyosin systems with different 'histories'. The shear rate is plotted against the shear stress to ascertain the various rheological features. In all plots the trends appear to be non linear even when the protein complex is formed with both native partners (fig. 5a). Moreover, in all plots the *up curve* and the *down curve* are not superimposable, and *transients* at critical values of velocity gradient are present. Consequently, all the protein systems exhibit non-Newtonian characters and tixotropy to different extents. Plasticity is present in all cases, except for the samples of the native material and those of actomyosin obtained by mixing native actin with myosin + Hp kept in the dark for 300 min. These systems turned out to be pseudoplastic both in the *up* and *down* curves. The rupture of an internal structure of the protein liquid is revealed by the appearance of the transients at critical values of shear gradient. The area included between the two curves (hysteresis area) gives the thixotropy due to the mechanical stress applied during the time of the

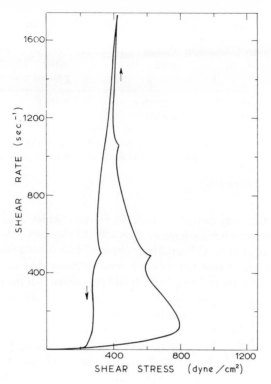

Fig. 7. Rheological behaviour of actomyosin system obtained by mixing photo-oxidized myosin for 300 min with native actin.

measurement. The actomyosin formed with dyed myosin kept in the dark (fig. 5b) shows trends for the *up curve* and *down curve* which are not appreciably dissimilar from those pertinent to the native system. Fig. 5c concerns dyed actin kept in the dark and shows qualitative differences, although the hysteresis area is similar to that of the control. The plots of fig. 6a and fig. 6b have been recorded for systems formed with dyed myosin kept in the dark for 30 and 300 min at 4°C (the exposure time for photo-oxidation was 300 min). The results show slight qualitative modifications and decrease of the hysteresis area.

Fig. 6c reveals the behaviour of the actomyosin formed with photo-oxidized actin. Here, some qualitative modification in the trends is evident; in addition, the presence of *creep* on the *up curve* for high value of gradient velocity is noticeable after the appearance of the two transients.

Fig. 7 concerns the actomyosin system formed with photo-oxidized myosin, and shows a marked modification in shape. Here the plastic viscosity, obtained by extrapolation from the *up curve*, is high, and the hysteresis area is very large.

The value of plastic viscosity of the different systems are reported in table 1.

Table 1

Plastic viscosity (U)* of actomyosin complex obtained with native and dyed proteins

Native complex	Complex formed with dyed myosin	Complex formed with dyed actin	Complex formed with native proteins + haematoporphyrin
0.1505	0.4345** 7.0002***	0.3907** 0.4404***	0.3011

*Defined by: $U = (\tau - f)/(dv/dx)$, where τ is the stress, f the so-called yield value (extrapolated from the linear part of trend on the τ axes), and dv/dx is the velocity gradient.
**In the dark.
***After 300 min of illumination.

On the whole, the study carried out with the Ferranti-Shirley apparatus clearly shows that myosin is much more sensitive than actin to photo-oxidation, and it confirms the data relative to ATP-sensitivity obtained with diluted protein solutions.

Apart from the help of photo-oxidation in demonstrating the remarkable differences in behaviour of actin and myosin, it should be recalled that the Ferranti-Shirley rheological method can reveal definite differences in the native materials. In our previous study we observed that for skeletal actomyosin systems, at a protein partner concentration corresponding to 6 mg/ml for myosin and 2 mg/ml for actin, the hysteresis area of skeletal actomyosin is larger than that of cardiac system and three transients are apparent at 383.02, 748.63, and 1549 sec^{-1}. On the contrary the cardiac system, for same protein concentrations, does not exhibit a real transient at 383.02 sec^{-1}, while some degree of strain is revealed in the trend in connection with an extensive interaction among the particles. Two other transients are present, but they correspond to a lower shear gradient (651.58 and 1349.27 sec^{-1}) in comparison to those of the skeletal system [4].

4. Theoretical considerations

The physico-chemical measurements carried out by Mueller et al. [8] with sedimentation methods on the size and shape of cardiac and skeletal myosin demonstrated that no significant differences were detectable within the range of experimental error. On the other hand, the chemical measurements carried out by Barany et al. [1] on the cystein content of cardiac and skeletal meromyosins showed that these proteins are not identical.

Our physico-chemical measurements, especially the determinations of the rheological characteristics with the Ferranti-Shirley apparatus, demonstrated that a physico-chemical study of the two contractile protein systems was profitable when it originates from proper theoretical considerations. In the living state these proteins constitute dense phases; therefore, one should try to produce these physical states experimentally and to apply suitable methods of investigation. To fulfill these requirements we used rather concentrated protein solutions and the Ferranti-Shirley apparatus which permits rheological characteristics to be determined in rather dense materials. The use of our concentrated protein solutions satisfied the experimental requirement relative to the living state according to the following considerations.

The acqueous protein systems obtained under our experimental conditions are such that the mixture of solutions of actin and myosin, prepared in molecular form, leads to a complexation with remarkable increase in viscosity. This is connected to (a) interactions between actin and myosin, (b) interactions with the medium constituents, and (c) other important interactions occurring among the formed particles, as expected by the force of London—Van der Waals fluctuation of charge, as formulated by Jehle [5]. The interaction with the constituents of the dispersion medium are of the kind of cooperative processes.

At this point, it is worthwhile paying attention to more general theoretical considerations, stressing the fact that in our experimentation the physico-chemical differences between the two contractile protein systems were properly revealed by the use of mechanical perturbations. The latter were such that even weak interactions were detected to both qualitative and quantitative extents. As previously reported [10], we think that the occurrence of all interactions described above in the complexation phenomenon fits in with a physical concept of the living state of matter, namely a particular form of aggregation whose definition, besides chemical information of the 'isolated macromolecules', includes, necessarily, the characteristics of the 'medium' where the macromolecules are embedded. Therefore we have to deal with the physical concept of 'field' since the medium is affected by the action of the macromolecules and at the same time the former has own effects on the macromolecules.

Accordingly, it does not make sense any more to refer to the *solute* and to the *solvent* as in the conventional terms of the traditional physical chemistry, which employs methods of the diluted solutions to extrapolate at infinite dilution the molecular parameters of the solute. Here it appears convenient to deal with *concentrated systems*. Several methods have been proposed for studying these systems but from time to time particular difficulties and limitations arise. However, the results so far obtainable by means of these methods are leading to a new physical chemistry, that of the so-called 'concentrated systems'. Indeed, the data relative to this kind of experimentation have shown the importance of the *medium* in determining a particular conformation of the *solute* and the influence that this may exert on the intrinsic association capacity of the *solvent*; as a consequence equilibria are originated, responsible for a peculiar aggregation of the matter that characterizes the *native state*. We may then understand why the physical picture of the denaturation problem has not been adequately approached by means of an elegant and precise technique such as X-ray diffraction which, on the other hand, is being applied with increasing success for describing the fine structure of the macromolecules. At first glance the meaning of such a situation appears to be a paradox; actually this is merely connected to the fact that X-ray diffraction analysis is really powerfull on molecular grounds, and, in this connection, it represents the best technique for the spatial resolution of each atom within the whole molecular assembly. On the other hand, when we deal with living matter as such, we need at the same time to account for the macromolecules and the medium where they are embedded, so that X-ray diffraction becomes unapplicable owing to the complexity of the involved parameters. In fact the macromolecules are not any more isolated entities arranged in the lattice of a pure crystal; here the physical requirements impose the need to consider a further structure where all the fundamental constituents are interplaying: salts, water, and macromolecules.

This structure is statistically formed and could be investigated by methods capable of perturbing the equilibria responsible for the overall aggregational system; a very elegant treatment of the structure of the condensed systems was given by Eyring and coworkers in their theory of relaxation of transport phenomena [9]. As examples of perturbations we may recall the frequencies of absorption and emission spectra, the T-jump, the P-jump, the application of stress for getting viscous flow, etc. In our study we selected a rheological method that allows the recording of the applied stress over velocity gradient range, since condensed phases, such those under our experimental conditions, consist of patches of molecules vibrating about a minimum of free energy well. The applied stress causes such patches of molecules to jump over the barrier energy of their equilibrium positions. Some of the applied stress is stored as potential energy and it is released when the stress is released so that it is not delivered as thermal energy; this occurs when elasticity is associated to viscous flow.

Actually the strain present in some portions of the trends recorded by the Ferranti-Shirley apparatus reveals qualitatively a degree of viscoelasticity for our systems, although in this connection a quantitative account could be properly obtained by means of special rheometers as the Weissemberg rheogoniometer.

In conclusion, experimental approaches dealing with protein systems reproducing the dense phases of the living state and adequate methods for their study do appeal to physiologists in the effort to explain functional processes at a molecular level. At the same time, the obtained results often provoke new horizons in theoretical developments which lead to a better understanding of the physical state of the living matter which arises by complex interactions among all the components of the basic system. Nevertheless, it should be still acknowledged that the conventional use of aqueous solutions of highly purified proteins, even with characteristics far from those pertinent to ideal solutions, does permit extensive physico-chemical characterizations at a molecular level, provided definite values of pH and ionic strength are used. This was recently proved by Wu [14] who has been able to prove differences between skeletal chicken myosin obtained from breast and leg muscles.

5. Summary

Definite differences in sensitivity to photo-oxidation between actin and myosin extracted from cardiac muscle are demonstrated using capillary viscometry. Extensive rheological analyses of the actomyosin complex formed with actin and myosin alternatively photo-oxidized in the presence of haematoporphyrin were carried out. The results, made possible by the use of rather concentrated solutions, are discussed in terms of theoretical concepts of the physical state of living matter. The importance of using protein systems reproducing dense phases pertinent to the living state is discussed. Adequate methods are needed for the study of functional processes at a molecular level.

References

[1] M. Barany, E. Gaetjens, K. Barany and E. Karp, Comparative studies on rabbit cardiac and skeletal myosins. Arch. Biochem. Biophys. *106* (1964) 280–293.

[2] G. Granato Corigliano, R. Santamaria and F. Comite, Photodynamic process of skeletal muscle proteins. Enzymologia *35* (1968) 303–315.

[3] G. Granato Corigliano, G. Pagnini, R. Santamaria, F. Di Carlo and E. Genazzani, Influenza della k-strofantina sul comportamento reologico di miosina ed actina di muscolo scheletrico e cardiaco. Boll. Soc. It. Biol. Sper. *44* (1968) 1965–1968.

[4] G. Granato Corigliano, R, Santamaria, G. Pagnini and E. Genazzani, Differenza nel comportamento biofisico dei sistemi di actomiosina di muscolo scheletrico e cardiaco. Boll. Soc. It. Biol. Sper. *44* (1968) 1962–1965.

[5] H. Jehle, Amino acid sequence in protein synthesis. Proc. Nat. Acad. Sci. U.S. *45* (1959) 1360–1384.

[6] A.M. Katz, I. Doris, C.T. Repke, B. Bonnie and B.A. Rubin, Adenosinetriphosphatase activity of cardiac myosin, Circ. Res. *19* (1966) 611–621.

[7] R. McKennel, Instr. Manual (Ferranti, Monston, Manchester, 1960).

[8] M. Mueller, G. Franzen, R.V. Rice and R.E. Olson, Characterization of cardiac myosin from the dog. J. Biol. Chem. *239* (1964) 1447–1456.

[9] T. Ree and M. Eyring, The relaxation theory of transport phenomena. Rheology Theory and Application, ed. F.R. Eirich, Vol. 2 (Academic Press, New York, 1958) pp. 83–144.

[10] R. Santamaria, Fondamenti per una teoria fisica dello stato vivente della materia. Rend. Atti Acc. Sci. Med. Chir. Napoli *119* (1965) 283–294.

[11] D.S. Smith, The organization and function of the sarcoplasmic reticulum and T-system of muscle cell. Progr. Biophys. Mol. Biol., eds. J.A.V. Butler and H.E. Huxley, Vol. 16 (Pergamon Press, London, 1966) pp. 110–142.

[12] T.C. T'sao and K. Bailey, The extraction, purification, and some chemical properties of actin. Biochim. Biophys. Acta *11* (1953) 102–113.

[13] H.M. Weber and H. Portzehl, Muscle contraction and fibrous muscle proteins. Adv. Prot. Chem., eds. M.L. Anson, K. Bailey and J.T. Edsall, Vol. 7 (Academic Press, New York, London, 1952) pp. 161–252.

[14] C.S. Wu Chung, Comparative studies on myosin from breast and leg muscles of chicken. Biochemistry *8* (1969) 39–48.

SECTION III

GENETIC DAMAGE BY PHOTODYNAMIC ACTION

CHAPTER 15

INTRODUCTORY REMARKS

M.R. ZELLE
Colorado State University

Progress in radiation biology during the past 25 years has been remarkable. These rapid advances stem not only from greatly increased interest because of the development of nuclear energy but also from the even more spectacular progress in the biological sciences. In turn, development of the biological sciences has been based largely upon the revolutionary advances in genetics, especially molecular genetics, which have had far reaching implications for many other areas of biology such as biochemistry, cell biology, radiation biology, photobiology, and medicine and in addition, have provided new approaches and new insights to the study of evolution [9].

Although a number of investigators were working in the field of microbial genetics during the 1930's, the publication of Luria and Delbruck's work on mutation to bacteriophage resistance in 1943 was a mile-stone marking the beginning of modern bacterial and bacteriophage genetics which has had tremendous impact on the subsequent development of genetics [20]. The pneumococcal transformation principle was identified as deoxyribonucleic acid (DNA) at about the same time by Avery and his associates [2] and, along with the demonstration by Hershey and Chase [12] that it was primarily the phage DNA that entered the host cell, not the protein constituents, caused the realization that DNA, not protein, was the carrier of genetic information. With the establishment of the double helical structure of DNA by Watson and Crick in 1953 [33], the foundations of the revolution in genetics were completed. Since that time, the genetic code has been elucidated (see Yčas [37] for review) and great progress in the knowledge of mutagenesis, and in the genetic control of protein structure and specificity, for example, has followed.

In radiation biology, D.E. Lea's book, the first real attempt at a synthesis in the field, was published in 1946 [18]. Indirect effects of X rays were already known at that time, in large part from Lea's own work with viruses. But renewed interest in the oxygen effect [30], studies of chemical protection and electron spin resonance studies of free radical production by X rays, along with significant advances in radiation chemistry including discovery of the hydrated electron (see Schwarz [26] for review) and the development of new sources including accelerators making possible more critical studies of linear energy transfer have combined to add greatly to our understanding of the physical and chemical events initiating radiation dam-

273

age to living systems. The nature of the lesions responsible for lethal damage in cells from ionizing radiations is beginning to emerge. Quantitative studies of the production by X rays of single and double strand breakage in DNA and of their repair indicate that double strand breakage may be one of the most important lethal lesions in bacteriophage and bacterial cells [19]. In eukaryotic cells, the production of chromosome aberrations has long been known to be a major mechanism leading to cell death and significant advances in the understanding of the production and repair of chromosome breaks have been made (see Wolff [36] for review). Studies of breakage of DNA molecules are now being extended to eukaryotic cells and there is evidence that the basic lesion leading to chromosome breakage and aberration production by ionizing radiations may be breakage of DNA strands [21].

Although X rays and other ionizing radiations have long been known to produce point mutations, much remains to be known of the basic kinds of damage and the mechanisms of point mutation production by ionizing radiation.

Even greater progress has been made in the understanding of the basic lesions and mechanisms underlying the effects of ultraviolet (UV) radiation. Interest in the nucleic acids dates from 1928 [10] when Gates published the first nucleic acid type action spectrum for the UV inactivation of bacterial cells though I have long been intrigued by Henri's work published in 1914 [11] which had no real impact upon the field but which was a deliberate and successful attempt to induce heritable changes in *Bacillus anthracis* cells surviving UV radiation. He was led to make this attempt by his conclusion from crude action spectrum studies utilizing filters that UV lethal effects resulted from absorption of UV in the nucleus and hence, surviving cells might display permanent genetic damage. During the 1930's and 1940's a number of action spectra implicating nucleic acids in both the lethal and mutagenic effects of UV in a variety of organisms was demonstrated (see Zelle and Hollaender [38] for review). Progress has been especially rapid since about 1949 when Roberts and Aldous [23] published their observations on the effect of post-irradiation treatment in modification of the lethal effects of UV and Kelner [16] discovered photoreactivation. This marked the beginning of the study of the repair of UV damage and the isolation and utilization of a number of mutant strains of especially *Escherichia coli* and coliphages has led to the discovery and at least partial understanding of the mechanisms involved in several different kinds of repair of UV damage to nucleic acids. (see Howard-Flanders [13], Rupert and Harm [24], and Adler [1] for reviews.) These advances were concomitant with and contingent upon the demonstration that UV produces thymine dimers in DNA both in vitro [3] and in bacterial cells [32]. With the possible exception of strand breakage by ^{32}P decay in singled stranded DNA of phage ΦX174 [29], pyrimidine dimer formation between adjacent pyrimidines in the DNA chain, their monomerization by enzymatic photoreactivation, their photodestruction by shorter wavelengths, and their repair by the excision, repair-synthesis process is the clearest and best understood example of radiation produced damage of biological significance in both cell survival and mutagenesis (see Setlow [27] for review). Although it is clear that pyrimidine dimers are of major importance in UV cell killing and that most UV induced mutations result from dimer formation [34] it is also evident that pyrimidine dimer formation is not responsible for all UV lethal and mutagenic effects [27, 34].

Study of the repair of UV damage has been extended to mammalian cells by a number of investigators. Cleaver [7, 8] has shown that whereas normal human fibroblasts can remove pyrimidine dimers from DNA by the excision, repair-synthesis mechanism, fibroblasts from patients suffering from xeroderma pigmentosum, an autosomal recessive genetic disease in which sunlight induces skin cancer, are unable to repair the UV damage. These studies are being extended to other skin disorders.

With the elucidation of the genetic code which has been comprehensively reviewed by Yčas [37], and by the use of a variety of chemical mutagens of known mode of action on DNA, it is now possible in certain biological systems, most notably the coat protein of tobacco mosaic virus and the tryptophane synthetase of *E. coli* to relate specific changes in the amino acid sequences of proteins to corresponding changes in the base composition of the DNA. Such knowledge of the molecular basis of mutation coupled with the repair of UV damage has made possible significant advances in the understanding of UV mutagenesis. It is impossible to review all of this important work here but certain general conclusions can be mentioned. The photoreversibility of many mutations shows clearly that pyrimidine dimers are involved in the induction of most mutations but there is also evidence that certain other mutations, for example the mutation to the inability to ferment lactose in *E. coli* is not photoreversible and hence derives from some other photoproduct than pyrimidine dimers. In UV resistant strains, premutational damage is repaired in the dark by the excision, repair-synthesis process and in UV sensitive mutants lacking the ability for excision-repair, the yield of mutations is much greater. With certain mutations, the fraction of premutational lesions not normally repaired includes pyrimidine dimers but for other mutations, suppressors leading to mutation from auxotrophy to prototrophy, the normally non-repaired fraction does not seem to include dimers. This unique fraction of premutational lesions is subject to 'mutation frequency decline' when protein synthesis is inhibited after irradiation and, although photoreversible, the photoreversibility is of the indirect type and does not involve the enzymatic monomerization of pyrimidine dimers [34]. Recently, Witkin [35] has postulated that most UV-induced mutations result from errors in the recombinational repair of daughter strand gaps which occur opposite unexcised dimers [25] in the first postirradiation DNA replication. This is supported by markedly lower frequencies of UV induced mutations in recombination deficient mutants. Though much remains to be known concerning UV mutagenesis, the use of genetic mutants differing in UV sensitivity due to the lack of the normal repair capabilities has made possible significant progress.

The investigation of the biological effects of photodynamic action with which this meeting is concerned began in 1900 when Raab observed that irradiation with visible light greatly accelerated the killing of paramecia by acridine dyes and that the effective wavelengths where those absorbed by the dye. The involvement of oxygen in the effect was suggested by the work of Ledoux-Lebard in 1902 and proven in 1905 by Jodlbauer and von Tappeiner in studies of the erythrosin-sensitized inactivation of invertin and the killing of bacteria. In 1904, Straub made comparative studies of the eosin-sensitized oxidation of iodide ions with dye-sensitized effects in biological systems and concluded that both were dye-sensitized photooxidations involving molecular oxygen. It has been shown since that significant biological damage follows exposure to visible light of many types of biological systems or components of biological systems in the presence of oxygen and a wide range of photosensitizing substances including dyes, plant pigments, and carcinogenic hydrocarbons for example. Blum [4] has reviewed the early development of the field including discussion of a number of diseases of animals and man caused by light. Later developments have been considered in a number of quite recent reviews (see Spikes and Livingston [28]).

The lethal effects of photodynamic action have been studied extensively in a variety of species. The kinetics resemble those of UV and X rays. The genetic effects have also been investigated (see Zelle this volume for review) and not only point mutations but also chromosome aberrations are produced by photodynamic action. There are many similarities but also differences between photodynamic action and UV and X ray effects.

In general, photodynamic damage is not photoreversible which indicates pyrimidine dimers are not formed. Recently however, Lamola [17] has shown that pyrimidine dimers are formed efficiently in DNA by acetophenone sensitization and that the yield of thymine dimers is at least 30 times that of any other pyrimidine dimer or other photoproduct.

Böhme [5] isolated two mutants of *Proteus mirabilis* which display increased sensitivity to ethylmethanesulfonate (EMS), a monofunctional alkylating agent, UV, and photoinactivation following sensitization with thiopyronin. One of these mutants is of the recombinationless type. Böhme's results parallel the results obtained with *E. coli* mutants lacking normal repair mechanisms for UV damage and indicate that certain cells, at least, may normally possess mechanisms which can function in repair of certain kinds of photodynamic (and EMS) damage. Uretz [31] reported no difference in sensitivity to acridine sensitized photoinactivation of the extremely UV-sensitive *E. coli* B_{s-1} strain which lacks both the excision repair and recombination repair mechanisms as compared to *E. coli* B and B/R. Ito and Hieda [15] report that diploid yeast display no liquid holding recovery for either inactivation or forward mutation at the adenine locus following photosensitization with acridine orange and exposure to light. Their strain displays such recovery following UV irradiation, X rays, and nitrogen mustard and their observations confirm earlier results of Patrick et al. [22] for photosensitized inactivation of yeasts. In a recent study utilizing near UV irradiation following sensitization with 8-methoxypsoralen Igali et al. [14] studied the induction of true reverse mutations and suppressor mutations of three auxotrophic mutations in a series of mutant *E. coli* strains differing in the ability for enzymatic photoreactivation, excision-repair, and recombination repair. 8-Methoxypsoralen is believed to form a 1 : 1 adduct with thymine and other pyrimidines with the formation of a cyclobutane ring. Both premutational and potentially lethal damage were largely excisable and were not photoreactivable. Recombinationless mutants (both rec- and exr-) were more sensitive to the lethal effect of 8-methoxypsoralen plus near UV just as for far UV radiation but no induced mutations could be detected. The mechanism of mutagenesis thus appears similar to that of UV radiation [35].

The production of free radicals in DNA by photosensitized oxidation has been studied by Calberg-Bacq et al. [6] (see also Delmelle, this volume) who related the efficiency of radical production as measured by electron spin resonance (ESR) to the biological effect observed with a series of sensitizing dyes. Inactivation and induction of rapid lysis mutants in phage T4B where the biological indicators. In general, the inactivating efficacy of 5 acridine dyes paralleled the efficiency of energy transfer to the phage DNA as measured by ESR. Methylene blue, the strongest inactivating sensitizer, behaves differently. All of the dyes were about equally effective in mutation induction.

In summary, it is evident that there are many similarities and differences between photodynamic action and UV or ionizing radiation. The similarities include among others, the kind of effects produced, the kinetics of their production, the production of free radicals, the involvement of DNA damage, and some similarity in responses of mutant strains of bacteria and other organisms differing in their ability to repair UV induced damage in DNA. The differences include certain differences in the ability of various organisms to repair photosensitized damage to DNA which undoubtedly reflects differences in the kinds of damage produced by different photosensitizers. In view of the rapid progress in radiation biology and UV photobiology of the past 25 years, with identification of some of the important lethal and mutagenic lesions in DNA, and the increasing understanding of the various repair mechanisms and of the mutagenic process itself, the time seems ripe for an increased effort in research in photodynamic action. Comparative studies of the ability of known repair mechanisms to repair photosensitized damage to DNA and a search for mutants with greater sensitivity to photooxidized inactivation and hence perhaps lacking normal repair mechanisms seem promising. With increasing knowledge of the specific nature of the binding of different sensitizers to DNA and of the chemical nature of the photodynamic lesions produced which in general are of a more specific nature than the DNA lesions resulting from UV or ionizing radiation, future research gives promise not only of a fuller understanding of photodynamic effects and the mechanisms whereby they are produced but also of contributing to a more complete understanding of radiation effects in general and of the consequences, both lethal and mutational, to living cells of damage to DNA.

References

[1] H.I. Adler, The genetic control of radiation sensitivity in microorganisms. In: Advances in Radiation Biology, eds. L.G. Augenstein, R. Mason and M.R. Zelle (Academic Press, New York and London, 1966) Vol. 2, pp. 167–191.

[2] O.T. Avery, C.M. MacLoed and M. McCarty, Studies on the chemical nature of the substance inducing transformation of pneumococcal types. Induction of transformation by a desoxyribonucleic acid fraction isolated from pneumococcus type III. J. Exp..Med. *79* (1944) 137–158.

[3] R. Beukers and W. Berends, Isolation and identification of the irradiation product of thymine. Biochem. Biophys. Acta *41* (1960) 550–551.

[4] H.F. Blum, Photodynamic Action and Diseases Caused by Light (Reinhold, New York, 1941).

[5] H. Böhme, Absence of repair of photodynamically induced damage in two mutants of *Proteus mirabilis* with increased sensitivity to monofunctional alkylating agents. Mutation Res. *6* (1968) 166–168.

[6] C.M. Calberg-Bacq, M. Delmelle and J. Duchesne, Inactivation and mutagenesis due to the photodynamic action of acridines and related dyes on extracellular bacteriophage T4B. Mutation Res. *6* (1968) 15–24.

[7] J.E. Cleaver, Defective repair replication of DNA in xeroderma pigmentosum. Nature *218* (1968) 652–656.

[8] J.E. Cleaver, DNA damage and repair in light sensitive human skin disease. J. Invest. Dermatol. *54* (1970) 181–195.

[9] W.M. Fitch and E. Margoliash, Construction of phylogenetic trees. Science *155* (1967) 279–284.

[10] F.L. Gates, On nuclear derivatives and the lethal action of ultraviolet light. Science *68* (1928) 479–480.

[11] V. Henri, Etude de l'action metabiotique des rayons ultraviolettes. Productions de formes de mutation de la bacteridie charbonneuse. Compt. Rend. *158* (1914) 1032–1035.

[12] A.D. Hershey and M. Chase, Independent functions of viral protein and nucleic acid in growth of bacteriophage. J. Gen. Physiol. *36* (1952) 39–56.

[13] P. Howard-Flanders, DNA repair. Ann. Rev. Biochem. *37* (1968) 175–200.

[14] S. Igali, B.A. Bridges, M.J. Ashwood-Smith and B.R. Scott, Mutagenesis in *Escherichia coli*. IV. Photosensitization to near ultraviolet light by 8-methoxypsoralen. Mutation Res. *9* (1970) 21–30.

[15] T. Ito and K. Hieda, Absence of apparent dark recovery from the photodynamically-induced mutational damages in yeast. Mutation Res. *5* (1968) 184–186.

[16] A. Kelner, Effect of visible light on the recovery of *Streptomyces griseus* conidia from ultraviolet injury. Proc. Nat. Acad. Sci. U.S. *35* (1949) 73–79.

[17] A.A. Lamola, Specific formation of thymine dimers in DNA. Photochem. Photobiol. *9* (1969) 291–294.

[18] D.E. Lea, Actions of radiations on living cells (Cambridge University Press, London, 1946).

[19] J.T. Lett, I. Caldwell and J.G. Little, Repair of X ray damage to the DNA in *Micrococcus radiodurans*: The effect of 5-bromodeoxyuridine. J. Mol. Biol. *48* (1970) 1746–1759.

[20] S.E. Luria and M. Delbruck, Mutations of bacteria from virus sensitivity to virus resistance. Genetics *28* (1943) 491–511.

[21] G.J. Neary, Chromosome aberrations, cell killing and the molecular basis of relative biological effectiveness of ionizing radiations. In: Radiation Research, ed. G. Silini (North-Holland, Amsterdam, 1967) pp. 445–454.

[22] M.H. Patrick, R.H. Haynes and R.B. Uretez, Dark recovery phenomena in yeast. I. Comparative effects with various inactivating agents. Radiation Res. *21* (1964) 144–163.

[23] R.B. Roberts and E. Aldous, Recovery from ultraviolet irradiation in *Escherichia coli*. J. Bacteriol. *57* (1949) 363–375.

[24] C.S. Rupert and W. Harm, Reactivation after photobiological damage. In: Advances in Radiation Biology, eds. L.G. Augenstein, R. Mason and M.R. Zelle (Academic Press, New York and London, 1966) Vol. 2, pp. 1–81.

[25] W.C. Rupp and P. Howard-Flanders, Discontinuities in the DNA synthesized in an excision-defective strain of *Escherichia coli* following ultraviolet irradiation. J. Mol. Biol. *31* (1968) 291–304.

[26] H.A. Schwarz, Recent research on the radiation chemistry of aqueous solutions. In: Advances in Radiation Biology, eds. L.G. Augenstein, R. Mason and H. Quastler (Academic Press, New York and London, 1944) Vol. 1, pp. 1–32.

[27] J.K. Setlow, The molecular basis of biological effects of ultraviolet radiation and photoreactivation. In: Current Topics in Radiation Research, ed. M. Ebert and A. Howard (North-Holland, Amsterdam, 1966) Vol. II. pp. 195–248.

[28] J.D. Spikes and R. Livingston, The molecular biology of photodynamic action: sensitized photooxidations in biological systems. In: Advances in Radiation Biology, eds. L.G. Augenstein, R, Mason and M.R. Zelle (Academic Press, New York and London, 1969) Vol. 3, pp. 30–121.

[29] I. Tessman, Some unusual properties of the nucleic acid in bacteriophages S13 and ΦX174. Virology *7* (1959) 263–275.

[30] J.M. Thoday and J. Read, Effect of oxygen on the frequency of chromosome aberrations produced by X rays. Nature *160* (1947) 608–610.

[31] R.B. Uretz, Sensitivity to acridine sensitized photoinactivation in *Escherichia coli* B, B/r, and B_{s-1}. Radiation Res. *22* (1964) 245.

[32] A. Wacker, H. Dellweg and P. Jacherts, Thymin-dimerisierung und Überlebensrate bei Bakterien. J. Mol. Biol. *4* (1962) 410–412.

[33] J.D. Watson and F.H.C. Crick, Molecular structure of nucleic acid. A structure for deoxyribose nucleic acid. Nature *171* (1953) 373–738.

[34] E.M. Witkin, Radiation-induced mutations and their repair. Science *152* (1966) 1345–1353.

[35] E.M. Witkin, The role of DNA repair and recombination in mutagenesis. Proceedings of the XII Int. Congress of Genetics. Tokyo, Vol. III (1968) pp. 225–245.

[36] S. Wolff, Mechanisms of dose rate effects: insights obtained from intensity and fractionation studies on chromosome aberration induction. Jap. J. Genet. *40* (1965) Suppl. pp. 38–48.

[37] M. Yčas, The Biological Code (North-Holland, Amsterdam-London, 1969).

[38] M.R. Zelle and A. Hollaender, Effects of radiation on bacteria. In: Radiation Biology, ed. A. Hollaender (McGraw-Hill, New York, 1955) Vol. II. pp. 365–430.

CHAPTER 16

GENETIC EFFECTS OF PHOTODYNAMIC ACTION

M.R. ZELLE
Colorado Stata University

Although study of the biological effects of photodynamic action dates from 1900 when Raab observed that irradiation with visible light greatly accelerated the the killing of paramecia by acridine dyes, studies of the genetic effects are more recent with most of the available data having been published in the last decade. Genetic effects including both point mutations and chromosome aberrations have been reported in a number of species ranging from viruses and bacteriophages to higher plants and drosophila. It is the purpose of this discussion to review briefly the data concerned with the genetic effects of photodynamic treatment and to relate it to radiation genetic effects and mutagenesis in general.

1. Early studies of mutation induction

Doring [15] was the first to observe induced mutations following photodynamic treatment. Morphological mutations in *Neurospora crassa* were studied following sensitization with a 10^{-4} concentration of eosin. All presumed mutations were tested in genetic crosses and only those showing normal segregation were counted as mutations. Five mutations were found in 4297 control isolations yielding a spontaneous mutation rate of 0.116 ± 0.052. Among eosin-treated conidia not exposed to light, 2 mutations in 2124 isolations were observed yielding an eosin-dark control rate of 0.094 ± 0.020 which is not different from the spontaneous rate. By contrast, 12 mutations were found in 2084 eosin-sensitized conidia which were subsequently exposed to visible light. The corresponding mutation rate, 0.577 ± 0.116 is significantly higher than the eosin-dark controls or the spontaneous rates. Thus Doring's careful experiments involving statistical analysis, appropriate controls, and genetic verification of the mutations clearly demonstrated the induction of mutations in Neurospora following photodynamic treatment with eosin.

The first evidence of the photodynamic induction of mutations in bacteria was provided by Kaplan [26, 27] who showed that dwarf colony mutations in *Bacterium prodigiosum* could be produced in high proportions, up to 30% of the surviving clones, by exposure to visible light following sensitization with 1/40,000 aqueous erythrosin. Although the survival curve was exponential, the dose-mutation curve for the dwarf mutations was of a two or three hit character. The genetic nature of the dwarf mutation is not fully known and Kaplan [31] discussed the finding that many of the presumed dwarf mutations were only temporary modifications and not heritable, perhaps being due to a delay in cell fission. Hence, only an unknown fraction of the observed dwarf colonies can be considered as mutants and there is, therefore, some doubt that the photodynamic treatment did, in fact, increase the mutation rate. In later papers, however, Kaplan [28] showed a significant increase in mutation rates to T7 bacteriophage resistance in *E. coli* B and to histidine independence [29] following erythrosin sensitization and exposure to visible light. An increase in absolute number of histidine-independent cells was observed in the photodynamically treated cultures. Considered together, the studies on T7-phage resistance and back mutation to histidine independence do provide convincing evidence that the photodynamic treatment did increase the mutation rate significantly.

Kaplan [30] extended his studies of photodynamic induced mutations following sensitization with erythrosin to colony-morphology mutations in *Penicillium notatum*. High yields of morphological mutants were observed following both UV and photodynamic treatment of conidia. Similar kinetics were observed for both UV and photodynamic treatment. Mutation induction increased linearly with exposure whereas inactivation as measured by loss of colony forming ability followed a 2 or 3 hit survival curve.

Complex inactivation and mutation induction kinetics were observed by Kaplan [31] in extensive studies of S-mutations in *Serratia marcescens*. S-mutations occur in high frequency following irradiation and are evidenced by white (occasionally pink) sectors in the normally red pigmented colonies. Comparative studies were made utilizing X rays, 254 nm (nanometers) ultraviolet with and without photoreactivating light exposure, near ultraviolet (310—400 nm) and photodynamic treatment with erythrosin as sensitizer. Following photodynamic treatment, exponential survival curves were obtained but S mutations increased approximately linearly to a maximum and then decreased. A significant increase in S mutations was observed following exposure to near ultraviolet light. Various models to account for the complex inactivation and mutation-dose curves are discussed.

Convincing evidence of mutagenic activity of photodynamic treatment was obtained by Bohme and Wacker [7] who made comparative studies of thiopyronine and methylene blue in order to see if the greater efficacy of thiopyronine as photosensitizer in oxidative destruction of guanine residues in DNA was paralleled in inactivation and mutation induction. Two strains of *Proteus mirabilis* were used. One was streptomycin dependent and was used to test for mutation to streptomycin independence by plating on unsupplemented nutrient agar plates. The second strain required methionine and phenylalanine and was used in tests of mutation to phenylalanine independence by plating on minimal agar supplemented with 20 μg/ml methionine. With both strains, exposure to the photosensitizing dye or visible light alone produced no inactivation or mutations. Photodynamic treatment resulted in rapid killing of both strains with sigmoidal survival curves and with both strains, thiopyronine

was a more effective photosensitizer. Mutations to streptomycin independence
were observed, the mutation-dose curves reaching a plateau of about 5×10^2 mutants per 10^8 survivors for both dyes, there being no significant difference between
them. For mutations to phenylalanine independence, however, essentially linear
mutation-dose curves were obtained with thiopyronine being a more effective photosensitizer thus paralleling the results for inactivation and guanine destruction.

2. Genetic effect of photosensitization with acridines

Some of the acridine dyes have been of especial interest in studies of photodynamic mutagenesis in microorganisms, viruses and bacteriophages because of
their high affinity for DNA [48] and their property of intercalation between successive bases in the DNA helix [40]. Other effects of acridines contributing to this
interest include a high mutagenecity in the T-even bacteriophages [14, 47], the
interference with photoreversal of UV mutagenic damage in auxotrophic *E. coli*
strains [58], the decrease in excision repair of UV damage in *E. coli* [21, 22], the
protection against UV inactivation [2], and the antimutagenic effect during vege-
tative growth of *E. coli* [56] and yeast [41].

2.1. *Micro-organisms*

Webb and Kutitschek [56] were the first to demonstrate the induction of mutations following photosensitization with acridine orange. They studied the induction of mutations to resistance to bacteriophage T5 in *E. coli* B/r/l, try⁻ in chemostat cultures limited with glucose in which the cells were grown in a minimal medium supplemented with tryptophane. Acridine orange was added in concentrations
ranging from 1 to 5×10^{-6} M. At a concentration of 10^{-6} M acridine orange, a mutation rate of about 1×10^{-8} mutants per day per bacterium per foot candle was
found. At the dye concentration and levels of illumination used, there was no detectable killing of the cells. Acridine orange plus light thus is a very efficient mutagenic treatment in this system. Interestingly, Webb and Kubitschek observed an
antimutagenic effect of acridine orange in the absence of light since the spontaneous mutation rates observed were significantly lower than those observed in the absence of the dye. A similar, approximately 3-fold reduction in mutation rates was
observed when acridine orange was added to chemostat cultures in which caffein
was the mutagen.

The antimutagenic effect of acridines has been confirmed in yeast by Magni et
al. [41] and Puglisi [49] in studies of mutation to canavanine resistance and to
histidine independence in mitotically reproducing, haploid strains of yeast.

In studies utilizing a tryptophane-requiring strain of *E. coli* K-12 (λ⁻);W3623,
Nakai and Saeki [46] showed that oxygen was required for the photodynamic induction of mutations to prototrophy. The mutation rate was an increasing function
of the light exposure and increased with increasing concentration of acridine up to
about 15 μg/ml where saturation levels were reached.

In more extensive and detailed studies of acridine orange photosensitized mutation induction utilizing essentially the same methods as in the earlier Webb and Kubitschek investigations, Kubitschek [36] showed that mutations to T5 bacteriophage resistance were expressed with no segregational division following the mutagenic treatment. Similar results were obtained in later experiments [37] in which *E. coli* B$_{s-1}$ and *E. coli* WP2 hcr⁻ were used. Both of these strains are known to lack the ability for excision repair of UV damage to DNA [23] . Since one segregational cell division would be expected if only one of the DNA strands were mutated and if both strands code for the synthesis of complementary strands during DNA replication, Kubitschek postulates on the basis of his observation of the absence of a segregational division that one strand of the DNA molecule codes for both daughter strands.

Further evidence of the induction of mutations by visible light following sensitization with acridines has been published by Zampieri and Greenberg [62] who studied the induction of lac⁺ revertants in a lac⁻ strain of *E. coli* S which lacks the ability to ferment lactose. Lac⁺ revertants were observed following photosensitization with both acridine orange and proflavine. In addition, these authors observed a significant increase in lac⁺ mutants following treatment of the lac⁻ cells by both acridine orange and proflavine in the dark. These results appear somewhat at variance with the antimutagenic effect of acridine orange reported by Webb and Kubitschek [56] . However, Zampieri and Greenberg used much higher concentrations of dye sufficient to kill two decades of the bacteria in 10 minutes whereas Webb and Kubitschek used non-lethal dye concentrations. Furthermore, different mutation systems were utilized and Witkin [59] has shown that UV-induced premutational damage leading to lac⁻ mutations in lac⁺ *E. coli* strains is not photoreversible which suggests the possibility that somewhat different mechanisms may be involved in the mutation of genes controlling fermentation reactions in *E. coli*.

The induction of forward mutations at the adenine locus of yeast by acridine orange and visible light treatment has been demonstrated by Ito and Hieda [25] who also showed that there was no decrease in the lethal or mutagenic effects (liquid holding recovery) of photodynamic treatment upon delayed plating after holding the treated cells in distilled water for periods up to 25 hr. UV killing and induced mutations, however, do exhibit liquid holding recovery.

Extensive studies of growth inhibition and mutation of *Aspergillus nidulans* have been reported by Ball and Roper [4] . They utilized a number of different acridines and studied their effects both in the dark and after exposure to visible light. Mutants differing in sensitivity to the various dyes were employed and both morphological mutants and reversions to methionine independence (suppressor mutations at a number of loci) in strains carrying the *meth 1* mutation were studied. Their data on growth inhibition and killing will not be reviewed here beyond noting their general conclusion that the effectiveness of any one acridine depends on interaction of genotype and conditions of treatment such as temperature, pH, treatment medium, and light intensity; that mutant alleles which confer growth resistance to acriflavine are selective in their actions towards other acridines, differ in their dominance with different acridines, and differ with regard to the conditions under which they confer resistance to acriflavine. Interestingly, different pairs of acridines used together showed additive effects, potentiation, or sometimes complete annulment by one member of the pair of the inhibition shown by the other. Only two stable morphological variants, both due to gene mutations, were found among some 1300 colonies tested. Significant increases in revertant mutants were observed after treatment with acriflavine in the dark and the frequency was not dependent on the presence of oxygen. The frequency of reversions was somewhat higher following acriflavine plus visible light treatment and oxygen was required for the induction of reversion by photodynamic treatment. The acriflavine concentration used in the photodynamic experiments was only 1 percent of that used in the dark mutation tests. The frequency of reversions either by acriflavine in the dark or with exposure to light was not appreciably affected by the Acr 1 mutation which protects significantly against the growth inhibition and killing effects both in the dark and in the light. Ball and Roper [4] emphasize that in their tests for acriflavine induced mutations in the dark, conidia were treated under conditions such that genetic recombination was excluded.

2.2. *Viruses and bacteriophages*

Ritchie [50] was the first to demonstrate the production of mutations in bacteriophages by photodynamic treatment. He studied the induction of rapid lysis mutants (*r* mutants) in phage T4 following sensitization with proflavine. The spontaneous frequency of *r* plaques in his *r*$^+$T4 phage was 0.01%. Treatment by proflavine or light alone each resulted in a mutant plaque frequency of 0.1%, a tenfold increase over the spontaneous frequency. Photodynamic treatment of proflavine plus light yielded 0.4% of mutant plaques, a twofold increase over the sum of the frequencies of the light and proflavine controls. The increase over the spontaneous frequency from proflavine is not surprising since Brenner et al. [9] have observed *r* mutations following growth of T4 phage in the presence of proflavine. The significant increase in *r* mutants following the light exposure is, however, somewhat unexpected. The light source was a Hanovia Alpine 10 sun lamp which emits about 50% of its energy between 300 nm and 450 nm. The light was filtered through a glass dish containing 5% copper sulphate. Thus, far UV wavelengths were excluded but Ritchie in a later paper [51] points out that long wavelength UV could be responsible for the mutations induced by the light treatment. Ritchie isolated 12 *r*II mutants from the photodynamic survivors, all of which reverted spontaneously, and tested their reversion rates after growth in the presence of 5-bromodeoxyuridine (BD) and proflavine. Five were found to revert after BD treatment and were considered as base-analogue mutants [19], four after proflavine treatment and were considered acridine-type mutations, and three did not revert after either treatment. These three were considered to be base analogue mutants in which the mutant base pair was adenine—thymine. Ritchie postulated that photodynamically

induced mutants were of the base-analogue type and that the four acridine-type
mutants resulted from the proflavine treatment alone.

In an extension of these studies, Ritchie [51] classified a number of proflavine-
induced mutants (P mutants), light-induced mutants (L mutants) and photodynamically-induced mutants (PL mutants) into the 3 phenotypic groups rI, rII, and rIII,
as defined by Benzer [5]. Of the P and L mutants, the great majority, 75% and
90% respectively, were rII mutants. Of the PL mutants, half were rII mutants and
the rest were about equally divided between the rI and rIII classes. No good explanation is known for the disparate distributions of the P and L mutants. More extensive studies were made of the reversion properties of rII mutants following treatment with the base analogue mutagens, 5-bromodeoxyuridine (BD) and 2-amino-
purine (AP), and with proflavine. BP and AP are thought to induce base pair transitions [19] whereas acridine mutagenesis probably results from base pair additions
or deletions [12] which displace reading of the DNA code. Only 2 of 19 P or L mutants could be induced to revert with BD or AP and not with proflavine whereas
the other 17 were induced to revert only after proflavine treatment. It was concluded that the P and L mutations are of the acridine type and the two tested mutants which reverted after BD and AP treatment were of spontaneous origin. Among
15 tested PL mutants, 8 were of the acriflavine type and reverted only after proflavine treatment with the other 7 being of the base-analogue type and reverting
only after BD or AP treatment. Ritchie concludes that photodynamic treatment
produces mutations of the base-analogue type and that the nearly half of the presumed PL mutants of the acriflavine type are in reality either P or L mutants since
the combined mutant frequency following the latter two mutagenic treatments is
about half of that found after photodynamic treatment. The conclusion that photodynamically induced mutations are of the base-analogue type is supported by the
observation of Simon and Van Vanakis [53], among others, that several photosensitizing dyes selectively attack guanine residues in DNA, and by the observation of
Singer and Fraenkel-Conrat [54] that changes in the molar ratio of only guanine
resulted from the photodynamic treatment of tobacco mosaic virus RNA with
either thiopyronine or proflavine.

Similar but somewhat more extensive analyses of photodynamically-induced
rapid lysis mutants in phage T4 have been published by Drake and McGuire [16].
Thiopyronine and psoralen were the photosensitizing dyes and a 15 watt white
fluorescent bulb was the light source. With both sensitizers, inactivation occurred
with single hit kinetics but thiopyronine was about 110 times more effective on a
molar basis. Mutations were induced at rates of 3.3×10^{-4} and 4.6×10^{-4}, per
lethal hit for thiopyronine and psoralen respectively. These rates are similar to that
observed by Ritchie [50] 4.3×10^{-4} with proflavine as the photosensitizer. Although there were some small differences in the frequencies of rI, rII, temperature
sensitive or highly-reverting rII, and rIII mutants isolated following photosensitization with the two photosensitizers, more detailed mapping and reversion analysis
showed no significant differences and the two samples of mutants were not con-

sidered to differ. Ninety-one psoralen-induced and 47 thiopyronine-induced rII mutants were genetically mapped and were found to be essentially randomly distributed throughout the rII locus. By contrast, about half of spontaneously occurring rII
mutants are found to be located in two highly mutable sites [6]. The map distribution along with differences in patterns of induced revertibility show that the mutations were induced by the photodynamic treatment and not merely due to selection
of spontaneous mutations. One hundred five rII mutants were subjected to reversion
analysis by treatment with base analogues (2-aminopurine and 5-bromodeoxyuri-
dine) or proflavine. Fifty-one percent were induced to revert by base analogues and
were considered to have resulted from base pair substitutions. Twenty-two percent
reverted after proflavine mutagenic and were considered as sign or frame shift mutations and 27 percent did not revert after either base analogue or proflavine treatment. The nature of these latter remains unknown but since they all reverted spontaneously, it is unlikely that they contain multiple mutational lesions. Further, the
authors point out that it is unlikely they are frame shift mutations since proflavine-
induced mutations considered to be of this kind invariably are induced to revert
with proflavine and similarly, that it is unlikely that they are base pair transitions
since such mutations induced by base analogues are uniformly induced to revert by
base analogue mutagenic treatment. The authors speculate that these 27 percent of
the mutants may be transversions, base pair substitutions in which the purine—pyrimidine orientation is reversed. The authors further subdivided the 51 percent of
mutants which reverted following base analogue treatment by tests of reversion
with hydroxylamine which is a cytosine-specific mutagen. Nearly half responded
to hydroxylamine and were therefore considered to have guanine-hydroxylmethyl-
cytosine base pairs at the mutated sites.

The results of Drake and McGuire [16] and of Ritchie [50, 51] are in essential
agreement and are the most detailed studies of the molecular nature of photodynamic induced mutations thus far.

Three different plaque type mutations have been induced in the *Serratia marcescens* phage kappa following photodynamic treatment with methylene blue as sensitizer [8]. The mutation dose curves obtained had exponents of 2.3, 1.6, and 3.0
respectively for the clear, narrow, and pale mutant plaque types studied. Cross reactivation experiments showed that much of the lethal effect is due to DNA damage but neither the lethal or mutagenic effects were photoreactivable. The authors
made the interesting observation that about $\frac{1}{3}$ of the lethal lesions could be repaired
by host-cell reactivation which did not influence the rate of induced mutations.
Since methylene blue preferentially destroys guanine residues in DNA [53], this
finding would seem to indicate that damage to DNA other than pyrimidine dimers
may also be repaired by the excision repair mechanism.

A series of five acridines along with three structurally related dyes (pyronin,
acridine red, and methylene blue) have been compared in studies of inactivation
and r mutant induction in phage T4 and of free-radical production in the phage
DNA following prolonged intensive illumination at $-196°C$ [11]. Inactivation in

all cases was exponential but the efficiency of the different photosensitizers varied widely. Among the acridines, acridine orange was most effective followed by proflavin, acriflavin, acridine yellow, and 9-aminoacridine. Electron spin resonance (ESR) studies showed that the absorbed energy is transferred to the phage nucleic acid with efficiencies parallel to the lethal efficiency. Pyronin and acridine red exhibited little biological activity and little ESR signal. Methylene blue was the most effective inactivating dye but its action seems to differ in several aspects from the acridines. The induction of *r* mutants by proflavine sensitization was linearly related to light exposure and all of the acridines and methylene blue displayed about the same mutagenic efficiency as proflavine except 9-aminoacridine which was significantly less effective.

Studies of the genetic effects of photodynamic action have been extended to RNA viruses. Thus, Gendon [20] found that whereas treatment of the intact polio virus particle with proflavine and visible light produced no effect, similar treatment of the isolated RNA caused a loss of infectivity and the production of mutations. The mutational changes, which affected several traits including pathogenicity and the 'AG' marker which determines the intratypic antigenic specificity of the virus, were stable after passage in tissue culture.

Singer and Fraenkel-Conrat [54] studied the effects of photosensitization of infectious tobacco mosaic virus RNA with thiopyronine and proflavine. Both dyes caused loss of infective activity at least initially by first order kinetics, but thiopyronine was much more effective as a photosensitizer since the same rate of inactivation obtained by a molar ratio of thiopyronine to nucleotide of 0.001 required a molar ratio of proflavine of about 1. Studies of the molar ratios of the bases in treated RNA showed that both dyes catalyzed the destruction only of guanine and that the number of guanine residues destroyed exceeded the number of lethal events by a ratio of about 20 varying from 18 to 28 for different levels of inactivation. Sedimentation analysis indicated that the photosensitized inactivation of the RNA did not lead to degradation of the molecule since samples retaining only 0.5–0.1% of the initial infectivity corresponding to about 6 lethal events contained the same amount of 30S material as the untreated samples. The frequency of mutations causing necrotic lesions on inoculated leaves of *Nicotiana sylvestris* was increased 3- or 4-fold either by photodynamic treatment of the RNA or by growing the virus in the presence of the photosensitizing dye and light. Several amino acid exchanges were found in the coat proteins of a small sample of mutants analyzed but these failed to be explainable in all cases by codon changes involving guanine.

3. Mutagenesis with near ultraviolet and visible light

Incidental to studies of photosensitized mutation induction, several investigators have observed the induction of mutations by near UV and visible light. In studies of photodynamic induced *r* mutations in bacteriophage T4, Ritchie [50] observed approximately an 8-fold increase in mutant plaque frequency when the phage was exposed to his light source without any added photosensitizer. His source was a Hanovia Alpine 10 sun lamp emitting about 50% of its irradiation between 300 nm and 450 nm and it was filtered through a glass dish 1 cm thick apparently [51] containing 5% copper sulphate. This filter would remove the infrared and most of the UV wavelengths but as Ritchie points out some long UV radiation might be present and responsible for the mutations. More critical evidence for the induction of mutations by near UV light (320–400 nm) has been published by Kubitschek [38] who observed significantly increased mutation rates to T5 phage resistance when *E. coli* WP2 HCR⁻ was exposed to near visible light either in continuously growing chemostat culture or in static cultures. Induced mutation rates in chemostat cultures were not growth rate dependent and were 10–20 times the spontaneous rates which were growth rate dependent. Induced mutation rates under anaerobic conditions were also higher than the spontaneous rates but only about 60% as high as in aerobic conditions. This indicates at least two mechanisms are involved, one of which requires oxygen and may therefore be a photodynamic effect involving an endogenous photosensitizer. Kubitschek used two light sources: (i) two 4-watt black light fluorescent bulbs (Westinghouse F4T51BLB) with a largely continuous spectrum in the region or (ii) a single 100-watt 'black light' mercury vapor lamp (Magnaflux CH-4) emitting most of its energy in a sharp line at 366 nm. A Corning No. 5860 filter that transmits less than 1 percent of the incident irradiations at wavelengths below 310 nm was also employed. It appears therefore that almost all of the mutants must have been induced with light near 365 nm.

Kubitschek's findings have been confirmed by Webb and Malina [57] who also observed significant increases over the spontaneous mutation rate to T5 phage resistance in glucose-limited chemostat cultures of *E. coli* B/r/1 try continuously exposed to visible light of wavelength greater than 408 nm. A Corning filter, No. 3389 with a 37% cutoff at 424 nm and only 1 percent transmission at 408 nm was used to eliminate any near UV wavelengths from the visible light source. The mutation rates for both 'black' and visible light were proportional to the intensity and were 18 times the spontaneous rate for the highest visible light intensity. They also confirmed Kubitschek's findings that both an oxygen requiring and an oxygen independent mechanism was involved in 'black' light mutagenesis. However, with visible light under anaerobic conditions, no increase over the spontaneous mutation rate was observed. This strongly suggests a photodynamic mechanism for the visible light mutagenesis but as they point out, the amount of chromophore(s) present under anaerobic conditions could be greatly reduced and thus could account for the results. In later studies, Webb [55] extended the studies of visible light mutagenesis to induction of streptomycin resistant mutations with identical results and showed that the induction of T5 phage resistant mutations by visible light in chemostat cultures is independent of growth rate.

4. Photodynamic genetic effects in higher organisms

Although most of the studies of the genetic effects of photodynamic treatment have utilized microorganisms or bacteriophages, a number of investigators have made such studies in higher organisms. Although the data are not extensive, the available evidence indicates clearly that chromosome aberrations can be induced by visible light following photosensitization with acridine orange or methylene blue and there is some evidence for point mutation induction following 8-methoxy-psoralen and acridine photosensitization.

Altenberg [3] utilizing the Muller 'sifter' technique reported an increase in re-cessive lethal mutations in Drosophila following treatment of the polar cap cells of the developing egg with 8-methoxypsoralen and near ultraviolet light.

Somewhat more extensive experiments on point mutation induction in plants utilizing acridines as photosensitizers have yielded somewhat conflicting data. D'Amato [13] reported a significant increase in chlorophyll mutations in progenies of barley plants grown from seeds treated with acriflavine, acridine orange, and 9-aminoacridine. Ehrenberg [17] and his co-workers were not able to confirm D'Amato's observations, however. The mutagenic effect of acridine orange treat-ment of tomato seeds has been studied by Buiatti and Ragazzini [10]. Four treat-ments were utilized. Seeds treated with 1 g/l or 20 g/l acridine orange were divided into two parts, one of which was subsequently exposed to visible light in the pre-sence of oxygen. All four lots were then planted in the field and M_1 plants grown. The exposure to visible light, therefore, was sunlight during growth for two lots and sunlight plus post-dye treatment with visible light of seeds for the other two. Seeds were collected separately from each of the M_1 plants and 100 control plants and M_2 progenies of an average of about 250 plants grown from each M_1 and con-trol plant. The M_2 progenies were scored for chlorophyll mutations which were found in 8 progenies. The frequencies of progenies with mutations were: 2 of 107 progenies from the 1 g/l treatment; 4 of 88 progenies from the 1 g/l with post-light treatment; 2 of 97 progenies from the 20 g/l treatment. No mutations were found among 100 control progenies or in 96 progenies from the 20 g/l with post-light treatment. The total frequency of progenies with mutations from all experimental treatments is 8/388 which, compared to the control frequency of 0/100 yields a value of Chi-square corrected for small expected numbers of 1.013 which is not significant. Thus, although Buiatti and Ragazzini conclude their data demonstrate the photodynamic induction of chlorophyll mutations in the tomato and thus con-firm D'Amato's results, there is some question as to the statistical validity of this conclusion. These authors also observed a very high frequency of M_2 plant prog-enies, 80–98%, which showed rogue-like characteristics. However, M_3 progenies grown from seeds of the M_2 rogue-like plants were all normal in growth habit and it was concluded that the rogue-like phenotype induced by the acridine-orange treatment was not due to genetic change but rather was a kind of Dauermodification.

The evidence for the induction of chromosome aberrations in higher plants is by contrast, completely convincing. This was first shown by Kihlman [32—35] in studies of chromosome aberrations utilizing conventional cytological methods in *Vicis faba* root tips following acridine-orange sensitization. Significantly higher frequencies of both chromatid and chromosome aberrations were observed following photodynamic treatment, the frequencies observed following certain exposures being comparable to those following 100 *r* of X ray treatment and as high as can be conveniently and accurately scored. Appropriate experiments demonstrated that acridine orange alone, visible light alone, and visible light plus acridine orange treatment in the absence of oxygen resulted in no aberration production. The mechanism, therefore, appears to be of a typical photodynamic nature. In the course of these investigations, Kihlman observed that visible light and acridine-orange treatment could also produce chromosome aberrations in the presence of cupferron and nitric oxide in the absence of oxygen. Detailed consideration of these aspects of this work is not germane to this discussion but Kihlman suggests that nitric oxide can replace oxygen in the photodynamic mechanism and that the effect of cupferron is due to its decomposition following the visible light plus acridine-orange treatment since nitric oxide is known to be one of its decomposition products.

Extensive studies of chromosome aberration production by visible light following methylene-blue and acridine-orange photosensitization of *Vicia faba* and barley seeds have been carried out by Kumar and Natarajan [39]. Seeds of barley were soaked in water, decoated and treated with 3.31×10^{-5} M acridine orange and 0.31×10^{-5} M methylene blue, the highest concentrations which were non-toxic. Following dye treatment, the seeds were exposed to 1000 foot candles of visible light in oxygen saturated water. *Vicia faba* seeds were treated by quite similar methods. Conventional cytological preparations were made of root tips following germination. Appropriate experiments demonstrated the absence of aberration production in the absence of oxygen or by the light or dye treatments alone. In both species, the photodynamic treatment produced significant increases in both chromatid and chromosome aberrations. Both dyes caused mitotic inhibition. The distribution of chromosome breaks following visible light and methylene blue was found to be non-random with the centromeres and some other sites being affected preferentially. The authors speculate that this may be due to the higher guanine-cytosine content of the centromeric DNA. These authors also tested the effect of post-treatment of dry barley seeds with methylene blue and acridine orange in the dark without oxygen following X-ray irradiation. Both dyes increased the yield of X-ray induced aberrations by a factor of about 1.4. Interestingly, similar tests of the effect of post-X ray treatment with the dyes on chlorophyll mutation production in M_2 progenies of M_1 plants grown from the treated seeds showed about a five-fold increase in mutation frequency with methylene blue whereas acridine orange had no effect.

5. Mutation induction with near ultraviolet and 8-methoxypsoralen

The photosensitization reactions resulting from treatment with 8-methoxypso-ralen (8-MOP) differ from the usual photodynamic reactions in that oxygen is not required and in fact may have some inhibitory effect, presumably by destruction of some of the 8-MOP [42]. The most effective wavelengths lie in the near ultra-violet region (NUV) of the spectrum [45]. 8-MOP forms a 1 : 1 adduct with the pyrimidine bases of DNA or RNA by formation of a cyclobutane ring upon irradia-tion with NUV [18, 45, Musajo this volume, Wacker this volume]. Purines seem not to be affected.

Reference has been made earlier to Altenberg's [3] observation of recessive lethal mutation induction in Drosophila following treatment of the polar cap cells of developing fertilized eggs treated with NUV plus 8-MOP. The studies of Drake and McGuire [16] of rapid lysis mutations in phage T4 induced by visible light and either thiopyronine or psoralen have also been discussed earlier.

Mathews [42] was the first to demonstrate the induction of mutations in bac-teria by 8-MOP plus NUV treatment. She observed a potent mutagenic effect as measured by penicillin resistant mutants in both pigmented and mutant colorless strains of *Sarcina lutea*. No protection against either the lethal or mutagenic effects of 8-MOP plus NUV treatment was afforded by the normal carotenoid pigments as was found to be the case for photodynamic inactivation by visible light and tolui-dine blue [43].

The induction by 8-MOP plus NUV treatment of mutations to prototrophy in strains of *E. coli* auxotrophic at three different loci has been studied by Igali et al. [24]. Strain WP2 *trp* mutates to prototrophy either by true reverse mutations at the tryptophane locus or by *ochre* suppressor mutations. Strain WP2 Hcr⁻ is a mu-tant which is deficient in excision-repair [23]. Strain IV-5 is an Exr⁻ threonine auxotroph and strain V-5 is an Exr⁺ derivative of it; both mutate to prototrophy by the formation of *amber* suppressors. Strain H/r 30-R is an arginine auxotroph with the B/r phenotype which mutates to prototrophy by the formation of *amber* suppressors. Several mutants of H/r 30-R varying in UV sensitivity and ability to repair DNA were also used in the experiments. NUV light was obtained from a Phillips 125-watt mercury discharge black-light lamp emitting in the region of 300–400 nm with a peak at 365 nm. Photoreactivating light was from a 250-watt photographic flood lamp from which wavelengths below 380 nm were removed by a Kodak Wratten filter No. 85. Neither NUV nor photoreactivating light alone had any detectable lethal or mutagenic effect at the intensities and exposure times em-ployed. 8-MOP was dissolved in ethanol and diluted in saline to a final concentra-tion of 100 μg/ml which had no detectable effect in the absence of NUV.

With strain WP2 *trp*, 8-MOP + NUV treatment resulted in sigmoidal survival curves and a somewhat more than linear increase in induced mutations with in-creasing NUV exposure time. The supplementation of the glucose-salts plating medium with broth instead of tryptophane had no effect on survival and very little effect on mutation induction. This is in sharp contrast to the results after far ultraviolet irradiation where there are many more mutations on broth-enriched plates, the increase being entirely due to suppressor mutations. In the present experiments, approximately equal numbers of suppressor and true reverse mutations were induced by the 8-MOP+NUV treatment. Strain WP2 Hcr⁻ is a mutant which has lost the ability to excise pyrimidine dimers from its DNA and is about 16 times more sensitive to the lethal and mutagenic effects of ultraviolet. Replicated, comparative tests showed WP2 Hcr⁻ to be about 10 times as sensitive for lethality and about 15 times as sensitive for mutation-induction following 8-MOP plus NUV treatment. Essentially similar results were obtained in comparative tests with strains H/r 30-R (Hcr⁺) and H/s 30-R (Hcr⁻). It appears, therefore, that damage produced by 8-MOP plus NUV is also excisable. Strains H/s 30-R (Phr⁺) and H/s 30-S (Phr⁻) which differ presumably only in the ability of the former to repair UV damage to DNA by enzymatic photoreactivation were used in two kinds of experiments to test if 8-MOP plus NUV damage were photoreactivable. Neither the lethal nor mutagenic damage could be photoreactivated. Incidental to the tests of photoreactivation was the observation that wavelengths longer than 380 nm were ineffective in 8-MOP sensitized cells.

Pyrimidine dimers which are not excised may pass through the DNA replication point after some delay but the newly synthesized strands have discontinuities opposite the dimers [52]. These gaps may be repaired by a post-replication repair mechanism, the normal functioning of which requires the wild type alleles of the *rec* loci which are also essential for genetic recombination [23]. In Rec⁻ strains which are unable to effect post-replication repair, each unexcised pyrimidine dimer is lethal and such strains are not mutable by ultraviolet [44, 61]. Comparative tests with strain H/r 30-R (Rec⁺) and its derivative NG30 (Rec⁻) showed that the Rec⁻ strain is much more sensitive than the Rec⁺ strain to the lethal effects of 8-MOP plus NUV, the ratio of sensitivity being comparable to that observed after ultraviolet irradiation [24]. The induced mutation rate in the Rec⁻ strain did not exceed the spontaneous rate at the low doses of NUV necessary in order to have sufficient survivors to make an accurate assay of mutation. The data suggest but do not prove that Rec⁻ strains may be unmutable by 8-MOP plus NUV.

Witkin [60] has shown that the gene product of the Exr⁺ allele in *E. coli* B is necessary for normal post-replication repair of ultraviolet DNA damage and that Exr⁻ strains are less efficient in carrying out post-replication repair and are essentially unmutable by ultraviolet. Comparative tests of strains V-5 (Exr⁺) and IV-5 (Exr⁻) following 8-MOP plus NUV treatment yielded results essentially parallel to those following ultraviolet irradiation [24]. The Exr⁻ strain was significantly more sensitive to the lethal effects and no increase over the spontaneous mutation rate was observed at doses where the Exr⁺ strain yielded a significantly higher mutation rate.

Summarizing these important studies of Igali et al., the damage to DNA produced by 8-MOP plus NUV treatment affects *E. coli* cells in essentially the same fashion as ultraviolet damage with respect to excision-repair, post-replication repair, and mutation induction. The only difference is that it is not enzymatically photoreversible. The authors point out that these properties of 8-MOP plus NUV damage are consistent with the known photoproducts. Finally, these studies by their effective utilization of mutant strains which differ in their capability to repair DNA damage produced by ultraviolet by the several known repair mechanisms and in their mutagenic response can serve as model studies which, applied to other systems may contribute materially to our understanding of the basic mechanisms and effects of photodynamic action and of mutagenesis in general.

Alderson and Scott [1] have studied the induction of suppressor mutations in a methionine-requiring strain of *Aspergillus nidulans* by 8-MOP plus NUV treatment of conidia. They used the same light source and dye concentrations as Igali et al. [24]. The suppressor mutations are known to occur in at least 5 loci and can be divided phenotypically into 3 classes. Treatment of conidia by NUV alone, 8-MOP alone, visible light alone, visible light plus 8-MOP, or NUV followed by 8-MOP had no lethal or mutagenic effect. Inactivation and mutation induction required the presence of 8-MOP at the time of NUV exposure and a prior treatment of 8-MOP for about 5 min was more effective. 8-MOP plus NUV induced mutations at a rate essentially linearly related to NUV exposure time. The frequencies of the 3 classes of suppressor mutations were about equal after 8-MOP plus NUV. This is in contrast to the distribution of spontaneous suppressor mutations where from 60 to 85% of the mutants are of one type and to previously reported results with chemical mutagens. 8-MOP plus NUV is an efficient mutagen, some 12 times as many mutations being induced than by diethyl sulphate at corresponding levels of survival.

6. Conclusions

The available evidence clearly demonstrates that photodynamic treatment utilizing a variety of photosensitizing dyes can induce point mutations in a number of species including viruses, bacteriophages, bacteria, yeasts, and fungi. The evidence for point mutation induction in Drosophila and higher plants is less convincing which reflects not only the small amount of data but also the technical difficulty in achieving a significant photodynamic treatment of the DNA in the reproductive cells in such species. Chromosome aberrations are, however, produced in plant root tip cells by photodynamic treatment in frequencies comparable to those observed following X-ray radiation.

Analysis of the reverse mutability of photodynamically induced mutations at the *r*II locus of bacteriophage T4 indicates that they are base-pair subsitutions, both transitions and transversions, and not frame shift mutations. Recent studies utilizing mutant strains of *E. coli* which differ in their ability to repair UV damage to DNA indicates that lethal and premutational damage produced by 8-methoxypsoralen sensitization and exposure to near ultraviolet is also repairable by the excision-repair and post-replication repair processes, is not photoreversible by enzymatic mediated photoreactivation, and acts in mutagenesis in recombination-deficient and Exr⁻ mutant strains in a manner analogous to UV-induced damage. The extension of such investigations to other photodynamic systems utilizing different photosensitizers, along with increasing knowledge of the specific photodynamic effects produced in DNA, promises to add significantly to our understanding of the basic mechanisms responsible for photodynamic induced genetic change and of mutagenesis in general.

References

[1] T. Alderson and B.R. Scott, The photosensitizing effect of 8-methoxypsoralen on the inactivation and mutation of aspergillus conidia by near ultraviolet light. Mutation Res. *9* (1970) 569–578.

[2] T. Alper and B. Hodgkins, Excision repair and dose modification: questions raised by radiobiological experiments with acriflavine. Mutation Res. *8* (1968) 15–23.

[3] E. Altenberg, Studies on the enhancement of the mutation rate by carcinogens. Texas Rept. Biol. Med. *14* (1956) 481.

[4] C. Ball and J.A. Roper, Studies on the inhibition and mutation and *Aspergillus nidulcns* by acridines. Genet. Res. Camb. *7* (1966) 207–221.

[5] S. Benzer, Fine structure of a genetic region in bacteriophage. Proc. Nat. Acad. Sci. U.S. *41* (1955) 344–354.

[6] S. Benzer, On the topography of the genetic fine structure. Proc. Nat. Acad. Sci. U.S. *47* (1961) 403–415.

[7] H. Bohme and A. Wacker, Mutagenic activity of thiopyronine and methylene blue in combination with visible light. Biochem. Biophys. Res. Comm. *12* (1963) 137–139.

[8] M. Brendel and R.W. Kaplan, Photodynamische Mutationauslosung und Inkativierung beim *Serratia*-Phagen x durch Methylenblau und Licht. Mol. Gen. Genet. *99* (1967) 181–190.

[9] S. Brenner, S. Benzer and L. Barnett, Distribution of proflavine-induced mutations in the genetic fine structure. Nature *182* (1958) 983–985.

[10] M. Buiatti and R. Ragazzini, The mutagenic effects of acridine orange in tomato (*Lycopersicum esculentum*). Mutation Res. *3* (1966) 360–361.

[11] C.M. Calberg-Bacq, M. Delmelle and J. Duchesne, Inactivation and mutagenesis due to the photodynamic action of acridines and related dyes on extracellular bacteriophage T₄B. Mutation Res. *6* (1968) 15–25.

[12] F.H.C. Crick, L. Barnett, S. Brenner and R.J. Watts-Tobin, General nature of the genetic code for proteins. Nature *192* (1961) 1227–1232.

[13] F. D'Amato, Mutazioni chlorofilliane nell'orzo indotte da derivati acridinici. Caryologia *3* (1950) 211–220.

[14] R.I. DeMars, Chemical mutagenesis in bacteriophage T2. Nature *172* (1953) 964.

[15] H. Doring, Photosensibilisierung der gene? Naturwissenschaften *26* (1938) 819–820.

[16] J.W. Drake and J. McGuire, Properties of *r* mutants of bacteriophage T4 photodynamically induced in the presence of thiopyronin and psoralen. J. Virol. *1* (1967) 260–267.

[17] L. Ehrenberg, Chemical mutagenesis: biochemical and chemical points of view on mechanism of action. In Chemische Mutagenese, ed. H. Stubbe (Akademie Verlag, Berlin, 1959) pp. 124–136.

[18] S. Farid and C.H. Krauch, Photochemical and biological reactions of the furocoumarins. In: Radiation Biology, Proceedings of the Third International Congress of Radiation Research, ed. G. Silini (North-Holland, Amsterdam, 1967) pp. 869–886.

[19] E. Freese, The difference between spontaneous and basa-analogue induced mutations of phage T4. Proc. Nat. Acad. Sci. U.S. *45* (1959) 622–633.

[20] Y.Z. Gendon, Vopr. Virusol. *8(5)* (1963) 542–547. (Abstract: Y.Z. Gendon, Mutations in polio virus induced by direct action of proflavine on viral ribonucleic acid (RNA). Chemical Abstracts *60* (1963) 4505d).

[21] W. Harm, The role of host-cell repair in liquid-holding recovery of UV-irradiated *Escherichia coli*. Photochem. Photobiol. *5* (1966) 747–760.

[22] W. Harm, Differential effects of acriflavine and caffeine on various ultraviolet irradiated *Escherichia coli* strains and T1 phage. Mutation Res. *4* (1967) 93–110.

[23] P. Howard-Flanders, DNA Repair. Ann. Rev. Biochem. *37* (1968) 175–200.

[24] S. Igali, B.A. Bridges, M.J. Ashwood-Smith and B.R. Scott, Mutagenesis in *Escherichia coli*. IV. Photosensitization to near ultraviolet light by 8-methoxypsoralen. Mutation Res. *9* (1970) 21–30.

[25] T. Ito and K. Hieda, Absence of apparent dark recovery from the photodynamically-induced mutational damages in yeast. Mutation Res. *5* (1968) 184–186.

[26] R.W. Kaplan, Auslosung von Mutationen durch sichtbares Licht in vitalgefarbten *Bacterium prodigiosum*. Naturwissenschaften *35* (1948) 127.

[27] R.W. Kaplan, Mutations by photodynamic action in *Bacterium prodigiosum*. Nature *163* (1949) 573–574.

[28] R.W. Kaplan, Auslosung von Phagenresistenzmutationen bei *Bacterium coli* durch Erythrosin mit und ohne Belichtung. Naturwissenschaften *37* (1950) 308.

[29] R.W. Kaplan, Mutation und Keimtotung bei *Bact. coli* histidinless durch UV und Photodynamie. Naturwissenschaften *37* (1950) 547.

[30] R.W. Kaplan, Photodynamische Auslosung von Mutationen in den Sporen von *Penicillium notatum*. Planta *38* (1950) 1–11.

[31] R.W. Kaplan, Dose-effect curves of S-mutation and killing in *Serratia marcescens*. Arch. Mikrobiol. *24* (1956) 60–79.

[32] B.A. Kihlman, Induction of structural chromosome changes by visible light. Nature *183* (1959) 976–978.

[33] B.A. Kihlman, Studies on the production of chromosomal aberrations by visible light: The effects of cupferron, nitric oxide, and wavelength. Exp. Cell Res. *17* (1959) 590–593.

[34] B.A. Kihlman, The production of structural chromosome changes in the presence of light and acridine orange. Radiation Botany *1* (1961) 35–42.

[35] B.A. Kihlman, Biochemical aspects of chromosome breakage. Advan. Genet. *10* (1961) 1–59.

[36] H.E. Kubitschek, Mutation without segregation. Proc. Nat. Acad. Sci. U.S. *52* (1964) 1374–1381.

[37] H.E. Kubitschek, Mutation without segregation in bacteria with reduced dark repair ability. Proc. Nat. Acad. Sci. U.S. *55* (1966) 269–274.

[38] H.E. Kubitschek, Mutagenesis by near-visible light. Science *155* (1967) 1545–1546.

[39] S. Kumar and A.T. Natarajan, Photodynamic action and post-irradiation modifying effects methylene blue and acridine orange in barley and *Vicia faba*. Mutation Res. *2* (1965) 11–21.

[40] L.S. Lerman, Structural considerations in the interaction of DNA and acridines. J. Mol. Biol. *3* (1961) 18–30.

[41] G.E. Magni, R.C. Von Borstel and S. Spora, Mutagenic action during meiosis and antimutagenic action during mitosis by 5-aminoacridine in yeast. Mutation Res. *1* (1964) 227–230.

[42] M.M. Mathews, Comparative study of lethal photosensitization of *Sarcina lutea* by 8-methoxypsoralen and by toluidine blue. J. Bacteriol. *85* (1963) 322–328.

[43] M. Mathews and W.R. Sistrom, The function of the carotenoid pigments of *Sarcina lutea*. Arch. Mikrobiol. *35* (1959) 139–146.

[44] A. Miura and J. Tomizawa, Studies on radiation sensitive mutants of *E. coli*. III. Participation of the Rec system in induction of mutation by ultraviolet irradiation. Mol Gen. Genet. *103* (1968) 1–10.

[45] L. Musajo, Photochemical interaction between skin-sensitizing furocoumarins and DNA. In: Radiation Research, Proceedings of the Third International Congress of Radiation Research, ed. G. Silini (North-Holland, Amsterdam, 1967) pp. 803–812.

[46] S. Nakai and T. Saeki, Induction of mutation by photodynamic action in *Escherichia coli*. Genet. Res. Camb. *5* (1964) 158–161.

[47] A. Orgel and S. Brenner, Mutagenesis of bacteriophage T4 by acridines. J. Mol. Biol. *3* (1961) 762–768.

[48] A.R. Peacocke and J.N.H. Skerrett, The interaction of aminoacridines with nucleic acids. Trans. Faraday Soc. *52* (1956) 261–279.

[49] P.P. Puglisi, Mutagenic and antimutagenic effects of acridinium salts in yeast. Mutation Res. *4* (1967) 289–294.

[50] D.A. Ritchie, Mutagenesis with light and proflavine in phage T4. Genet. Res. Camb. *5* (1964) 168–169.

[51] D.A. Ritchie, Mutagenesis with light and proflavine in phage T4. II. Properties of the mutants. Genet. Res. Camb. *6* (1965) 474–478.

[52] W.D. Rupp and P. Howard-Flanders, Discontinuities in the DNA synthesized in an excision-defective strain of *Escherichia coli* following ultraviolet irradiation. J. Mol. Biol. *31* (1968) 291–304.

[53] M.I. Simon and H. Van Vunakis, The photodynamic reaction of methylene blue with deoxyribonucleic acid. J. Mol. Biol. *4* (1962) 488–499.

[54] B. Singer and H. Fraenkel-Conrat, Dye-catalyzed photoinactivation of tobacco mosaic virus. Biochemistry *5* (1966) 2446–2450.

[55] R.B. Webb, Photomutagenesis in chemostat cultures of bacteria. Argonne National Laboratory, Biological and Medical Research Division Annual Report; ANL-7535 (Dec. 1968) p. 15–17.

[56] R.B. Webb and H.E. Kubitschek, Mutagenic and antimutagenic effects of acridine orange in *Escherichia coli*. Biochem. Biophys. Res. Comm. *13* (1963) 90–94.

[57] R.B. Webb and M.M. Malina, Mutagenesis in *Escherichia coli* by visible light. Science *156* (1967) 1104–1105.

[58] E.M. Witkin, The effect of acriflavine on photoreversal of lethal and mutagenic damage produced in bacteria by ultraviolet light. Proc. Nat. Acad. Sci. U.S. *50* (1963) 425–430.

[59] E.M. Witkin, Radiation-induced mutations and their repair. Science *152* (1966) 1345–1353.

[60] E.M. Witkin, Mutation-proof and mutation-prone modes of survival in derivatives of *Escherichia coli* B differing in sensitivity to ultraviolet light. Proceedings of Symposium on Recovery and Repair Mechanisms in Radiobiology, Brookhaven Symposia in Biology, No. 20 (1967) pp. 17–53.

[61] E.M. Witkin, The mutability towards ultraviolet light of recombination deficient strains of *Escherichia coli*. Mutation Res. *8* (1969) 9–14.

[62] A. Zampieri and J. Greenberg, Mutagenesis by acridine orange and proflavine in *Escherichia coli* strain S. Mutation Res. *2* (1965) 552–556.

CHAPTER 17

GENETIC DAMAGE BY PHOTODYNAMIC ACTION AND RECOMBINATION OF BACTERIOPHAGE*

Nobuto YAMAMOTO

Fels Research Institute and Department of Microbiology,
Temple University School of Medicine,
Philadelphia, Pa. 19140, USA

1. Introduction

The photosensitizing action of dyes on biological systems is commonly designated as "photodynamic action". This subject has been well summarized in a monograph by Blum [2].

The photodynamic action on viruses can be recognized by loss of viability of viruses: photodynamic inactivation of viruses. Earlier studies on photodynamic inactivation of bacteriophages were made in the early 1930's by Clifton [8] and by Perdrau and Todd [23]. Significance of the photodynamic inactivation of bacteriophages was greatly broadened by the similar findings with animal viruses by Perdrau and Todd [24] and later by Hiatt et al. [17, 18], and with a plant virus by Oster and McLaren [22].

Our earlier studies (1956) on the photodynamic inactivation of bacteriophages led us to the conclusion that the photodynamic inactivation of bacteriophages is due to the damage of phage DNA [31, 34].

In this communication we discuss genetic damages of bacteriophage by photodynamic action.

2. Historical background

2.1 *Reaction sites in bacteriophage*

Although considerable studies have been performed on the photodynamic inactivation of viruses, as yet relatively little is known regarding the biological sites involved in this type of reaction. This problem undoubtedly merits closer attention since there is a considerable degree of specificity involved both as regards the sensitizer and the acceptor for the photosensitized inactivation of enzymes and proteins [20].

*Supported in part by Grants from the U.S. Public Health Service (NIH AI-06429) and National Science Foundation (NSF GB-25098).

In 1933, Burnet demonstrated that serologically related phages exhibit similar susceptibility to visible light in the presence of a given dye, but that there are marked differences in sensitivity from one serological grouping to another. Hence, the relative sensitivity of phage strains to photodynamic inactivation was of taxonomic significance [5].

Quantitative studies of the kinetics of photodynamic inactivation of bacteriophage by Welsh and Adams (1954) confirmed Burnet's conclusion that the relative sensitivity of phage strains to photodynamic inactivation is correlated with their serological classification [28]. This indicates that loss of infectivity is due to photodynamic damage of the protein components. We studied photodynamic inactivation of T-group bacteriophages belonging to four serological groups as shown in table 1 [31]. The results were in accord with the conclusions of Burnet [5] and Welsh and Adams [28] regarding a correlation between serological grouping of phages and photodynamic sensitivity. In 1956 we suggested that the relatively high resistance of T-even bacteriophages is attributed to their supposedly smaller pore-sizecoat, since they are more susceptible to osmotic shock than the other T-odd phages; this would make penetration of dye molecules to the nucleic acid components more difficult [31, 33, 34].

Photodynamic sensitivity: T5, T3, T7 > T1 ≫ T2, T4, T6

Osmotic shock sensitivity: T5, T3, T7 < T1 ≪ T2, T4, T6

Table 1

Photodynamic activity of various dyes on the T-series of coliphages [31]; the velocity constant (min^{-1}) of serological groups

Dye	Strain						
	T1	T2	T4	T6	T3	T7	T5
Methylene blue	0.24	0.01	0.02	0.01	0.55	0.72	0.77
Toluidine Blue	0.12	0.00			0.36	0.57	0.40
Brilliant cresyl blue	0.07	0.00			0.18	0.33	0.26
Neutral red	0.01	0.01	0.01		0.06		0.01
Crystal violet		0.00			0.04		0.06
Methyl green		0.00					0.01*
Pyronine		0.01			0.02		0.05
Acridine orange	0.06	0.00	0.00	0.00	0.97	1.15	0.35
Acriflavine	0.04	0.01			0.29	0.46	0.28
Rivanol		0.01					0.10
No dye		0.02					

The buffered saline: pH 7.4
Dye concentration 1.5×10^{-5} M. Phage concentration: 5×10^7 particles/ml.
*pH: 8.0.

Furthermore we have observed that, when photodynamic sensitivity of T-even phage after infection to *E. coli* was measured by loss of infectious centers, T-even phages became sensitive a few minutes after infection, resistant at 7 min and later sensitive again at about 20 min (unpublished observation). Phage maturation should begin at about 20 min after infection. Thus, the above observation indicates that the naked DNA after phage infection and before maturation is sensitive to photodynamic action, supporting the theory that the nucleic acid of bacteriophage is a target for photodynamic inactivation of bacteriophage [31, 34].

Helprin and Hiatt (1959) reported some evidences in favor of such a conception in a study of the inactivation of $T2r^+$ and T3 coliphage by photodynamic action of toluidine blue, which demonstrated these phages differ widely in their rate of uptake and tenacity of binding of photosensitizing dyes. The inactivation of $T2r^+$ was found to be very markedly dependent on the time of pre-incubation with the dye in the dark whereas that of T3 was independent on the time of pre-incubation. In agreement with our hypothesis, it was concluded that association of viruses and dye was necessary prerequisite for photodynamic inactivation of bacteriophage [16].

Pre-incubation of T2 with dye at 37°C resulted in a higher degree of inactivation, suggesting the binding of a larger amount of dye. The high temperature coefficient of this process, a Q_{10} of about 38, is similar to that for diffusion across a membrane. If the dye—phage mixtures were suddenly diluted, the sensitivity of T3 phage decreased, as might be expected; but $T2r^+$ under such treatment was still inactivated at an appreciable rate, in agreement with the hypothesis that dye was bound at the interior of the phage from which its removal by diffusion was restricted [16]. Thus, they supported our conclusion that T2 was much less readily photoinactivated than T3 because the T2 was not as permeable to the dye than was T3.

Cummings and Kozloff (1960) have shown that intact T2 phage could exist in two interconvertible morphological forms classified with respect either to head length or to sedimentation rate. They found that the transition between the two forms of T2 was markedly dependent upon the temperature as well as the pH. At the particular pH the fast sedimenting form (short head form) was transformed into the slow sedimenting form (long head form) as the temperature of the solution was increased [10].

In view of the fact that the T2 head membrane can exist in two forms, the possibility was examined that these two forms differed greatly in their permeability. The permeability of T2 to methylene blue was measured from the rate of photodynamic inactivation of T2 after incubation in methylene blue solution (1×10^{-5} M). In a typical experiment the phage was incubated with methylene blue for 1 hr at a given pH at various temperatures. The rate of photodynamic inactivation was dependent on the preincubation temperature. The rate of inactivation was approximately logarithmic and first-order inactivation constants (k) ranged from 0.006 min^{-1} for preincubation at 9°C to 0.37 min^{-1} for preincubation at 44°C.

There was a good correlation between the amount of methylene blue inside the head, presumably in contact with the DNA, and the photodynamic inactivation rate constant. From the observation, Cummings and Kozloff concluded that the long head form is 20 to 30 times as permeable as the short head form to methylene blue. Thus, particles having the long head, slow sedimenting form are more permeable to passage of methylene blue through the head membrane than those having the short head, fast sedimenting form [10].

2.2. Protoplast infecting agent

In confirmation of the results reported by Yamamoto [31, 34] and by Hiatt [16, 17, 18], Fraser and Mahler found that T2 is relatively resistant to 'immediate' photodynamic inactivation by dyes but is rendered susceptible by a prolonged period of pre-incubation [13]. The earlier investigators [16, 17, 18] have postulated that the necessity for pre-incubation is imposed by a restricted permeability of the T2 (T-even) head membrane. Fraser and Mahler (1961) have shown that the protoplast-infecting agent, π (a phage degraded by 6 M urea), is accessible without lag to molecules even as large as trypsin or DNAse. They expected, therefore, that also should exhibit no lag to photodynamic inactivation. This was found to be the case: the rate of inactivation is immediately equal to that shown by T2 only after complete equilibration with the dye [13].

2.3. Guanine is destroyed by photodynamic action

Simon and Van Vunakis (1962) studied the photodynamic action of methylene blue on the DNA bases, nucleosides and nucleotides, and analyzed by spectral, chromatographic and monometric techniques. A rapid photodynamic reaction leading to the loss of ultraviolet absorbance accompanied by the uptake of one mole O_2/mole derivative occurred with the guanine derivatives. Thymine compounds reacted very slowly. The base composition found for photodynamically treated DNA indicated that guanine was preferentially destroyed [25]. The DNA photodynamically treated with methylene blue was analyzed by immunological techniques. The reaction led to partial denaturation of the DNA. Heating degraded the photodynamically treated DNA, indicating that photodynamic action on DNA produces many single strand breaks. The results of T_m and sedimentation measurements agreed with the interpretations of the immunological data in the degree of photodynamic destruction of DNA. Moreover, the photodynamic reaction was more rapid with denatured than with native DNA [25].

2.4. Active dyes for photodynamic inactivation of bacteriophage

We have investigated the chemical structure necessary for photodynamic inactivation of bacteriophage [31, 34]. Among many dyes tested thiazine dyes, oxazine dyes and acridine dyes are exceedingly active in photodynamic inactivation of bacteriophage (see fig. 1). These photodynamically active dyes for bacteriophages have common structure [31, 34]: (a) These active dyes are all basic. (b) These dyes are all hetero-tricyclic compounds. (c) These hetero-tricyclic compounds have two chargeable nitrogen atoms situated at comparable positions on both sides of the skeleton.

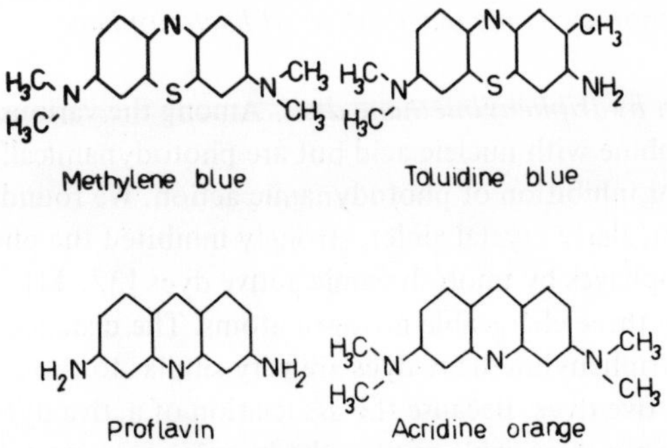

Fig. 1. Representatives of photodynamic active dyes for bacteriophages.

(d) The distances of these two nitrogens are almost equal in all active dyes, and conjugated double bond systems lie between these two atoms (see fig. 2). (e) All of these dyes posses an affinity for the nucleic acids and the association is of an ionic nature [7].

These active dyes belonging to the three groups can all be derived from the fundamental skeleton by suitable variation in X and Y (X=N or C; Y=S, O, or N). The cations of these dyes are a resonance hybrid of the structure (fig. 2). The cationic charge is located either on one of the two nitrogen atoms, or on the hetero-atom in the heterocyclic ring. The cationic charge may associate with an anionic charge of the phage (nucleic acid). The uncharged nitrogen atom on the side-chain can form a coordination bond with a cationic charge or an empty orbital via the lone pair of electrons. Therefore, these active dyes may associate with nucleic acid of phage through a bridge-like structure [31, 34].

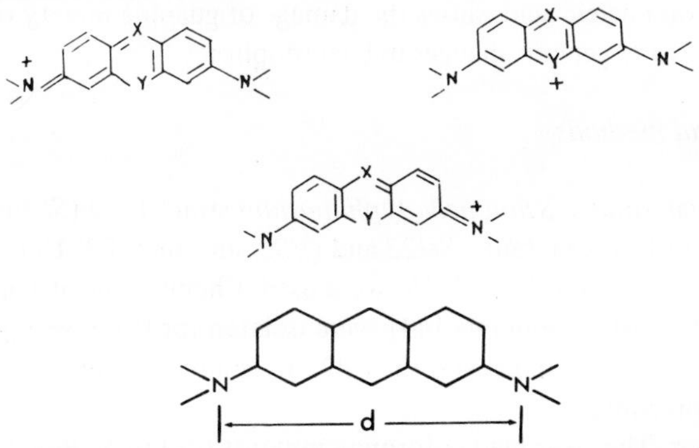

Fig. 2. The common structures of active dyes. I. (top left, top right, and center): resonance structures. II. (bottom): The distance (*d*) of the two nitrogens are almost equal in all active dyes. It is suggested that these two nitrogens of the dyes may associate with DNA through bridge like structure [31].

2.5. Inhibition of photodynamic inactivation of bacteriophage

2.5.1. Inhibition by triphenylmethane dyes. Among the various dyes we tested, some dyes can combine with nucleic acid but are photodynamically inactive. These dyes were tested for inhibition of photodynamic action. We found that triphenyl-methane dyes, particularly crystal violet, strongly inhibited the photodynamic inactivation of bacteriophages by photodynamic active dyes [32, 34]. The triphenyl-methane dyes carry three chargeable nitrogen atoms. The distances between the nitrogen atoms of triphenylmethane dyes are very similar to those of the two nitrogen atoms of the active dyes. Because the association of active dyes with the bacterio-phage was of reversibly dissociable nature, the kinetics on enzymology has been applied to the active dye—phage system. Application of the kinetics to the inhibition of the photodynamic action has revealed that this inhibition is a competitive one [32, 34].

2.5.2. Inhibition of photodynamic action by polyamines. Fraser and Mahler [13] have shown a pronounced stabilizing effect of various diamines on the thermal inactivation of π, presumably by interaction with DNA. Thus, they investigated effect of diamines on photodynamic inactivation of bacteriophage by proflavine, and they found that photodynamic inactivation of bacteriophage was competitively inhibited in the presence of aliphatic diamines with no particular specificity of chain length [13].

3. Recent genetic studies

From the accumulated evidences, it is concluded that the damage of phage genome is responsible for the photodynamic inactivation of bacteriophage. Since the photo-dynamic action on nucleic acid causes the damage of guanine moiety of nucleic acids, it is desirable to study genetic changes in bacteriophages.

3.1. Materials and methods

3.1.1. Bacterial Strains. Salmonella typhimurium strain LT-2 (*St* for short), *Q*; their P22 phage resistant mutants, *St/22* and *Q/22*; and their P221b lysogens, *St(P221b), Q(P221b)* and *Q/22(P221b)* were used. Chemical agent sensitive mutants [40, 44] and UV-sensitive mutants [44] were isolated for this investigation. The chemical agent sensitive mutants and their P22-resistants mutants lysogenic for P221b [45] were also prepared.

Bacteriophages. The clear plaque-forming mutants (c_2) of *Salmonella* bacterio-phage P22 and P221dis_1 [38, 39] were used for this study. P221bc+ [45] was used for preparing the lysogenic strains. Although P22 and P221 serologically are unrelated [43], they have a genetic homology containing the clear plaque-forming makers, $c(c_1, c_2$, and $c_3)$ [43]; the color indicator markers, g and h_{21} [35]; the prophage-

Table 2

Host ranges and immunity relationships of phages P22 and P221

Phage	St	St/22	Q	Q/22	St(P22)	St(P221b)	St/22(P221b)	Q(P22)	Q(P221b)	Q/22(P221b)
P22	+	−	+	−	−	+	−	−	+	−
P22h	+	(+)[a]	+	−	−	+	(+)[a]	−	+	−
P221b	+	+	+	+	−	−	+	−	−	−
P221dis₁	(±)[b]	+	(±)[b]	+	−	+	+	−	+	+

[a] (+) indicates formation of very faint plaques [43].
[b] (±) indicates formation of extremely faint plaques. Thus, for the study of P221dis₁, P22-resistant mutants and their P221b lysogens were used.

bacterium attachment homology region, att^{ϕ} and prophage integration function, int [42]. P22 and P221dis_1 can superinfect P221b lysogens [38, 39]. Immunity relations and host ranges of these phages were summarized in table 2.

3.1.2. *Media.* Regular nutrient broth containing of 8 g Difco nutrient broth and 5 g sodium chloride per liter of distilled water was used for bacterial aeration culture. For phage assay, we used hard agar containing 23 g Difco nutrient agar and 5 g sodium chloride per liter, and soft nutrient agar containing 7.5 g Difco bacto-agar, 5 g sodium chloride and 8 g of Difco nutrient broth per liter were used [1]. Phosphate buffered saline contained M/15 phosphate in 0.1 M NaCl at pH 7.0.

3.1.3. *Agents.* Methylene blue was purchased from Matheson, Coleman and Bell, Cincinnati, Ohio. A potent mutagen, N-methyl-N'-nitro-N-nitrosoguanidine (NG) was obtained from Aldrich Chemical Co., Inc., Milwaukee, Wisconsin. A potent carcinogen, 4-nitroguinoline 1-oxide (4QNO) was kindly sent by Dr. Y. Tagashira. β-Propiolactone (BPL) was purchased from Eastman Kodak Company. Dimethyl sulfonate (DMSO) was used for solubilizing 4NQO [44]. It is well established that 4NQO reacts with guanine or adenine moiety of nucleic acid [21, 27] and BPL reacts with guanine moiety of nucleic acid [9], thus destroying biological activity of the genome [14, 44].

3.1.4. *Isolation of 4NQO-sensitive bacterial mutants by N-methyl-N'-nitro-N-nitrosoguanidine (NG).* It is established that photodynamic action on bacteriophages causes damage of guanine moiety of the phage genome. Since 4NQO also damages guanine, bacterial mutants sensitive to 4NQO may be useful for understanding mechanisms of photodynamic damage and repair of the genome.

NG is in widespread use as a mutagen for bacterial and phage mutation and appears to be the most potent chemical mutagen yet found. The strains used for mutation studies were a proline-requiring mutant of *Q1* and histidine-requiring mutants of *St.*

Washed cell suspension of the above auxotroph mutants in phosphate buffered saline were incubated in the presence of 50 μg/ml NG in phosphate buffered saline for 15 min at 37°C. Then, cells were washed by filtration with Millipore filter HA to remove residual NG molecules. The washed cells were suspended in nutrient broth and cultured for 3 hr at 37°C to allow segregation. After segregation the cultures were streaked on nutrient agar and incubated overnight. Colonies were transferred on nutrient agar plates by replica planting. A group of the replica plates were exposed to UV for identification of UV-sensitive mutants. The other groups of the replica plates contained 4NQO (0.4 μg/ml). Those which can not grow on the 4NQO-plates were designated 4NQO-sensitive mutants [44].

3.1.5. *Test for recombination ability of bacterial strains.* About 25 NQO- or UV-sensitive strains carring a auxotroph marker were streaked by tooth picks on a nu-

trient agar plate. After incubation for 4 hr at 37°C, the bacterial strains were transferred by replica plating method onto minimal agar plates previously streaked with
a male Hfr strain of *S. typhimurium*. After 24 hr incubation of these mating plates
at 37°C, recombination ability of these 4NQO- or UV-sensitive strains was analyzed.
Those strains which form colonies on the mating plate were considered to be recombination positive whereas those strains which can not form colonies to be recombination deficient (*rec⁻*) mutants [44].

3.1.6. *Determination of excision activity of bacterial strains.* Excision activity of
bacterial strains was determined by Dr. H. Takebe with a somewhat modified technique of Boyce and Howard-Flanders [4, 44].

3.1.7. *Inactivation kinetics of bacteria with 4NQO or BPL.* All reactions and
manipulations were performed at room temperature (25°C). Bacteria (10^7 cells/ml)
were treated with 4NQO in nutrient broth or BPL in buffered saline. Samples were
withdrawn at various intervals, diluted and assayed for inactivation of colony forming ability of bacteria [14, 44].

3.1.8. *Prophage induction of P221b lysogen.* Log-phase lysogenic cells ($10^7 - 10^8$
cells/ml) were treated with 4NQO in nutrient broth or BPL in buffered saline and
diluted 100 to 1000-fold in fresh nutrient broth. After 30 or 60 min incubation at
37°C, samples were plated on *St/22 Sm^r* with soft agar containing dihydrostrepto-
mycin (200 μg/ml for *rec⁺hcr⁺ (P221b)* and 25 μg/ml for *hcr⁻ (P221b)* and *rec⁻*
(P221b)). The streptomycin kills lysogenic cells on the plates, but does not inhibit
phage production of cells previously induced [26]. Therefore only cells that are
induced by 4NQO or by BPL give P221b infectious centers on *St/22 Sm^r* [14, 44].

3.1.9. *UV-irradiation of bacterial strains.* UV-irradiations were performed with
Gates Raymaster tubular 8-watt discharge bulbs, 12 inches long $\times \frac{5}{8}$ inch diameter,
purchased from Arthur H. Thomas Co., Philadephia. The lamp was fixed at a distance of 15 inches above the bacteria or phage suspensions to be irradiated. The intensity of UV at this distance was 21 erg/mm^2/sec for phage inactivation. The intensity of 2 erg/mm^2/sec was obtained by partially covering the bulb and used for
bacteria. Irradiation was carried out with shaking, in a glass dish in which the liquid
level did not exceed 1.0 mm. During irradiation, 0.05 ml samples were periodically
withdrawn by pipette and diluted 100-fold in the phosphate buffered saline. Samples
were diluted further, and 0.1 ml aliquots of each diluted sample were streaked on
nutrient agar. The number of colonies following 24 hr incubation at 37°C served as
the measure of bacterial inactivation. Similarly, for phage inactivation, the number
of plaques was counted after overnight incubation at 37°C.

3.1.10. *Procedure for photodynamic irradiation (PD).* Methylene blue at a final
concentration of 2.5×10^{-5} M was added to the phage suspension in phosphate

buffered saline at pH 7.0. A white GE fluorescent tubular 15-watt discharge bulb at a distance of 15 cm was used for irradiation. Irradiation was carried out by shaking the phage suspension in a glass dish in which the liquid level did not exceed 1.0 mm. During irradiation, 0.05 ml samples were periodically withdrawn by pipette and diluted 100-fold in the phosphate buffered saline. The samples were then further diluted, mixed with 0.1 ml of indicator bacteria and plates by overlay agar method [1, 37].

4. Recent observation

4.1. *Repair mechanism for damaged purine moiety of DNA*
Since photodynamic action on nucleic acid damages guanine moiety, the repair mechanism for the damage of genome induced by purine damaging agents was investigated.

4.1.1. *Sensitivity of the 4NQO-sensitive mutants to UV and other agents.* The 4NQO-sensitive mutants were tested for sensitivity to UV and BPL as described in the method section. All the 4NQO-sensitive mutants were sensitive to not only another guanine specific agent BPL but also UV [44]. Thus, the repair mechanism for damaged purine can repair also UV-induced pyrimidine dimers, suggesting that this is the dark repair mechanism for a variety of DNA damage and not specific for

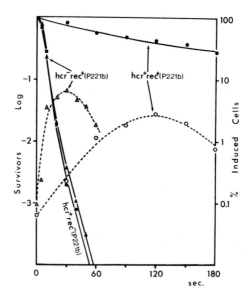

Fig. 3. UV-inactivation of P221b lysogens and UV-induction of the propage 221b. Solid lines indicate survivals. Broken lines indicate prophage induction.

Table 3

Characteristics of Salmonella typhimurium strains and their 4 NQO-sensitive mutants

Bacterial strain[a]	4NQO-sensitivity	UV-sensitivity	BPL-sensitivity	Bact. recombination	Excision activity	Host cell reactivation of UV-treated P22 phage	Phage recombination[b] P22 × P221b (UV-enhancement)
St	−	−	−	+	+	+	+ (+)
Q	−	−	−	+	+	+	+ (+)
rec⁻leu197[c]	+	+	+	−	+	+	+ (+)
hcr⁻ # 1-2	+	+	+	+	−	−	+ (+)
hcr⁺ # 1-3	+	+	+	+	−	−	+ (+)
rec⁻ # 22-1	+	+	+	−	+	+	+ (+)
hcr⁻ # 42-2	+	+	+	+	−	−	+ (+)
hcr⁻ # 712-1	+	+	+	+	−	−	+ (+)

a Strains St and Q are wild type, rec⁺hcr⁺.
b Recombination frequency was tested by mixed infection with P22 and P221b using a two point cross. (+) indicates UV-enhanced recombination by mixed infection with UV-treated phages.
c Kindly supplied by Dr. H.O. Smith [29, 30].

damaged purine. The results are summarized in table 3. Recently, five UV-sensitive mutants were isolated by screening with UV. All the UV-sensitive mutants were also sensitive to both 4NQO and BPL [44]. These observations agree with the Harm's observation that *E. coli* dark repair mechanism for UV damage repair photo-dynamic lesions in the bacterial DNA though the extent at which such damage is repaired is far less than in the case of UV damage [15]. UV-sensitivities of P221b lysogens of the above mutants were similar to those of the non-lysogens of the above mutants. The results are shown in table 4 and fig. 3.

4.1.2. *Characterization of 4NQO-sensitive mutants.* Since all the 4NQO-sensitive mutants were UV-sensitive, they were tested for recombination ability and activity to excise thymine dimers. As can be seen in table 3, the 4NQO-sensitive mutants were subdivided into two groups: recombination deficient (*rec⁻*) mutants and excision negative (*hcr⁻*) mutants**.

4.1.3. *Induction of the prophage from 4NQO-sensitive mutants.* Induction of bacteriophage formation in lysogenic *E. coli* by a potent carcinogen, 4-nitroquino-line 1-oxide and its derivative was previously reported by Endo and his coworkers [12].

Since all the 4NQO-sensitive mutants are also UV-sensitive and it is known that UV induction of UV-sensitive cell lysogenic for P221b is more efficient than that of the wild type cell lysogenic for P221b [40], induction by 4NQO of the prophage P221b from 4NQO-sensitive mutant may be very efficient. Indeed, it is found that induction of the prophage P221b from 4NQO-sensitive strains is far more efficient than that from the wild type cells. However, in agreement with our previous observation on UV induction [40], the P221b lysogens of *rec⁻* host were not inducible by 4NQO. These results were shown in fig. 4 and table 4. Similar data for UV-induction of these lysogens were shown in fig. 3.

4.2. *Photodynamic inactivation of bacteriophages.*

When bacteriophage P221*dis₁* was treated with PD, rapid inactivation of phage was observed. If the treated phage was assayed simultaneously on *hcr⁻* and wild type (*rec⁺hcr⁺*) strains, the number of survivors found on *hcr⁻* was slightly lower than those on *rec⁺hcr⁺* strains. The differences in phage survival must be ascribed to dark repair in the host cells although the degree of repair is small. A typical kinetics of inactivation is illustrated in fig. 5. Similar observations were also made by Harm [15] and by Uretz (personal communication).

**Since excision negative mutants can not reactivate UV-damaged phage [11, 19], the abbreviation *hcr⁻* for host cell reactivation negative was used. The abbreviations: *rec⁺hcr⁺* for wild type, *hcr⁻* for *rec⁺hcr⁻*, and *rec⁻hcr⁺*. All the *rec⁻* mutants studied in this paper were *rec A* mutants.

Table 4

Characteristics of P221b lysogens

Bacterial Strain lysogenic for P221b[a]	4NQO-sensitivity	UV-sensitivity	BPL-sensitivity	Phage induction		Recombination between superinfecting phage and prophage P221b[b]
				UV	4NQO	
St(P221b)	−	−	−	+	+	+
Q(P221b)	−	−	−	+	+	+
rec⁻leu 197(P221b)	+	+	+	−	−	−
hcr⁻ #1-3(P221b)	+	+	+	+	+	+
rec⁻ #22-1(P221b)	+	+	+	−	−	−
hcr⁻ #42-2(P221b)	+	+	+	+	+	+
hcr⁻ #72-1(P221b)	+	+	+	+	+	+

[a] Strains, St(P221b) and Q(P221b) are wild type, rec⁺hcr⁺ (P221b).

[b] + indicates PD-enhanced recombination between superinfecting phage and the prophage.
− indicates lack of PD-enhancement of the recombination.

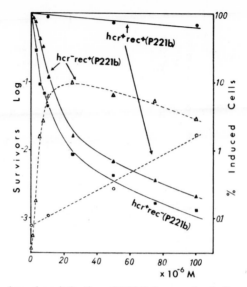

Fig. 4. Inactivation and prophage induction of P 221b lysogens by 4-nitroquinoline 1-oxide (4NQO). Solid lines indicate survivors. Broken lines indicate prophage induction.

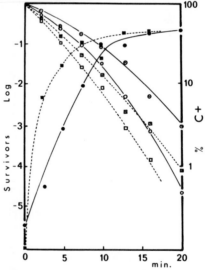

Fig. 5. Photodynamic inactivation of P221 $dis_1 c_2$ and formation of $c+$ plaques on P221b lysogens. Photodynamically treated samples were assayed simultaneously on a P221b lysogen and a non-lysogen of rec^+hcr^+ which was Q/22, and a P221b lysogen and a non-lysogen of rec^+hcr^- which was a P22 resistant hcr^- #1-3, hcr^- #1-3/22. The symbols: survivors (——○——) on rec^+hcr^+, survivors (——⊙——) and frequency of $c+$ plaque formation (——●——) on rec^+hcr^+(P221b), survivors (- - -□- - -) on rec^+hcr^-, survivors (- - -⊡- - -) and frequency of $c+$ plaque formation (- - -■- - -) on rec^+hcr^- (P221b).

4.3. *Photodynamic inactivation and recombination between superinfecting phage and prophage*

When P221dis_1 was treated with PD and assayed simultaneously on hcr^-, rec^-, rec^+, hcr^+ strains and their P221b lysogens, the number of survivors found on a P221b lysogen of hcr^- or rec^+hcr^+ was higher than those on the corresponding non-lysogen (see fig. 5). However there was no difference in survivors on rec^- and a P221b lysogen of rec^- strain (see fig. 6). As we reported previously [36, 37, 40], the slight reactivation in the former cases must be due to recombination between the PD-damaged P221dis_1 and the prophage P221b. Thus the lack of reactivation in the latter cases may be due to inability of recombination between the P221dis_1 and the prophage P221b.

When clear plaque-forming mutants (c_2) of P221dis_1 phage were treated with PD and assayed on P221b lysogens of rec^+hcr^+ or hcr^-, very high frequencies of turbid plaques ($c+$) were observed among the survivors. However no $c+$ plaques were found on non-lysogens. Thus, the formation of $c+$ plaques on the lysogenic strains must be a consequence of recombination between the PD-damaged P221dis_1c_2 and the prophage P221b$c+$. Since P221b and P221dis_1 are homologous phages [38, 39], it is evident that recombination leads to reactivation. A typical experiment on photodynamic inactivation of phage P221dis_1c_2 and $c+$ plaques formation is shown in fig. 5. The $c+$ formation increased greatly as survivors decreased. At optimal conditions $c+$ formation reached about 40% in comparison with 0.1–0.4% in untreated controls.

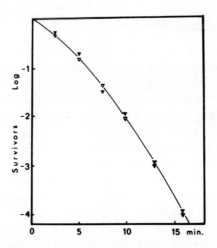

Fig. 6. Photodynamic inactivation of P221 dis_1c_2. Photodynamically treated samples were assayed simultaneously on a P221b lysogen (▼) and a non-lysogen (▽) of rec^-hcr^+ which was a P22-resistant rec^-leu 197, rec^-leu 197/22.

4.4. High frequency of recombination on hcr⁻ (P221b)

When PD-treated P221dis_1c_2 samples were assayed on hcr^- (P221b), a greatly
increased frequency of $c+$ formation was observed though the host does not have
hcr activity. The maximal $c+$ formation on hcr^- (P221b) was about 40%. As shown
in fig. 5, at small PD-dose recombination frequency on hcr^- (P221b) approached a
very high level whereas that on rec^+hcr^+ (P221b) rose relatively slowly. This suggests
that more PD-lesions are available for recombination in hcr^- (P221b) than in
rec^+hcr^+ (P221b) since the latter has the hcr mechanism for repair of PD-lesions.
The hcr mechanism repairs PD-damage resulting in an increase of non-recombination
type survivals. Therefore, at small PD-dose the recombination frequency on rec^+hcr^-
(P221b) is lower than that on hcr^- (P221b). At heavier PD-dose genomes contain
multiple hit damages. If the rec^+ function and the hcr mechanism of rec^+hcr^+ (P221b)
simultaneously act on multiply damaged phage genomes, viable recombinants should
increase. In contrast, since hcr^- (P221b) does not carry hcr function, many non-
viable recombinants in hcr^- (P221b) may be formed, resulting in fewer actual recom-
binant counts. However, the frequencies of recombination in these hosts reached to
a close value (see fig. 5).

4.5. Lack of PD-enhancement of recombination in rec⁻ (P221b)

When PD-treated P221dis_1c_2 samples were assayed simultaneously on rec^+hcr^+
(P221b) and rec^- (P221b), the $c+$ recombination on rec^+hcr^+ (P221b) increased as
inactivation proceeded whereas no significant increase of $c+$ formation on rec^- (P221b)
was observed. Since rec^- mutants have the same hcr activity as wild type, this observation indicates that the hcr function does not contribute to the phage recombination mechanism. Thus, the bacterial recombination function plays an important
role for the recombination.

4.6. Recombination and reactivation of PD-damaged P22 and P221b lysogens

Since P221dis_1 forms small plaques, it is difficult to study genetic markers of
this phage. P22 forms relatively large plaques. Although P22 and P221b are serologically and morphologically unrelated [43] they carry the genetic homology for the
entire region containing c_1, c_2, c_3, g, h_{21}, att and int markers [35, 42, 43]. Therefore the PD-damaged P22 was studied for detailed genetic analysis.

When a clear plaque forming mutant (c_2) of P22 was treated with PD and assayed
simultaneously on a P221b lysogen and non-lysogen of rec^+hcr^+, survivors on the
lysogenic strain rec^+hcr^+ (P221b) were slightly higher than those on non-lysogen
rec^+hcr^+ and a high frequency (6%) of turbid plaques ($c+$) was observed among survivors on rec^+hcr^+ (P221b) whereas no $c+$ plaques were found on rec^+hcr^+ (fig. 7).

While studying $c+$ formation of P22 on rec^+hcr^+ (P221b), a very high frequencies
of heterozygote ($c+/c_2$) plaques containing both $c+$ and c_2 phages were observed. As
shown in fig. 7, small dose of PD gave mostly heterozygote plaques whereas counts
for homozygote $c+$ plaques containing only $c+$ phage remained very low. However,
homozygote $c+$ plaque formation outnumbered heterozygote plaque formation at

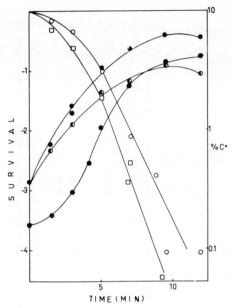

Fig. 7. Photodynamic inactivation of P22c_2 and formation of $c+$ plaques on a P221b lysogen. Photodynamically treated samples were assayed simultaneously on a P221b lysogen and a non-lysogen of rec^+hcr^+ which was Q1. The symbols: survivors (□) on rec^+hcr^+; survivors (○), total $c+$ frequency (●), frequency of $c+/c_2$ heterozygote plaque formation (○), and frequency of $c+$ ho-mozygote plaque formation (●) on rec^+hcr^+ (P221b).

heavy PD-dose level. This observation strongly suggests that single damage on one strand of a DNA molecule is responsible for the heterozygote plaque formation but damage of both strands by multiple hits (two or more) may be required for the formation of homozygote $c+$ plaques [37, 40].

4.7. Lack of PD-enhancement of recombination between the superinfecting phages P22 and the prophage P221b in rec^-

When PD-treated P22c_2 samples were assayed simultaneously on $rec^+hcr^+(P221b)$ and rec^- $(P221b)$, the $c+$ recombination on rec^+hcr^+ $(P221b)$ increased as inactivation proceeded whereas no significant increases of $c+$ formation on rec^- $(P221b)$ was observed. As mentioned previously, a similar observation was made between the PD-damaged P221dis_1 and the prophae P221b in rec^- strain. Moreover, observations, similar to the above data was observed between UV-damaged superinfecting phage and the prophage P221b in rec^- strain [40].

The rec^+hcr^+ $(P221b)$ and hcr^- $(P221b)$ strains spontaneously released substantial amounts of P221b phage (1.3 × 10^{-3} particles/cell and 6.0 × 10^{-4} particles/cell respectively) whereas rec^- $(P221b)$ released an extremely small amount of P221b (less than 2.5 × 10^{-8} particles/cell). As shown in table 4, UV and 4NQO induced the prophage P221b from rec^+hcr^+ $(P221b)$ and hcr^- $(P221b)$ but not from rec^- $(P221b)$.

Table 5

Induction of the prophage P221b by superinfection of P221b lysogens with UV-irradiated or unirradiated P22 and genomic masked particles carrying P221b genomes masked by P22 protein coat

Superinfecting UV-irradiated P22[a]	P221b lysogen								
	rec+hcr+ (P221b)			hcr- (P221b)			rec- (P221b)		
UV-dose (sec)	Inf. centers[b] / Inf. cell	Released[c] / Ind. cell	Masked Particles	Inf. centers / Inf. cell	Released / Ind. cell	Masked Particles	Inf. centers / Inf. cell	Released / Ind. cell	Masked Particles
0 (unirradiated)	0.05	15.3	0.81	0.043	13.7	0.72	0.038	4.5	0.039
20	0.41	13.6		0.20	11.4		0.26	3.9	
40	0.55	27.8	0.85	0.10	23.6	0.86	0.35	6.8	0.028
60	0.37	13.4		0.05	14.0		0.28	6.5	
120	0.10	12.1		0.025	9.6		0.18	14.6	

[a] For this study P22h was used because it is capable of infecting St/22 but forming only very faint plaques. Thus all P221b genomes irregardless of the capsid were scored [43].

[b] Frequency of induced cells was determined by P221b infectious centers per P22-infected cell.

[c] To calculate the number of P221b genomes released per induced cell, total released P221b genomes were divided by infectious centers.

[d] Genomic masked particles carrying P221b genome masked by P22 protein coat were detected by the method described previously [43].

If induction of the prophage is a prerequisite for recombination between the super-infecting phage and prophage, the inducibility may explain differences in recombi-nation frequency of P22 with the prophage P221b of these lysogens. In order to test this possibility, prophage induction of rec^+hcr^+ *(P221b)*, *hcr⁻ (P221b)* and *rec⁻ (P221b)* by superinfecting with P22 were examined for formation of infectious centers on *St/22* [40]. Unlike spontaneous and UV induction, superinfection of *rec⁻ (P221b)* with P22 induced about the same number of cells as P22 superinfection of rec^+hcr^+ *(P221b)* and *hcr⁻ (P221b)* (see table 5). Since the superinfection of *rec⁻ (P221b)* with P22 induces P221b, a mechanism for prophage induction may be due to de-struction or neutralization of repressor substance for the prophage P221b by P22. This may be supported by Wing's report [29, 30] that *rec⁻* cell has the machinery necessary for prophage excision since *rec⁻* cells lysogenic for a temperature-sensitive c_2 mutants *(ts c_2)* of P22 were heat inducible.

If in *rec⁺* lysogens there were a Borek–Ryan-like effect [3] that more P221b were indirectly induced by UV-irradiated or PD-treated P22 phage but not in *rec⁻* lysogens, this could account for enhancement of recombination in *rec⁺* lysogens and no enhancement of recombination in *rec⁻* lysogens. As shown in table 5, prophage inductions of both rec^+hcr^+ *(P221b)* and *rec⁻ (P221b)* were similarly enhanced by UV-damaged P22 phage and maximal frequencies of induction were about 55% for rec^+hcr^+ *(P221b)*, 20% for *hcr⁻ (P221b)*, and 35% for *rec⁻ (P221b)*. Therefore, the prophage inducibility can not explain no enhancement of recombination of P22 with the prophage P221b of *rec⁻ (P221b)*. Thus, we consider another possibility that *rec⁻* hosts may contain a repressor (or an inhibitor) for phage recombination. To test this possibility, mixed infection of *rec⁻* strain with P22 and P221b was stud-ied. The total recombination frequency for two markers (c_2 and h_{21}) in *rec⁻* strains was about 1.6% which is about the same as in rec^+hcr^+ hosts. When either rec^+hcr^+ or *rec⁻* strain was mixedly infected with UV (40 sec) irradiated P22c_1 and P221bc_2, about a five-fold increase in the frequency of *c+* recombination over unirradiated phages was observed in both rec^+hcr^+ and *rec⁻* strains. Thus, phage recombination was not repressed by *rec⁻* cells. From these observations, it may be concluded that the bacterial recombination mechanism plays an important role in the recombining the superinfecting phage P22 with the prophage P221b [40].

4.8. *Contribution of a superinfecting phage function to the recombination with the prophage*

When integration minus mutants of P22c_2, P22c_2*int⁻*, were treated with PD or UV and assayed on rec^+hcr^+ *(P221b)*, *hcr⁻ (P221b)* and *rec⁻ (P221b)*, no *c+* plaques were observed on these strains.

Thus, the recombination between superinfecting phage and the prophage P221b must be a consequence of combined action of the *int* function of the superinfecting phage and the bacterial *rec* function.

5. Discussion

5.1. Formation of recombination precursor molecules

The *int* function of the superinfecting phage can excise one of the *att* sites located at the ends of the prophage to form double lysogens with assistance of bacterial *rec* function or the *int* function can excise ('induced') the prophage out.

If the prophage is excised out, the bacterial *rec* function can join the superinfecting phage and the excised (or induced) phage. This would form circular or linear phage dimers (see fig. 8). Alternatively, if double lysogens are formed, they are unstable and may release the circular or linear phage dimers (see fig. 8).

If the superinfecting phage genome of the above dimers 'recombination precursor molecule' is damaged, recombinants of the superinfecting phage should be readily formed.

It should be mentioned that in addition to PD and UV, the frequency of recombination of the superinfecting phage is greatly increased by damage of the phage genome with a variety of agents: hydrogen peroxide [41], nitrogen mustard [36], 4-hydroxyaminoquinoline-1-oxide [44] and β-propiolactone [14]. The stimulation of recombination by a variety of phage inactivating agents suggest that damage of phage genome may be a prerequisite for genetic recombination in nature.

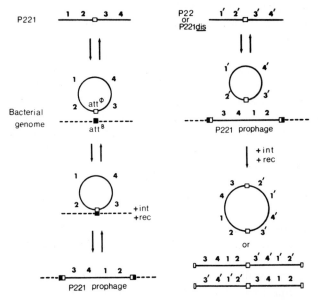

Fig. 8. Mechanisms of prophage integration and formation of recombination precursor dimers. (left): P221 integration according to the Campbell's model; (right): Superinfection and formation of recombination precursor dimers.

5.2. *Interaction between the superinfecting phage and the prophage genomes.*

Interaction between P22 and prophage P221b genomes was also studied by frequency of genomic masked particles [40, 43], carrying P221b genome masked by P22 protein coat, in the superinfection lysates. The method for detection of the genomic masked particles was previously described [43]. As shown in table 5, the majority (about 80%) of P221b genomes released from *rec⁺hcr⁺ (P221b)* or from *hcr⁻ (P221b)* by the superinfection with P22 was masked by P22 protein coat whereas only about 3% of P221b genomes in the superinfection lysate of *rec⁻ (P221b)* was masked by P22 coat. The frequency of the genomic masking in the latter case is comparable to the result obtained by mixed infection with P22 and P221b [43]. This suggests that in *rec⁻ (P221b)* strains induced P221b genomes interact with P22 genomes as in mixed infection. This seems to be supported by no significant UV-enhancement of recombination between the UV-treated superinfecting phage P22 and the prophage of *rec⁻ (P221b)* [40]. Furthermore, when *rec⁺hcr⁺ (P221b)* was superinfected with $P22c_2int^-$, no significant increase in frequency of the genomic masked particles was found.

The above observations suggest that in *rec⁺* hosts the superinfecting phage and the prophage are replicating in the same replicative pool whereas in *rec⁻* hosts they are replicating mainly in separate replicative pools. Therefore it seems likely that the recombination precursor dimer molecules are formed by combined action of the *int* function of the superinfecting phage and the *rec* function in *rec⁺* host, and replicate.

6. Conclusion

Our earlier studies showed that permeability of the bacteriophage protein coat to dyes seemed to be an important factor for photodynamic inactivation of bacteriophage. Therefore, it appeared that damage of the phage DNA is responsible for the photodynamic inactivation.

When clear plaque-forming mutants (c_2) of the phage P22 and $P221dis_1$ were photodynamically inactivated and assayed simultaneously on *Salmonella typhimurium* strains and their lysogenic for P221b, reactivation and greatly increased frequencies of turbid plaques (*c*+) were formed on P221b lysogens whereas no *c*+ plaques were observed on non-lysogens. The maximal *c*+ formation was about 6% for P22 and 40% for P221*dis*, compared with 0.1–0.4% in the untreated controls. Since no *c*+ plaques were found on non-lysogens, the high frequency of *c*+ plaque formation on the lysogenic strains must be a consequence of recombination between the damaged superinfecting phage and prophage P221b. Among *c*+ plaques, a very high frequency of heterozygotes $(c+/c_2)$ plaques were observed. The heterozygote plaque formation may be due to single strand damage by photodynamic action.

If the photodynamically damaged phages were plated on P221b lysogen of *rec⁻* hosts, no significant increase in the frequency of *c*+ plaques was found. Furthermore, when integration minus (*int⁻*) mutants of the phage were photodynamically

treated and assayed on P221b lysogens of wild type hosts, no significant increase in
c+ plaques were found. Thus, the recombination between the superinfecting phage
and the prophage P221b must be a consequence of combined action of the *int* function of the superinfecting phage and the bacterial *rec* function.

The above genetic modification presents the most crucial evidence that photodynamic actions acts directly on the phage DNA.

References

[1] M.H. Adams, Bacteriophages (Interscience, New York, 1959).
[2] H.F. Blum, Photodynamic action and diseases caused by light. (Rheinhold, New York,
    1941).
[3] E. Borek and A. Ryan, The transfer of irradiation-elicited induction in a lysogenic orga-
    nism. Proc. Nat. Acad. Sci. Wash. *44* (1958) 374–377.
[4] R.P. Boyce and P. Howard-Flanders, Release of ultraviolet light-induced thymine dimers
    from DNA in *E. coli* K-12. Proc. Nat. Acad. Sci. Wash. *51* (1964) 293–300.
[5] F.M..Burnet, The classification of dysentery-coli bacteriophages. III. A correlation of the
    serological classification with certain biochemical tests. J. Pathol. Bacteriol. *37* (1963)
    179–184.
[6] A. Campbell, Advan. Genet. *11* (1962) 101–145.
[7] E. Chargaff and J.N. Davidson, The nucleic acids, Vol. 2. (Academic Press, Inc., New York,
    1955) pp. 51–92.
[8] C.E. Clifton, Photodynamic action of certain dyes on the inactivation of staphylococcus
    bacteriophage. Proc. Soc. Exp. Biol. Med. *28* (1931) 745–746.
[9] N.H. Colburn, R.G. Richardson and R.K. Boutwell, Studies of the reaction of β-propiolac-
    tone with deoxyguanine and related compounds. Biochem. Pharmacol. *14* (1965) 113–118.
[10] D.J. Cumming and L.M. Kozloff, Various properties of the head protein of T2 bacterio-
    phage. J. Mol. Biol. *5* (1962) 50–62.
[11] S.A. Ellison, R. Feiner and R. Hill, A host effect on bacteriophage survival after ultraviolet
    irradiation. Virology *11* (1960) 294–296.
[12] H. Endo, M. Ishizawa and T. Kamiya, Induction of bacteriophage formation in lysogenic
    bacteria be a potent carcinogen, 4-nitro-quinoline 1-oxide and its derivatives. Nature *198*
    (1963) 195–196.
[13] D. Fraser and H.R. Mahler, Studies in partially resolved bacteriophage-host systems. VII.
    Diamines, dyes, empty phage leads, and protoplast-infecting agent. Biochim. Biophys.
    Acta *53* (1961) 199–213.
[14] S. Fukuda and N. Yamamoto, Effect of β-propiolactone on *Salmonella typhimurium* and
    bacteriophages. Cancer Res. *30* (1970) 830–833.
[15] W. Harm, Dark repair of acridine dye-sensitized photoeffects in *E. coli* cells and bacterio-
    phages. Biochem. Biophys. Res. Commun. *32* (1968) 350–358.
[16] J. Helprin and C.W. Hiatt, Photosensitization of T2 coliphage with toluidine blue. J. Bac-
    teriol. *77* (1959) 502–505.
[17] C.W. Hiatt, Photodynamic inactivation of viruses. Trans. New York Acad. Sci. *23* (1960)
    66–78.
[18] C.W. Hiatt, E. Kaufman, J.J. Helprin and S. Baron, Inactivation of viruses by the photody-
    namic action of toluidine blue. J. Immunol. *84* (1960) 480–484.
[19] P. Howard-Flanders, R.P. Boyce, E. Simon and L. Therior, A genetic locus in *E. coli* K-12
    that control the reactivation of UV-photoproducts associated with thymine in DNA. Proc.
    Nat. Acad. Sci. *48* (1962) 2109–2115.
[20] A.D. McLaren and D. Shugar, Photochemistry of Proteins and Nucleic Acids (Pergamon,
    New York, 1964).

[21] C. Nagata, M. Kodama, Y. Tagashira and A. Imamura, Interaction of polynuclear aromatic hydrocarbons, 4-nitroquinoline 1-oxide, and various dyes with DNA. Biopolymers *4* (1962) 409–427.

[22] G. Oster and A.D. McLaren, The ultraviolet light and photosensitized inactivation of tobacco mosaic virus. J. Gen. Physiol. *33* (1950) 215–228.

[23] J.R. Perdrau and C. Todd, The photodynamic action of methylene blue on bacteriophage. Proc. Roy. Soc. (London) *112* (1933) 277–287.

[24] J.R. Perdrau and C. Todd, The photodynamic action of methylene blue on certain viruses. Proc. Roy. Soc. (London) *112* (1933) 288–298.

[25] M.I. Simon and H. Van Vunakis, The photodynamic reaction of methylene blue with deoxyribonucleic acid. J. Mol. Biol. *4* (1962) 488–499.

[26] E. Six, The rate of spontaneous lysis of lysogenic bacteria. Virology *7* (1959) 328–346.

[27] T. Sugimura, H. Otake and T. Matsushima, Single strand scissions of DNA caused by a carcinogen, 4-hydroxylaminoquinoline 1-oxide. Nature *218* (1968) 392.

[28] J.N. Welsh and M.H. Adams, Photodynamic inactivation of bacteriophage. J. Bacteriol. *68* (1954) 122–127.

[29] J.P. Wing, Intergration and induction of phage P22 in recombination-deficient mutant of *Salmonella typhimurium.* J. Virol. *2* (1968) 702–709.

[30] J.P. Wing, M. Levine and H.O. Smith, Recombination-deficient mutant of *Salmonella typhimirum.* J. Bacteriol. *95* (1968) 1828–1834.

[31] N. Yamamoto, Photodynamic action on bacteriophage. I. The relation between chemical structure and the photodynamic activity of various dyes. Virus *6* (1956) 510–521.

[32] N. Yamamoto, Photodynamic action on bacteriophage. II. The kinetics of competitive inhibition of various dyes in photodynamic action. Virus *6* (1956) 522–530.

[33] N. Yamamoto, Studies on osmotic shock of bacteriophage. Nagoya J. Med. Sci. *19* (1957) 175–177.

[34] N. Yamamoto, Photodynamic inactivation of bacteriophage and its inhibition. J. Bacteriol. *75* (1958) 443–448.

[35] N. Yamamoto, An unusual hybrid of serologically unrelated phages P22 and P221. Science *143* (1964) 144–145.

[36] N. Yamamoto, Recombination: damage and repair of bacteriophage genome. Biochem. Biophys. Commun. *27* (1967) 263–269.

[37] N. Yamamoto, Photodynamic action on bacteriophage genome: Inactivation and genetic recombination of bacteriophage. Stud. Biophys. *3* (1967) 175–180.

[38] N. Yamamoto, The origin of bacteriophage P221. Virology *33* (1967) 545–547.

[39] N. Yamamoto, Genetic evolution of bacteriophage. I. Hybrids between unrelated bacteriophages P22 and *Fels 2.* Proc. Nat. Acad. Sci. *62* (1969) 63–69.

[40] N. Yamamoto, Damage, repair, and recombination. I. Contribution of the host recombination between the superinfecting phage and the prophage. Virology *38* (1969) 447–456.

[41] N. Yamamoto, Damage, repair, and recombination. II. Effect of hydrogen peroxide on the bacteriophage genome. Virology *38* (1969) 457–463.

[42] N. Yamamoto, Boundary structure between homologous and non-homologous regions in serologically unrelated bacteriophages P22 and P221. II. Physical and genetic structure. In preparation.

[43] N. Yamamoto and T.F. Anderson, Genomic masking and recombination between serologically unrelated phages P22 and P221. Virology *14* (1961) 430–439.

[44] N. Yamamoto, S. Fukuda and H. Takebe, Effects of a potent carcinogen, 4-nitroquinoline 1-oxide and its reduced form 4-hydroxyaminoquinoline 1-oxide on bacterial and bacteriophage genomes. Cancer Res. *30* (1970) 2532–2537.

[45] N. Yamamoto and M.L. Weir, Genetic relationships between serologically unrelated bacteriophages P22 and P221*b*. Virology *28* (1966) 168–169.

CHAPTER 18

PHOTODYNAMIC TAGGING OF POLIOVIRUS WITH PROFLAVINE

C.W. HIATT

*Department of Bioengineering, The University of Texas Medical
School at San Antonio, San Antonio, Texas 78229, USA*

1. Introduction

Following an observation by Schaffer [17] in 1960 to the effect that poliovirus incorporates proflavine into its ribonucleic acid (RNA) if the dye is present during intracellular maturation of the virus, we* speculated that combination with the dye in this way would render poliovirus susceptible to inactivation by exposure to light. If so, this would favor the hypothesis [6] that the resistance of poliovirus (under ordinary conditions) to photosensitizing dyes is due to impermeability of the protein coat, making it impossible for the dye to reach critical sites of combination with the viral nucleic acid. This explanation would be in accord with the earlier conclusion [4, 5, 16] that restricted dye permeation is a rate-controlling factor in photodynamic inactivation of T2 coliphage.

Our preliminary experiments were misleading in that we failed to observe any appreciable photosensitization when poliovirus was propagated in rhesus kidney cell cultures containing 5 μg/ml of proflavine. At this concentration, the dye was toxic to the cells, and virtually no virus multiplication occurred. The harvested virus consisted almost entirely of surviving virus from the inoculum, and any photosensitivity of the progeny was obscured.

Schaffer meanwhile determined [18] that poliovirus containing proflavine was actually quite photosensitive and Crowther and Melnick [3] reported qualitatively similar findings with other photodynamic dyes.

We have confirmed Schaffer's observation of proflavine-induced photosensitivity and have investigated the possible uses of this property as a phenotypic marker in tracer experiments.

---

*Most of the experimental data reported here were obtained by the author in collaboration with Dorothy E. Moore, at the Division of Biologics Standards, National Institutes of Health, Bethesda, Maryland.

2. Experimental results

Monolayer cultures of trypsinsized rhesus monkey kidney cells in medium 199 [15] containing 2% calf serum and 2 μg proflavine per ml were inoculated and incubated at 36°C in total darkness. After three serial passages, the harvested fluids from cultures of poliovirus type 1 (Mahoney) exhibited the photosensitivity shown in fig. 1. Poliovirus type 1 (Sabin) and type 2 (MEF) yielded harvests of virus of comparable photosensitivity, as shown in fig. 2. To obtain the survival curves displayed in these figures, aliquots of the virus were irradiated for graded times between two 350 watt Photoflood lamps at a face-to-face distance of 15 cm. A circulating cold water bath maintained temperature control during irradiation [6].

From an inspection of figs. 1 and 2, it may be seen that irradiation doses of 300 sec or more will result in survival ratios of 10^{-5}. When viewed as a tagging device, proflavine sensitization thus exhibits a very high degree of efficiency, surpassing that normally encountered in radioisotope labelling.

Under ordinary storage conditions, the photosensitivity is exceedingly stable, as

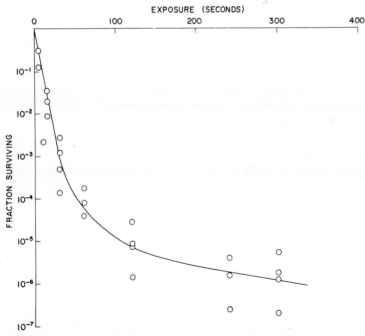

Fig. 1. Survival curve for poliovirus (type 1, Mahoney) sensitized by repeated serial passages in the dark in cell cultures containing 2 μg of proflavine per ml of fluid. Irradiation with polychromatic light as described in the text.

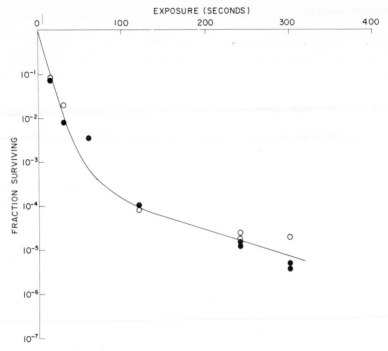

Fig. 2. Survival curves for poliovirus (type 1, Sabin) (●) and (type 2, MEF) (○) sensitized and irradiated under the same conditions described in fig. 1.

shown in table 1, making it possible and convenient to store sensitized virus suspensions in the dark at 4–8°C for prolonged periods for use as laboratory reagents for a variety of experiments.

2.1. *Differential inactivation*

To test the specificity of the proflavine label, we prepared a synthetic mixture

Table 1

Persistence of photosensitivity of poliovirus (type 1, Mahoney) after 1:100 dilution in pH 7.2 buffered salt solution (Hanks)

Time of storage in the dark at 4–8°C	Initial velocity constant at standard light intensity $k(\mathrm{sec}^{-1})$
24 hr	−0.28
(a) 50 days	−0.28
(b) 24 hr	−0.27
(c) 64 days	−0.26

Table 2
Diffusion inactivation of photosensitive poliovirus

| | Plaque-forming units (PFU)/0.2 ml | |
	Type 1[+]	Type 3
Photosensitive Type 1 + normal Type 3	$10^{5.4}$	$10^{2.0}$
Same, after irradiation	$<10^{0.0}$	$10^{2.0}$

[+]Assayed in the presence of Type 3 specific antiserum.

of photosensitive type 1 poliovirus and normal type 3 virus in a concentration ratio of about 2500:1. When this mixture was irradiated for 300 sec and assayed (with and without specific type 3 antiserum) we obtained the results shown in table 2, indicating that the photosensitive virus was reduced in concentration below the limit of detection while no change in concentration of the normal virus was observed.

2.2. *Growth curves*

If photosensitive poliovirus is inoculated into media containing no proflavine, all progeny virus would be expected to be normal, and the only photosensitive virus in the system would consist of particles surviving from the original inoculum. In experiments to confirm this expectation, we inoculated replicate monolayer (rhesus kidney cell) cultures at a multiplicity of 1 (i.e., a ratio of about 1 virus particle per cell), allowed 5 min for penetration, then washed the cell sheet to remove as much extracellular virus as could easily be dislodged. The cultures were then incubated in the dark at 36°C, and aliquot bottles were harvested at graded intervals by alternately freezing and thawing and low speed centrifugation. The supernatant fluid was sampled for virus assay before and after irradiation (in the apparatus previously described) for 300 sec.

A graph of a typical growth curve obtained under these conditions is shown in fig. 3, from which it may be seen that, by invoking the photosensitivity of the parent virus, it is possible to remove it from consideration and view only the new virus produced as a function of time. In the experiment illustrated, the first detectable progeny virus appeared $2\frac{1}{2}$ hr after inoculation and was clearly identified as new virus even though the parent virus surviving after that length of time was 1000 times as abundant.

The ability to detect small quantities of progeny virus in the presence of large quantities of parent virus would greatly facilitate investigations of the host range of poliovirus, or other viruses which could be similarly tagged with photosensitizing dyes. In a simple experiment to confirm the inability of poliovirus (type 1, Mahoney) to grow on rabbit kidney monolayers, we obtained the results displayed in fig. 4. The absence of any detectable progeny virus after six hours of incubation might be taken (by comparison with fig. 3) as unequivocal evidence that rabbit kidney cells

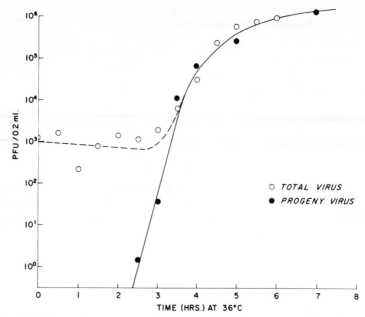

Fig. 3. Growth curve of poliovirus (type 1, Mahoney) in rhesus kidney cell monolayer cultures. After irradiation as described only the photoresistant progeny virus (solid circles) could be found in the culture harvests.

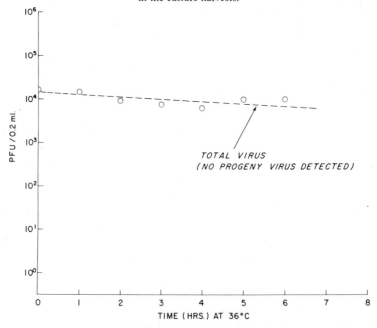

Fig. 4. Results of an attempt to propagate poliovirus on monolayers of primary rabbit kidney cells. Experimental conditions were the same as those described under fig. 3.

will not support the multiplication of poliovirus. When the experiment was extended, however, to longer periods of observation, we encountered small amounts of photo-resistant virus that seemed to signify a low level of multiplication in rabbit kidney cells and other cells of non-primate origin. These observations led us to a more intensive study of the phenomenon [8].

2.3. *Evidence for multiplication of poliovirus in cells of non-primate origin.*

When proflavine-sensitized poliovirus (type 1, Mahoney) was added to *rabbit kidney* monolayers, at a concentration of approximately 1 plaque-forming unit (PFU) per cell and poured off after 5 min at 25°C, the total residual virus recoverable by repeated freezing and thawing of the cell sheet was 10^3 to 10^4 PFU per 0.2 ml, as shown in table 3. In 4 of the 12 trials summarized in this table, a small but measurable virus was recovered after incubation for 24 hr or 48 hr in the dark at 36°C. This photoresistant virus did not increase in titer above the levels shown, and could not be cultivated in subsequent attempts at blind passage in rabbit kidney cells. Spent culture medium from cells of the same age did not increase the photo-sensitivity of the inoculum, so we were led to conclude, at least provisionally, that the photoresistant virus appearing in these cultures was *new* virus and that we were observing a low level of multiplication in supposedly non-susceptible host cells. The photoresistant virus did not invariably occur, and its appearance seemed to be associated with particular batches of cells; the same seed virus was used in all of these experiments. With a virus dose of 0.1 PFU per cell (i.e. by reducing the inoculum to 1/10th of its previous value) no photoresistant virus was detected.

On *chick-embryo* cell cultures the same type of response was obtained, even at 0.1 PFU/cell, as in the trials shown in table 4. In 8 of the 10 trials reported, some photoresistant virus was found after 24 hr.

In monolayers of second generation *mouse-embryo* cells (see table 5), photo-resistant virus was recovered in 7 out of 9 trials, all at an inoculation multiplicity of 0.1. In the second trial shown here, after 48 hr we found photoresistant virus

Table 3
Development of photoresistant virus in monolayer cultures of rabbit kidney

	Residual inoculum[+]	Photoresistant virus[+]		
		0 hr	24 hr	48 hr
4 successful trials	4.2	< 0	1.2	
	4.1	< 0	1.2	1.1
	3.9	< 0	1.4	0.8
	3.4	< 0	0.0	0.0
8 unsuccesful trials	2.0–3.5	< 0	< 0	< 0

[+] Log (PFU/0.2 ml).

Table 4
Development of photoresistant virus in monolayer cultures of chick embryo

	Residual inoculum[+]	Photoresistant virus[+]		
		0 hr	24 hr	48 hr
	2.7	< 0	0.6	0.3
	2.7	< 0	0.8	
	2.6	< 0	0.8	
8 succesful trials		< 0	1.1	
	2.0	< 0	0.4	
	2.8	< 0	0.8	0.5
	2.3	< 0	1.1	1.3
	3.8	< 0	1.1	0
2 unsuccessful trials	2.5, 3.0	< 0	< 0	<0

[+]Log (PFU/0.2).

at a titer of 1.9, which is the highest we have obtained. This is equivalent to 300 PFU in the 6 ml of fluid in the bottle.

The distribution of photoresistant virus among replicate bottles of cell lots in which the overall occurrence of the finding is low, as shown in table 6, is a source of valuable information. In 2 lots of 10 bottles each, photoresistant virus was found in 1 bottle in each lot. The entire contents of each of the negative bottles was tested; no photoresistant virus was found. The yield of about 200 particles in each of the two positive bottles suggests that 2 infective centers were present in a total of 20

Table 5
Development of photoresistant virus in monolayer cultures of mouse embryo cells

	Residual inoculum[+]	Photoresistant virus[+]		
		0 hr	24 hr	48 hr
7 successful trials	2.0	< 0	1.3	1.0
	3.0	< 0	1.4	1.9
	3.3	< 0	1.3	0.8
	2.7	< 0	0.8	0.5
	3.2	< 0	1.0	1.1
	3.1	< 0	< 0	0.1
	2.3	< 0	< 0	0.5
2 unsuccessful trials		< 0	< 0	< 0

[+]Log (PFU/0.2 ml).

Table 6
Distribution of infective centers in replicate bottles of mouse embryo cells

Bottle	New PFU/bottle[+]	Bottle	New PFU/bottle[+]
1A	0	2A	0
B	0	B	0
C	0	C	0
D	180	D	0
E	0	D	0
F	0	F	0
G	0	G	240
H	0	H	0
I	0	I	0
J	0	J	0

[+]Assayed in monkey kidney cells.

Table 7
Evidence that the photoresistant virus was new

1. Analogy with results on money-kidney cells.
2. Stability of the proflavine tag.
3. Occurrence of only a few infective centers.
4. Failure to occur when proflavine is present in the medium.
5. Blockage by fluorophenylalanine (40 μg/ml)

bottles. Each of these bottles contained about 900,000 cells at the time of inoculation. Thus the frequency of infective centers was about 1 in 9 million.

Clearly we were dealing with an event with low probability of occurrence. The event was either *multiplication*, producing virus which was photoresistant because it was new, or *desensitization*, some unknown cell-mediated process which eliminated photosensitivity of the seed virus. Evidence in favor of the belief that the photoresistant virus was new, and represented multiplication, is summarized in table 7. The growth-curve experiments with monkey-kidney cells showed clearly that new virus, when it appears, is readily detected. The demonstrated stability of the proflavine tag argues against desensitization as an explanation. The fact that the photoresistant virus is found in only a few infective centers is strong evidence that multiplication has occurred.

Two additional experiments were performed to elucidate the mechanism. In one of these, proflavine (2 μg/ml) was added to the cell culture medium. No photoresistant virus appeared, indicating that progeny virus, if it was formed, was photosensitive. In the second experiment, fluorophenylalanine, an inhibitor of protein synthesis, was added to the culture medium at a concentration of 40 μg/ml. No photore-

Table 8
Possible explanations for the limited yield of new virus

A. A few particles can infect any of the cells tested.
B. Infectious RNA goes through one cycle.
C. Only a few cells have specific receptor sites.

---

sistant virus developed, suggesting that protein synthesis (presumably synthesis of the protein coat) is essential to the appearance of photoresistant value.

With good reason to believe that low level multiplication did in fact occur in these non-primate cell cultures, we were presented with the three possible explanations summarized in table 8.

The possibility that our inoculum contained a previously undetected mutant virus capable of infecting non-primate cells can be ruled out because the new virus did not participate in more than one cycle of multiplication. It would certainly have continued to propagate itself if it had the inherent capability of doing so.

Since poliovirus ribonucleic acid (RNA) is known [10] to be able to carry out one cycle of multiplication in non-primate cells, it might be supposed that sub-viral infectious RNA in the inoculum resulted in the limited multiplocation observed. On quantitative grounds this possible explanation is not very plausible, since the efficiency of poliovirus RNA is quite low [19] and the amount present in the inoculum would have been very small. In an experiment bearing on this question, we determined that pretreatment of the seed virus with ribonuclease does not prevent the subsequent appearance of photoresistant virus. Thus we believe that sub-viral infectious RNA in the inoculum cannot explain our observations.

The third, and to us the most convincing explanation of our findings, is that a few cells in each of the species tested contain specific receptor sites for poliovirus. When these cells are exploited, no further multiplication can occur, and the infection terminates. Such cells, if they are present, would occur with the frequency of less than 10^{-5}, and could represent rare aberrations or conceivably contaminant cells of primate origin.

2.4. *In vivo tracer experiments*

It was a natural step to extend the proflavine-tagging techniques to studies of poliovirus multiplication in vivo. Shortly after the methodology was first described [7], experiments were undertaken by Black and Miller [1] to test a long-standing hypothesis that poliovirus is capable of low-level multiplication in the common house-fly. Virus recovered from the bodies of flies which had ingested fluids containing proflavine-tagged poliovirus was unchanged in photosensitivity, indicating that multiplication did not occur.

Cords and Holland [2] used proflavine sensitization of poliovirus (type 1) and Coxsackie B1 virus as a means of destroying the residual inoculum in tissues of mice which had been injected with mixed virus. Among the progeny virus particles detected by this method were some poliovirus particles which evidently contained poliovirus RNA enclosed in Coxsackie B1 capsid protein.

In an extensive study of the pathogensis of poliomyelitis in monkeys, Miller and Horstmann [14] fed proflavine-tagged poliovirus (type 1, Mahoney) to cynomolgus monkeys which were then sacrificed after 12, 18, 24 or 72 hr. Homogenates of various organs were assayed for virus content before and after irradiation with visible light. These experiments indicated that new virus appeared in the wall of the pharynx within 12 hr if the monkeys had not been previously immunized. Monkeys which had been immunized with oral vaccine developed no new virus in the pharynx, although small amounts could be found in the feces 24 hr after feeding.

2.5. Studies of intracellular events

By making it relatively easy to detect progeny virus in the presence of abundant particles from the parent generation, the proflavine tagging technique offers some . interesting possibilities for studying the sequence of events in the host cell after infection with poliovirus is initiated. Wilson and Cooper [23], using neutral red as a photosensitizer, instead of proflavine, found that the tagged infecting virus produced an infective center which went through a transient period of light resistance even when the dye was present in the cells.

Schaffer and Hackett [20] using acridine orange, neutral red, or proflavine as the photosensitizer for poliovirus inoculated into cultures of HeLa cells, found that the virus loses its sensitivity to light promptly upon entry into a susceptible cell and attributed this loss of photosensitivity to an opening of the protein coat allowing the bound dye to diffuse away from the viral RNA. Further experiments by Wilson and Cooper [24] seemed to indicate that this simple explanation of the loss of photosensitivity was not sufficient (at least in the case of virus sensitized with neutral red) since the loss of photosensitivity occurred at the same rate when the dye was present in the cell. Under these conditions the viral RNA would not lose its bound dye after removal of the capsid. Consequently some other explanation for the loss of photosensitivity was called for.

Hiatt and Moore [9], using proflavine-tagged poliovirus, demonstrated a total eclipse phase of the virus in rhesus kidney cells and were able to produce cell suspensions containing several thousand infective centers but no recoverable virus after cell disruption. A suggested explanation for the loss of photosensitity after entry of the virus into the cell was that the genetic information was transcribed quickly and therefore protected from degradative effects, photodynamic or otherwise, upon the RNA of the infecting particle.

3. Discussion

The remarkable resistance of mature poliovirus, as ordinarily prepared, to direct sensitization with photodynamically active dyes has been interpreted [6, 18, 3] as an indication that the protein capsid prevents access of dye to the RNA core of the virus, where the critical sites of combination are presumed to be situated. During maturation of the virus particle in its host cell, proflavine, acridine orange, or neutral red, if present within the cell, can combine with the RNA prior to encapsidation [18, 3, 13, 12]. The impenetrability of the viral capsid then appears to operate in both directions, since if effectively 'locks in' the dye and keeps it from disassociating or diffusing away from its sites of combination. The acridine dyes, such as proflavine and acridine orange, in fact have been shown to bind to double-stranded deoxyribonucleic acid (DNA) by intercalation [11] and it is possible that some similar mode of attachment to the single-stranded RNA of poliovirus accounts in part for the efficacy of these dyes as sensitizers. Nevertheless, the high degree of stability associated with proflavine-tagged poliovirus seems to result from a special property of the protein coat, since infectious RNA extracted from poliovirus suspensions loses its photosensitivity readily when it is allowed to equilibrate with diluents containing no dye [20].

Impenetrability of the capsid, however, is not an invariant property of poliovirus, for Wallis and Melnick [21] have demonstrated quite clearly that if mature poliovirus particles are freed of extraneous organic material and maintained at pH 8.0 in phosphate buffer containing proflavine, neutral red, or toluidine blue, the virus will combine with the dye and become photosensitive. The dye—virus complex formed under these conditions readily dissociates when the pH is lowered or the suspension is put into contact with cation exchangers. Stability of the proflavine tag therefore is more limited than originally supposed, and investigators using the technique for tracer experiments must avoid conditions under which poliovirus is deprived of the organic constituents of its normal environment. Interestingly, the presence of organic buffers, such as tris or glycine, is sufficient to prevent direct photosensitization of the virus.

Extension of the proflavine tagging method to other viruses is a matter of obvious interest, and we have seem from the report of Cords and Holland [2] that it works effectively with Coxsackie B1 virus. Other enteroviruses, which have capsids similar to that of poliovirus, would presumably be amenable to the same procedures. Wallis and Melnick [22] in studying a variety of animal viruses, found that echovirus 1, Coxsackie A9 and Coxsackie B3 were closely comparable to poliovirus in their reactivity with photosensitizing dyes, but that certain other viruses such as herpesvirus, measles, reovirus, and influenza could be irreversibly sensitized by direct exposure of the virus to dye under optimal conditions. Certain of these agents perhaps could be tagged directly with a suitable dye and thus made eligible for the variety of interesting applications that have been found for proflavine-tagged poliovirus.

References

[1] F.L. Black, Department of Epidemiology and Microbiology, Yale University School of Medicine (1963). Personal communication of findings in collaboration with D.G. Miller.

[2] C.F. Cords and J.J. Holland, Replication of poliovirus RNA induced by heterologous virus. Proc. Nat. Acad. Sci. *51* (1964) 1080–1082.

[3] D. Crowther and J.L. Melnick, The incorporation of neutral red and acridine orange into developing poliovirus particles making them photosensitive. Virology *14* (1961) 11–21.

[4] D.J. Cummings, Photooxidative elimination of the head form transition in T2 bacteriophage. Virology *19* (1963) 536–541.

[5] J.J. Helprin and C.W. Hiatt, Photosensitization of T2 coliphage with toluidine blue. J. Bacteriol. *77* (1958) 502–505.

[6] C.W. Hiatt, Photodynamic inactivation of viruses. Trans. N.Y. Acad. Sci. *23* (1960) 66–78.

[7] C.W. Hiatt and D. Moore, Tagging of poliovirus by photosensitization with proflavine. Fed. Proc. *21* (1962) 460.

[8] C.W. Hiatt and D. Moore, Evidence for multiplication of poliovirus in tissue cultures of non-primate origin. Fed. Proc. *22* (1963) 558.

[9] C.W. Hiatt and D.E. Moore, Virus-free infective centers produced by photodynamic inactivation of poliovirus in rhesus kidney cell suspensions. Proc. Soc. Exp. Biol. Med. *119* (1965) 203–205.

[10] J.J. Holland and B.H. Hoyer, Early stages of enterovirus infection. Cold Spring Harbor Symp. Quant. Biol. *27* (1962) 101–112.

[11] L.S. Lerman, The structure of the DNA-acridine complex. Proc. Nat. Acad. Sci. *49* (1963) 94–102.

[12] H.D. Mayor, Biophysical studies on viruses using the fluorochrome acridine orange. Prog. Med. Virol. *4* (1962) 70–86.

[13] H.D. Mayor and J.L. Melnick, Intracellular and extracellular reactions of viruses with vital dyes. Yale J. Biol. Med. *34* (1962) 340–358.

[14] D.G. Miller and D.M. Horstmann, The use of proflavine-tagged virus to study the pathogenesis of poliomyelitis in monkeys. Virology *30* (1966) 319–327.

[15] J.F. Morgan, H.J. Morton and R.C. Parker, Nutrition of animal cells in tissue culture. I. Initial studies on a synthetic medium. Proc. Soc. Exp. Biol. Med. *73* (1950) 1–8.

[16] D.A. Ritchie, Mutagenesis with light and proflavine in phage T4. Genet. Res. Camb. *5* (1964) 168–169.

[17] F.L. Schaffer, The nature of the non-infectious particles produced by poliovirus infected tissue cultures treated with proflavine. Fed. Proc. *19* (1960) 405.

[18] F.L. Schaffer, Binding of proflavine by and photoinactivation of poliovirus propagated in the presence of the dye. Virology *18* (1962) 412–425.

[19] F.L. Schaffer, Physical and chemical properties and infectivity of RNA from animal viruses. Cold Spring Harbor Symp. Quant. Biol. *27* (1962) 89–99.

[20] F.L. Schaffer and A.J. Hackett, Early events in poliovirus–HeLa cell interaction: acridine orange photosensitization and detergent extraction. Virology *21* (1963) 124-126.

[21] C. Wallis and J.L. Melnick, Photodynamic inactivation of poliovirus. Virology *21* (1963) 332–341.

[22] C. Wallis and J.L. Melnick, Irreversible photosensitization of viruses. Virology *23* (1964) 520–527.

[23] J.N. Wilson and P.D. Cooper, Photodynamic demonstration of two stages in the growth of poliovirus. Virology *17* (1962) 195–196.

[24] J.N. Wilson and P.D. Cooper, Aspects of the growth of poliovirus as revealed by the photodynamic effects of neutral red and acridine orange. Virology *21* (1963) 135–145.

SECTION IV

CELL REPAIR MECHANISMS OF LIGHT DAMAGE

CHAPTER 19

PHOTOREPAIR OF BIOLOGICAL SYSTEMS

Jane K. SETLOW

Biology Division, Oak Ridge National Laboratory,
Oak Ridge, Tennesse, USA

1. Introduction

Biological systems altered by ultraviolet irradiation may sometimes be partially or completely restored by further irradiation under different conditions, usually at a different wavelength. In all cases where something is known about the mechanism of light-induced repair in complex systems such as cells, it is nucleic acid, usually DNA, which has been found to be directly or indirectly restored by the light treatment. The involvement of DNA in photorepair of a biological system indicates that DNA damage is responsible for the alteration of that biological system by the ultraviolet radiation.

There are three main types of photorepair. The first involves only the interaction of photons and nucleic acid molecules. The best-known example of this type which clearly has a biological effect is called short-wavelength reversal. The other two types are more complex, in that at least one other macromolecule besides DNA is involved. Both are called photoreactivation. Photoreactivation is the reduction in response of a biological system to ultraviolet irradiation by posttreatment with radiation from about 310 to 500 nm. Types of biological properties which have been shown to be photoreactivable are: inactivation of colony formation [30, 31], delay in DNA synthesis [32], division delay [5, 35a], or filament formation [18], mutation [31] and virus infectivity [16]. In the case of viruses, the viral DNA inside the cell is the photoreactivable entity. Photoreactivation which results in DNA repair at the time of the light treatment is called direct photoreactivation. If the actual repair takes place in the dark following the light treatment, the photoreactivation is called indirect.

2. Short-wavelength reversal [73]

A biological system damaged by long-wavelength ultraviolet irradiation (such as at 280 nm), may be reactivated by further irradiation at a shorter wavelength (such as at 240 nm). An example of such a reaction is seen in fig. 1. Purified transforming DNA from cathomycin and streptomycin-resistant *Haemophilus influenzae* has

335

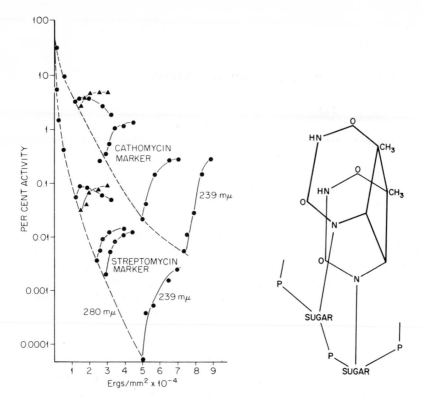

Fig. 1. Inactivation of *Haemophilus influenza* transforming DNA by 280 mµ radiation, ------, followed by reactivation at 239 mµ, ——— [73].

Fig. 2. Schematic diagram of thymine–thymine dimer in a polynucleotide [67a]. Copyright 1966 by the American Association for the Advancement of Science.

been irradiated with various doses at 280 nm, followed by 240 nm irradiation. The ability of this DNA to transform sensitive *H. influenzae* cells to cathomycin or streptomycin resistance was measured. The transforming activity of the DNA is decreased by the 280 nm irradiation, but some of this biological damage is eliminated by further irradiation at 240 nm. Similar results have been obtained with *Bacillus subtilis* DNA [58].

The molecular basis for the biological reactivation observed from such short-wavelength irradiation is that the long-wavelength inactivation was caused by the formation of cyclobutyl-pyrimidine dimers between any of two types of adjacent pyrimidines in the transforming DNA (see fig. 2), and the short-wavelength irradiation splits many of these dimers. The formation of pyrimidine dimers is a reversible reaction (fig. 3), but the wavelength dependencies and the quantum yields of the forward and back reactions are different [5, 29, 67, 69, 71, 84]. At long ultraviolet

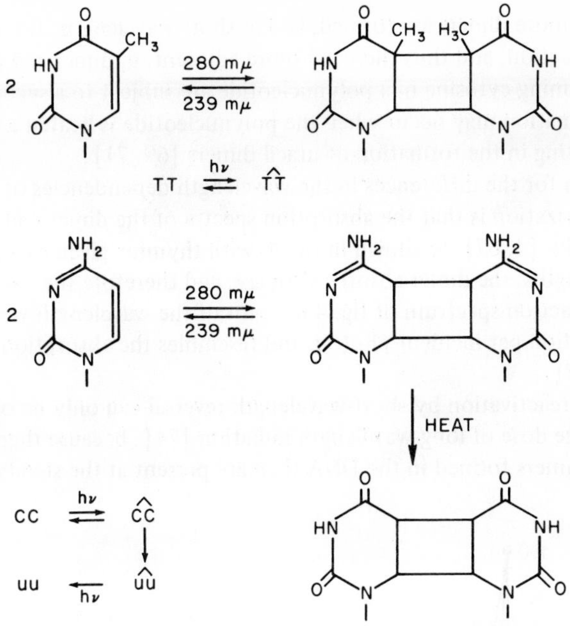

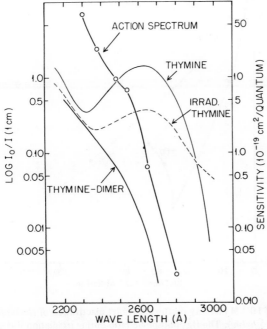

Fig. 3. The dimerization of pyrimidines. T: thymine; C: cytosine; U: uracil.

Fig. 4. The absorption spectra of thymine and thymine dimer and the action spectrum for the monomerization of dimer by ultraviolet radiation [67].

wavelengths, more dimers are formed, and at short wavelengths there is relatively more back reaction, and thus there are more adjacent, undimerized pyrimidines. Dimers containing cytosine in a polynucleotide are subject to a secondary reaction, in that deamination may occur when the polynucleotide is heated after irradiation (fig. 3), resulting in the formation of uracil dimers [69, 71].

The reason for the differences in the wavelength dependencies of dimerization and monomerization is that the absorption spectra of the dimer and the pyrimidine differ markedly [4, 67], as shown in fig. 4, with thymine given as an example. At short wavelengths, the dimer absorbs strongly, and therefore the back reaction is favored. The action spectrum of fig. 4 is a plot of the wavelength versus the rate of the back reaction per incident photon, and resembles the absorption spectrum of the dimer [67].

Biological reactivation by short-wavelength reversal can only be observed following a very large dose of long-wavelength radiation [74], because there must be many more dimers formed in the DNA than are present at the steady state value

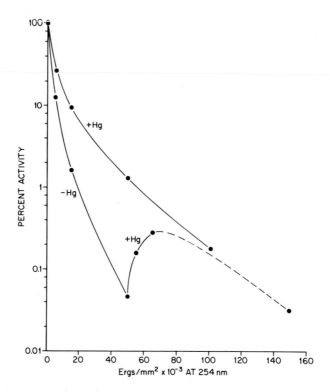

Fig. 5. The effect of 10^{-4} M mercury on the ultraviolet inactivation of *Haemophilus influenzae* transforming DNA (5 μg/ml). The Hg : phosphate ratio during irradiation was greater than 1. The mercury was dialyzed away before the biological assay [40].

for short-wavelength radiation, if the net effect of the short-wavelength irradiation is to *reduce* the fraction of pyrimidine present as dimer. Very few biological systems are suitable, because there must be measurable biological activity after a dose of about 10^4 erg/mm^2 at 280 nm, which sterilizes most bacterial cultures and even virus suspensions. Transforming DNA, however, is sufficiently resistant to ultraviolet radiation to show the short-wavelength reversal effect (fig. 1).

3. Photochemical reactions analogous to short-wavelength reversal

Any agent which decreases the rate of the forward dimer reaction relative to that of the back reaction can have an effect similar to that of short-wavelength reversal. Thus it is possible to reactivate transforming DNA by further irradiation at the same wavelength by addition of such an agent [66]. An example of this type of reactivation [40] is shown in fig. 5. After a large dose at 254 nm to transforming DNA, mercury ions were added to the DNA solution and irradiation was continued, resulting in an *increase* in the biological activity of the DNA. Similarly, proflavine decreases the forward reaction rate [3, 70], apparently by an energy transfer mechanism [78]. At long wavelengths, dimers containing cytosine may be selectively monomerized by irradiation in the presence of proflavine [70]. Use has been made of this phenomenon to show that cytosine-containing dimers are also involved in the inactivation of transforming DNA [66].

4. Evidence that pyrimidine dimers are important in biological inactivation

In order to understand the mechanisms of the various types of photorepair, it is necessary to review some of the evidence that pyrimidine dimers are important contributors to many types of biological damage. The first evidence was the short-wavelength reversal experiment of fig. 1. Pyrimidine dimers in DNA have unique photochemical properties with respect to the kinetics of their formation and monomerization at different wavelengths. The kinetics of biological inactivation and reactivation of transforming DNA are similar to the dimerization-monomerization kinetics. Therefore it has been concluded that pyrimidine dimers cause most of the inactivation [65, 73].

There are other methods of altering the number of pyrimidine dimers in DNA relative to the number of other photoproducts. One method is to change the temperature of irradiation [41]. Such an experiment is shown in fig. 6. Transforming DNA, labeled with tritiated thymidine, was given a single dose of 254 nm radiation at many different temperatures, varying from -180 to $25°$C. The number of dimers formed at each temperature was measured by counting the radioactivity in them (see Chapter 5). The yield of dimers varied greatly over the temperature range used, falling to low levels at the lowest temperatures. The biological inactivation was also measured, as a decrease in the number of transformations resulting from the irradiated DNA. There is a dramatic correlation between the formation of dimers and the

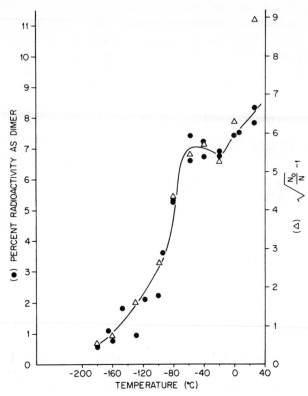

Fig. 6. Thymine-containing dimer formation and inactivation of *Haemophilus influenzae* transforming DNA as a function of temperature during irradiation [41] for a fixed dose of 10^4 erg/mm^2 at 254 nm. The DNA was in a 50% ethylene glycol solution, which forms a clear glass at the lower temperatures. N and N_0 are the number of transformations produced by the irradiated and unirradiated DNA. $\sqrt{(N_0/N - 1)}$ is usually linearly related to dose [52, 64].

biological inactivation as the temperature is varied. Since other known photoproducts do not show this type of temperature dependence [41], this experiment provides strong support for the conclusion concerning the importance of dimers derived from the short-wavelength reversal evidence.

It is not as simple to show the biological importance of a particular DNA photoproduct in cells as it is in transforming DNA. The evidence for the role of pyrimidine dimers in cellular inactivation is indirect. Many cells, bacterial [68] as well as mammalian [45], have been found to have a mechanism for excising pyrimidine dimers from their DNA following ultraviolet irradiation. Mutant cells have been found which exhibit a failure to perform such excision. Examples of such mutants are: *Escherichia coli* strain B_{s-1} [68], *Haemophilus influenzae* strain DB112 [63], and cells from humans suffering from the genetically determined disease xeroderma pigmentosum [72]. The excisionless mutants are more sensitive to the lethal effects of ultraviolet radiation than the related cells which can excise dimers. It can be concluded that such sensitive cells cannot replicate because they cannot deal with the pyrimidine dimers in their DNA, which inhibit DNA synthesis [7, 79].

Information on the mechanism of direct photoreactivation provides further evidence that dimers are important in inactivation of some cells. Since pyrimidine dimers are apparently the only photoproducts eliminated in direct photoreactivation (see below), the inactivation observed must to some extent result from pyrimidine dimers.

5. Direct photoreactivation

Evidence for a particular mechanism for direct photoreactivation began with the discovery by Goodgal, Rupert, and Herriott in 1957 [20] that an extract from *Escherichia coli* plus light could reactivate ultraviolet-inactivated transforming DNA, and that the active agent in the extract was an enzyme. Rupert later found that an extract from baker's yeast was also effective [47]. The yeast photoreactivating enzyme has been studied much more thoroughly than that of *E. coli*.

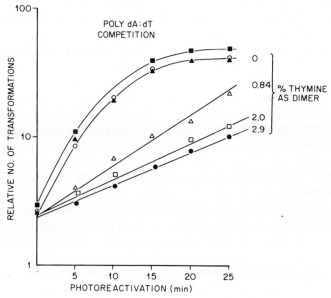

Fig. 7. Photoreactivation of transforming DNA from *Haemophilus influenzae* with yeast photoreactivating enzyme in the presence of competing synthetic polymer [61]. The DNA (0.1 μg/ml) has been exposed to 2400 erg/mm² at 254 nm. Yeast extract was at a concentration of 80 μg/ml protein. Tritium-labeled poly dA:dT* (1 μg/ml) was exposed to 3700 erg/mm² at 254 nm and photoreactivated for 0 (●), 10 (□), 20 (△), and 40 (○) min before addition of the transforming DNA. ■, No poly dA:dT*; ▲, unirradiated poly dA:dT*. The fraction of thymine in the poly dA:dT* converted to dimer was measured as radioactive counts following acid hydrolysis and chromatography to separate thymine from dimer.

Table 1

Dimer formation and competition for yeast photoreactivating enzyme in various irradiated deoxypolynucleotides

	Dimers	Competition
dA	−	−
dI	−	−
d(AT):(AT)	−	−
dC	$\widehat{CC}, \widehat{UU}$	+
dI:dC	$\widehat{CC}, \widehat{UU}$	+
dG:dC	$\widehat{CC}, \widehat{UU}$	+
dT	$\widehat{TT}$	+
dA:DT	$\widehat{TT}$	+
DNA	$\widehat{CC}, \widehat{TT}, \widehat{UT}, \widehat{CT}, \widehat{UU}$	+

The rate of the reaction in the light between the ultraviolet-irradiated transforming DNA and yeast photoreactivating enzyme may be decreased by adding substrates that compete for the enzyme. Certain irradiated polydeoxyribonucleotides act as competitors [48, 61]. An example of such a competition experiment is shown in fig. 7. The synthetic deoxypolymer dA:dT contains adenine in one chain and thymine in the other, and after various treatments is present in the mixture of irradiated transforming DNA and yeast enzyme. For example, the ultraviolet-irradiated polymer has been treated with yeast enzyme in the light for various times, resulting in the monomerization of thymine dimers in the dA:dT, before the transforming DNA was added. The longer the pretreatment of the polymer, the smaller the number of ultraviolet-induced dimers between the adjacent thymine residues, the smaller the number of substrate-competitors, and the faster the reactivation of the transforming DNA. The unirradiated polymer and the maximally photoreactivated polymer do not affect the rate of photoreactivation.

A wide range of synthetic polymers have been used in such experiments, as shown in table 1. Only polydeoxyribonucleotides in which pyrimidine dimers can be formed are able to compete [61]. Thus the alternating polymer d(AT):d(AT) does not compete, and ultraviolet irradiation does not induce dimers in it. Deoxypolymers containing adjacent cytosine residues form cytosine—cytosine dimers which deaminate on heating to form uracil—uracil dimers [69, 71]. All of these dimers in deoxypolymers are good competitors and in all cases the ability of an irradiated polymer to compete may be eliminated by treatment with photoreactivating enzyme plus light, with the same rate at which the pyrimidine dimers disappear from the polymers.

A competition experiment with irradiated oligodeoxynucleotides is shown in fig. 8. This experiment indicates that the irradiated deoxypolymer must be at least nine residues long in order to act as a substrate for the yeast photoreactivating en-

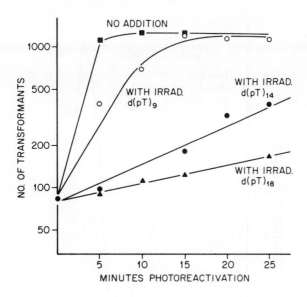

Fig. 8. Competition between *Haemophilus influenzae* transforming DNA (0.025 μg/ml, irradiated with a dose of 2400 erg/mm² at 254 nm) and deoxythymidylates of different lengths
(1.5 μg/ml, irradiated with a dose of 6300 erg/mm² at 254 nm) for yeast photoreactivating enzyme (6 μg/ml protein) [62].

zyme [62]. The oligomer containing five residues showed no competition, while
the eighteen-residue compound competed as well as polydeoxythymidylate containing about 600 residues. Thus the photoreactivating enzyme must recognize a
pyrimidine dimer as part of a deoxypolymer which is not too short. Irradiated ribo-
polymers do not compete [48].

All these experiments indicate that the substrate for the photoreactivating enzyme is a pyrimidine dimer in a polydeoxynucleotide, but do not divulge the chemical nature of the product of the reaction. In the case of the reaction with a polymer
containing radioactively-labeled cytosine as the only pyrimidine (such as poly dC or
dI:dC), the appearance of product may be readily measured. When the polymer is
heated after irradiation, the ultraviolet-induced cytosine dimers are converted to
uracil dimers [71]. The photoreactivating enzyme plus light then causes the appearance of a new component, uracil, in the polymer, which after hydrolysis is readily
separable from the great majority of the radioactive label that is still in cytosine.
Thus the photoreactivating enzyme apparently monomerizes uracil dimers, convert-
ing them to uracil. Demonstration of monomerization of thymine-containing dimers
in DNA is considerably more difficult, because the product, thymine, is indistin-
guishable from the great bulk of the radioactively-labeled thymine in the DNA.
Nevertheless, Cook has elegantly shown that disappearance of thymine-containing
dimers from DNA resulting from treatment with photoreactivating enzyme plus

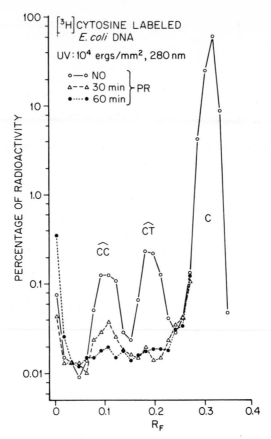

Fig. 9. Chromatogram of formic-acid-hydrolyzed DNA from *Escherichia coli* containing a ^{3}H-cytosine label, and irradiated with a dose of 10^4 erg/mm^2 at 280 nm and photoreactivated with yeast enzyme for 0 (○), 30 (△), and 60 (●) min before hydrolysis [75]. Descending chromatography in n-butanol, acetic acid and water, 80:12:30 by volume [76]. The high temperature of the hydrolysis converts the cytosine-containing dimers (ĈC and ĈT) to uracil-containing dimers, as shown in fig. 3.

light is accompanied by a stoichiometric increase in the amount of thymine in the DNA [10]. The product of the reaction between photoreactivating enzyme and DNA is therefore the same one obtained with short-wavelength reversal, namely residues in monomeric form. Unless the DNA is heated after irradiation, photoreactivation restores the dimerized bases to their original state in the DNA.

It is an important question whether *all* the dimers in DNA can be monomerized by photoreactivating enzyme. This is certainly true of the thymine-containing dimers [85]. Fig. 9 shows the result of an experiment in which DNA with a radioactive label in cytosine has been irradiated and photoreactivated for various times

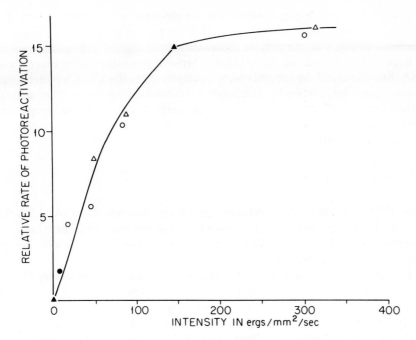

Fig. 10. Rate of photoreactivation versus intensity of illumination at 365 nm [59]. Dilution of yeast extract ● X 2, ○ X 4, △ X 20, and ▲ X 100. The rates were normalized by multiplying them by the ratios of the dilution factors.

[75]. The DNA was then hydrolyzed and chromatographed to separate the uracil dimers from the cytosine (the acid hydrolysis procedure having deaminated the cytosine dimers). The chromatograms show that the cytosine–thymine dimers and all but about 1.0% of the cytosine–cytosine dimers have been monomerized by the reactivating enzyme treatment. It is probable that this does not represent the completed reaction. However, even if it does, the fraction of cytosine–cytosine dimers remaining is considerably more than an order of magnitude too small to account for the inactivation of maximally photoreactivated transforming DNA which represent around ten per cent of the total damage in *Haemophilus influenzae* transforming DNA. The important consequence of these experiments is that most non-photoreactivable damage, at least in transforming DNA, must be non-dimer damage.

If the reaction between irradiated transforming DNA and photoreactivating enzyme is carried on at different radiation intensities, the rate of the reaction (increase in transforming ability of the DNA per unit time of photoreactivating illumination) varies as shown in fig. 10 [59]. At low intensities there is a linear relationship between rate and intensity, which is evidence that only one photon is required for the reaction. At higher intensities, the rate levels off, indicating that some component of the system other than light is limiting the reaction. Kinetic studies of the reaction have shown that this limiting component is the photoreactivating enzyme [49]. It can be seen from the data of fig. 10 that the linear portion of the curve as well the rate varies directly with the enzyme concentration.

Rupert [50] showed that in the reaction between photoreactivating enzyme and DNA the enzyme and the substrate form a complex in the dark, which protects the enzyme against inactivation by such agents as heat and certain metals. Illumination of the mixture of enzyme and substrate results in the dissolution of the complexes, as well as loss of the protection of the enzyme. The reaction scheme indicated by Rupert's experiments is:

$$E + S \underset{k_2}{\overset{k_1}{\rightleftharpoons}} ES \overset{h\nu}{\underset{k_3}{\rightarrow}} E + P,$$

where E is the enzyme, S is the substrate (pyrimidine dimers in the DNA), ES is the complex, P is the product (the set of two adjacent pyrimidines which have been monomerized), and k_1, k_2, and k_3 are the reaction constants. Only the second step in the reaction is light-dependent.

The kinetics of the reaction between enzyme and irradiated transforming DNA have been studied in detail with high-intensity light flashes, around one millisecond in duration [22]. The flash technique makes it possible to get a clear separation of the light-dependent and the light-independent steps, and thus to estimate k_1, k_2 and k_3. A set of complexes was allowed to form in the dark, and the increase in transforming ability resulting from a *single* flash was measured. The amount of photoreactivation from a single flash was dependent on the amount of enzyme relative to the amount of substrate. Increasing the flash intensity did not increase the amount of photoreactivation. Thus the biological effect observed was apparently the result of a photochemical reaction between complexes present at the time of the flash. By flash illumination at various times after mixing the enzyme and substrate, the rate of formation of the complex could be followed. For the case of the dilute reagents used in these experiments, complex formation required times of the order of minutes, which is considered to be due mostly to the time required for diffusion of enzyme and substrate, and therefore is temperature dependent within the range 2–37°C. Variation of the temperature at which the flash was administered, however, did not affect the reaction. Thus, k_3 is temperature-independent in this range.

The time required to form the first set of complexes in the reaction mixture was greater than that required for the next set, suggesting that once an enzyme has found a pyrimidine dimer on a DNA molecule, it can go more quickly to a neighboring dimer. This phenomenon has been interpreted in terms of DNA molecules representing local concentrations of pyrimidine dimers in the reaction mixture and not as sequential repair in which the enzyme slides along the polymer from one dimer to another. Evidence against completely sequential repair of dimers is that at 5°C the amount of repair due to the first three flashes is approximately the same whether the interval between flashes is ten minutes or thirty seconds, whereas the ten-minute interval provides more repair in the case of subsequent flashes. This experiment probably means that some lesions bind enzyme more readily than others. Moreover, the rate of monomerization by yeast enzyme of different kinds of dimers in DNA differs appreciably under conditions of excess of substrate where the dark reaction determines the rate [57 and fig. 9]. Thymine—thymine dimers are monomerized ten times faster than cytosine—cytosine dimers, and twice as fast as thymine—cytosine dimers.

At very low temperatures (down to −196°C), the repair efficiency of the light-dependent reaction drops markedly [21]. This is not because of dissociation or damage to the complexes, but is considered to be due to a change in configuration of the complex.

The enzyme—substrate complexes are not totally stable in the dark, as shown by an experiment in which irradiated transforming DNA was allowed to complex with enzyme and then mixed with either irradiated wild-type DNA or a synthetic deoxypolymer containing pyrimidine dimers [22]. The amount of photoreactivation from a single flash was measured as a function of time after this mixture was made, and was found to decrease somewhat, indicating that enzyme molecules are to a limited extent able to move in the dark from one site to another.

The flash photolysis method has permitted an estimate to be made of the number of active enzyme molecules in a preparation from baker's yeast. The initial number of pyrimidine dimers in the irradiated transforming DNA is known, and the number that are monomerized by a single flash, calculated from the transformation data, should be equal to the number of enzyme molecules available for forming complexes. Harm and Rupert [22] estimate that the proportion of photoreactivating enzyme to total protein in yeast is about 1 in 5×10^5. Purification of the photoreactivating enzyme to this extent should make it possible to observe some absorption by the enzyme or complex in the effective wavelength region for the reaction, with maximum effectiveness at 385 nm [59].

A completely purified enzyme would be highly desirable for a study of enzyme—substrate relations by physico-chemical techniques, since the enzyme—substrate complex could be studied separate from the products of the reaction, simply by keeping the enzyme and the substrate in the dark.

Extensive purification has been carried on by Muhammed [37], and later by the author, but the most highly purified material, estimated to be about as pure as required for complete homogeneity from the Harm and Rupert estimate [22], did not show any absorption in the region of 385 nm, either alone or in the presence of the substrate. This disappointing lack of success is not all understood.

One possible approach to purification which has not so far been used is to start with cells which have an enhanced concentration of the enzyme. Mutants of *Escherichia coli* have been found which have around six times the normal amount of enzyme [23]. Yeast cells in stationary phase contain about an order of magnitude more enzyme than cells in logarithmic growth [6].

6. Indirect photoreactivation

Some bacterial cells which do not contain photoreactivating enzyme may nevertheless be photoreactivated after ultraviolet irradiation [25]. This type of repair differs from direct photoreactivation in that pyrimidine dimers are not monomerized [26]. The amount of dark repair by the excision mechanism or the time at which it occurs relative to DNA initiation or cell division is apparently altered by the photoreactivation, possibly as a result of growth delay [28] which accompanies this type of photoreactivation. Therefore a dark repair system must be present in order for a cell to exhibit indirect photoreactivation. It is not known how the illumination of the cell causes a growth delay, nor is it certain that the growth delay is responsible for the added repair. It has been postulated that quinone destruction causes growth delay [27], but the evidence is so far not very convincing.

Indirect photoreactivation may be readily distinguished from direct photoreactivation by the following criteria: (1) it is relatively independent of the temperature during illumination; (2) the effective wavelength range is restricted to the shorter side of the near ultraviolet spectrum; and (3) the photoreaction is relatively independent of intensity [25].

The existence of indirect photoreactivation makes it dangerous to conclude that pyrimidine dimers are involved in a biological effect in cells on the basis that the effect has been found to be photoreactivable.

7. Distribution of photoreactivating enzyme in nature

Most microorganisms which have been investigated are photoreactivable, and probably most of these contain photoreactivating enzyme. An intriguing generalization has been made by Kelner [33, 34] that bacteria which are able to be transformed are not photoreactivable (see table 2). The only exception known is an unusual type of photoreactivation in non-competent *Bacillus subtilis,* which is oxygen-dependent, and is probably indirect photoreactivation [34]. In the case of two of the microorganisms listed in table 2, *Micrococcus radiodurans* [60] and *Haemophilus influenzae* [20], it has been shown that they do not contain photoreactivating enzyme. It is reasonable to assume that the other transformable species will also be found to be lacking in photoreactivating enzyme.

Table 2
Transformable bacteria which are known to be lacking in direct photoreactivation

Haemophilus influenzae	[20]
Diplococcus pneumoniae	[20]
Bacillus subtilis	[34]
Streptococcus sp.	[2, 8]
Micrococcus radiodurans	36, 60]
Micrococcus luteus (lysodeikticus)	[17, 35, 38]

Table 3

Plants and animals other than bacteria and yeast in which photoreactivating enzyme has been found present

Scientific name	Common name	References
Neurospora crassa	breadmold	[80]
Euglena gracilis		[15a]
Plectonema boryanum	blue-green alga	[83]
Phaseolus vuglaris	pinto bean, white navy bean, great northern bean	[53]
Phaseolus lunatus	large lima bean	[53]
Nicotiana tabacum	tobacco	[81]
Ginkgo biloba	ginkgo	[82]
Paramecium aurelia	paramecium	[77]
Tetrahymena pyriformis		[19]
Arbacia punctulata	sea urchin	[15]
Echinarachnius parma	sand dollar	[46a]
Anagasta kühniella	flower moth	[12]
Gecarcinus lateralis	land crab	[12]
Haemulon sciurus	blue-striped grunt	[12, 42]
Pimephales promelas	fathead minnow	[14]
Bufo marinus	cowflop toad	[12]
Xenopus leavis	African clawed toad	[43]
Rana pipiens	greenfrog	[12]
Terrapene carolina	box turtle	[44]
Iguana iguana	iguana	[44]
Gekko gecko	gekko	[44]
Gallus gallus	chicken	[12]
Didelphis masupialis	North American opossum	[13]
Caluromys derbianus	South American woolly opossum	[13]
Potorous tridactylis	Tasmanian rat-kangaroo (potoroo)	[13]

Plants and animals other than bacteria and yeast in which photoreactivating enzyme has been found absent

Phaseolus aureus	mung bean	[53]
Haplopappus gracilis		[81]
Oryctolagus cuniculus	rabbit	[12]
Mus musculus	mouse	[12]
Rattus novegicus	rat	[12]
Bos taurus	cow	[12]
Homo sapiens	man	[11]

There are mutants of *Escherichia coli* [24] and the yeast *Saccharomyces cerevisiae* [46] which contain no photoreactivating enzyme [46, 51, 56], although indirect photoreactivation may be present [25]. Transformation apparently does not occur as a result of losing the enzymatic activity.

The first metazoan tissue shown to contain photoreactivating enzyme was the egg of the sea urchin *Arbacia punctulata* [15]. Since then the enzyme has been

found to be extraordinarily widespread (table 3). The criteria used for the presence
of enzyme have been the ability of extracts of the organisms plus light (1) to monomerize dimers in ultraviolet-irradiated DNA, and (2) to increase the transforming
ability of irradiated transforming DNA [12]. In the case of some organisms, such as
Paramecia aurelia, the application of these criteria is technically very difficult because of nucleases on the extracts, so that a third criteria has been used, namely the
in vivo monomerization of dimers in the DNA of the organism by illumination [77].
In most of the animals and plants listed in table 3, it has not been shown that the
photoreactivating enzyme contributes to survival in vivo following ultraviolet irradiation.

One of the most curious aspects of the distribution of enzyme shown in table 3
is that placental mammals do not contain the enzyme, although marsupials do.
Cook [11] has pointed out that placental mammals are unique in being the only
class of organism incapable of complete regeneration of complex tissues. He men-
tions that nerve tissue has been shown to be involved in regeneration, and brain is
the vertebrate tissue with the highest photoreactivating activity (including that of
the marsupials). Thus there may be a connection between lack of regeneration and
the lack of photoreactivating enzyme in placental mammals.

The distribution of photoreactivating enzyme within the cell has been studied
in two very different biological systems: the bacterium *Escherichia coli* [9] and
frog liver [12]. There is a strain of *E. coli* which exhibits abnormal division properties, in that in addition to the production of replicas of itself, it also produces cells
one-tenth the normal value which do not contain DNA [1]. These small cells, called
minicells, also do not contain photoreactivating enzyme [9], unlike the normal
cells in the population. These data indicate that in these bacterial cells the photoreactivating enzyme is localized with the DNA. A similar result was obtained from
fractionation of frog liver cells [12], in that most of the photoreactivating activity
was found associated with the nucleus.

8. Speculation on the roles of the photoreactivating enzyme

The information which has been obtained on photoreactivating enzyme distribution may be used in some interesting speculation. The enzyme has remained
functional throughout a long evolutionary process, culminating in the marsupial
[13]. It is present in many tissues and organisms which are opaque to ultraviolet
radiation, and sometimes even to photoreactivating wavelengths. The enzyme is
also present in fish cells which have been in tissue culture for a number of years
[42]. Unessential functions are expected to be lost from such cells, and the photoreactivating activity certainly has not been essential to cell survival, because there
has been no need to monomerize pyrimidine dimers. Therefore one might postulate that the enzyme has an important function in addition to the monomerization
of dimers [12]. The localization of the enzyme with the DNA of cells suggests that
the hypothetical second function concerns DNA. This hypothesis could have an

important bearing on the problem of why transformable strains of bacteria are lacking in the enzyme.

A cell which can undergo transformation is a cell which allows DNA to enter its interior from the surrounding medium at some stage of its growth. In general, the only DNA which can transform the cell is DNA from the same or a related species, although there is little or no specificity with regard to the type of DNA which can enter the competent cell [54]. If this cell contains an enzyme which not only performs an essential function concerned with DNA but also has a special affinity for DNA containing dimers, the entrance of dimer-containing DNA into the cell might be expected to disrupt the essential function of the enzyme. Therefore the loss by the important enzyme of the special affinity for dimers might have survival value for the bacteria that permit DNA to enter them, potentially transformable bacterial species.

It is now widely believed that life began on our planet in shallow water exposed to considerable ultraviolet radiation because of the lack of the protective ozone which exists now [39]. When polynucleotides first began to evolve in these primaeval puddles, there was the possibility of crippling pyrimidine dimers. Therefore it is reasonable that DNA replication and dimer monomerization should be closely allied.

9. Summary

Photorepair can be a reaction between photons and nucleic acid only, or one or more additional macromolecules may be involved. Direct photoreactivation is a photoreaction between a photoreactivating enzyme complexed with ultraviolet-induced pyrimidine dimers in DNA, in which the dimers are monomerized. The photoreactivating enzyme is extraordinarily widespread in nature, being notably absent only in bacterial cells capable of transformation and in placental mammals. The enzyme is localized in cells with the DNA. It is postulated that the enzyme has another function besides monomerization of pyrimidine dimers, and that this function also concerns DNA.

Acknowledgements

This research was sponsored by the United States Atomic Energy Commission under contract with the Union Carbide Corporation.

I am grateful to J.S. Cook, J.M. Boyle and R.B. Setlow for helpful comments on the manuscript.

References

[1] H.I. Adler, W.D. Fisher, A. Cohen and A.A. Hardigree, Miniature *Escherichia coli* cells deficient in DNA. Proc. Nat. Acad. Sci. U.S. *57* (1967) 321–326.

[2] W.D. Bellamy and M.T. Germain, An attempt to photoreactivate ultraviolet inactivated Streptococci. J. Bacteriol. *70* (1955) 351–352.

[3] R. Beukers, The effect of proflavine on UV-induced dimerization of thymine in DNA. Photochem. Photobiol. *4* (1965) 935–937.

[4] R. Beukers and W. Berends, Isolation and identification of the irradiation product of thymine. Biochim. Biophys. Acta *41* (1960) 550–551.

[5] R. Beukers, J. Ylstra and W. Berends, The effect of UV-light on some components of the nucleic acids. V. Reversibility of 'the first irreversible reaction' under special conditions. Rec. Trav. Chim. Pays-Bas *78* (1959) 883–887.

[5a] H.F. Blum, G.M. Loos, J.P. Price and J.C. Robinson, Enhancement by 'visible' light of recovery from ultra-violet irradiation in animals cells. Nature *164* (1949) 1011.

[6] M.E. Boling and J.K. Setlow, Photoreactivating enzyme in logarithmic-phase and stationary-phase yeast cells. Biochim. Biophys. Acta *145* (1962) 502–505.

[7] F.J. Bollum and R.B. Setlow, Ultraviolet inactivation of DNA primer activity. I. Effects of different wavelengths and doses. Biochim. Biophys. Acta *68* (1963) 599–607.

[8] R.M. Bracco, M.R. Krauss, A.S. Roe and C.M. MacLeod, Transformation reactions between pneumococcus and three strains of streptococci. J. Exp. Med. *106* (1963) 247–258.

[9] A. Cohen, W.D. Fisher, R. Curtis III and H.I. Adler, The properties of DNA transferred to minicells during conjugation. Cold Spring Harbor Symp. Quant. Biol. *38* (1968) 635–641.

[10] J.S. Cook, Direct demonstration of the monomerization of thymine-containing dimers in UV-irradiated DNA by yeast photoreactivating enzyme and light. Photochem. Photobiol. *6* (1967) 97–101.

[11] J.S. Cook, Photoreactivation in Animal Cells. In: Photophysiology, ed. A.C. Giese, Vol. 5 (Academic Press, New York, 1970) pp. 191–233.

[12] J.S. Cook and J.R. McGrath, Photoreactivating enzyme activity in metazoa. Proc. Nat. Acad. Sci. US *58* (1967) 1359–1365.

[13] J.S. Cook and J.D. Regan, Photoreactivation and photoreactivating enzyme activity in an order of mammals (Marsupialia). Nature *223* (1969) 1066–1067.

[14] J.S. Cook and J.D. Regan, unpublished experiment.

[15] J.S. Cook and J.K. Setlow, Photoreactivating enzyme in the sea urchin egg. Biochem. Biophys. Res. Commun. *24* (1966) 285–289.

[15a] J. Diamond, J.A. Schiff and A. Kelner, Photoreactivating enzyme from *Euglena gracilis* var. *Bacillaris* and a mutant lacking chloroplast DNA. Plant Physiol. *44* Suppl. (1969) 9.

[16] R. Dulbecco, Reactivation of ultraviolet-induced bacteriophage by visible light. Nature *163* (1949) 949–950.

[17] R.L. Elder and R.F. Beers Jr., Nonphotoreactivating repair of ultraviolet light-damaged *Micrococcus lysodeikticus* cells. J. Bacteriol. *89* (1955) 1225–1230.

[18] M. Errera, Induction of filamentous forms in ultraviolet-irradiated *E. coli* B. Brit. J. Radiol. *27* (1954) 76–80.

[19] A.A. Francis and G.L. Whitson, Ultraviolet-induced pyrimidine dimers in *Tetrahymena pyriformis*. II. In vivo photoreactivation. Biochim. Biophys. Acta *179* (1969) 253–257.

[20] S.H. Goodgal, C.S. Rupert and R.M. Herriott, Photoreactivation of *Hemophilus influenzae* transforming factor for streptomycin resistance by an extract of *Escherichia coli* B. In: The Chemical Basis of Heredity, eds. W.D. McElroy and B. Glass (John Hopkins Press, Baltimore, Maryland, 1957) pp. 341–343.

[21] H. Harm, Analysis of photoenzymatic repair of UV lesions in DNA by single light flashes. III. Comparison of the repair effects at various temperatures between +37°C and −196°C. Mutat. Res. *7* (1969) 261–271.

[22] H. Harm and C.S. Rupert, Analysis of photoenzymatic repair of UV lesions in DNA by single light flashes. I. In vitro studies with *Haemophilus influenzae* transforming DNA and yeast photoreactivating enzyme. Mutat. Res. *6* (1968) 355–370.

[23] W. Harm, Analysis of photoenzymatic repair of UV lesions in DNA by single light flashes. IV. Mutations affecting the number of photoreactivating enzyme molecules in *E. coli* cells. Mutat. Res. *8* (1969) 411–415.

[24] W. Harm and B. Hillebrandt, A non-photoreactivable mutant of *E. coli* B. Photochem. Photobiol. *1* (1962) 271–272.

[25] J. Jagger and R.S. Stafford, Evidence for two mechanisms of photoreactivation in *Escherichia coli* B. Biophys. J. *5* (1965) 75–88.

[26] J. Jagger and R.S. Stafford, Evidence that indirect photoreactivation does not split dimers. Abstract, Pacific Slope Biochemical Conference, Los Angeles, California, Sept. 2–3, 1965.

[27] J. Jagger and H. Takebe, On the possible role of quinones in ultraviolet-induced growth delay and photoprotection in bacteria. Abstract, Fifth International Congress on Photobiology, Hanover, New Hampshire, August 26–31, 1968.

[28] J. Jagger, W.C. Wise and R.S. Stafford, Delay in growth and division induced by near ultraviolet radiation in *Escherichia coli* B and its role in photoprotection and liquid holding recovery. Photochem. Photobiol. *3* (1964) 11–24.

[29] H.E. Johns, S.A. Rapaport and M. Delbrück, Photochemistry of thymine dimers. J. Mol. Biol. *4* (1962) 104–114.

[30] A. Kelner, Effect of visible light on the recovery of *Streptomyces griseus* conidia from ultraviolet irradiation injury. Proc. Nat. Acad. Sci. US *35* (1949) 73–79.

[31] A. Kelner, Photoreactivation of ultraviolet-irradiated *Escherichia coli* with special reference to the dose-reduction principle and to ultraviolet-induced mutation. J. Bacteriol. *58* (1949) 511–522.

[32] A. Kelner, Growth, respiration and nucleic acid synthesis in ultraviolet-irradiated and in photoreactivated *Escherichia coli*. J. Bacteriol. *65* (1953) 252–262.

[33] A. Kelner, Correlation between genetic transformability and nonphotoreactivability in *Bacillus subtilis*. J. Bacteriol. *87* (1964) 1295–1303.

[34] A. Kelner, Nature of photorestoration in the genetically transformable *Bacillus subtilis* Sb-1. Radiat. Res. *25* (1964) 205.

[35] I. Mahler and L. Grossman, Transformation of radiation sensitive strains of *Micrococcus lysodeikticus*. Biochem. Biophys. Res. Commun. *32* (1968) 776–781.

[35a] A. Marshak, Recovery from ultra-violet light-induced delay in cleavage of Arbacia eggs by irradiation with visible light. Biol. Bull. *97* (1949) 315–322.

[36] B.E.B. Moseley and J.K. Setlow, Transformation in *Micrococcus radiodurans* and the ultraviolet sensitivity of its transforming DNA. Proc. Nat. Acad. Sci. US *61* (1968) 176–183.

[37] A. Muhammed, Studies on the yeast photoreactivating enzyme. I. A method for the large scale purification and some properties of the enzyme. J. Biol. Chem. *241* (1966) 516–523.

[38] S. Okubo and H. Nakayama, Evidence of transformation in *Micrococcus lysodeikticus*. Biochem. Biophys. Res. Commun. *32* (1968) 825–830.

[39] A.I. Oparin, A.G. Pasynskii, A.E. Braunshtein and T.E. Pavlovskaya, eds. The Origin of Life on the Earth. Moscow, 1957 (Pergamon, New York, 1959).

[40] R.O. Rahn and J.K. Setlow, unpublished experiment.

[41] R.O. Rahn, J.K. Setlow and J.L. Hosszu, Ultraviolet inactivation and photoproducts of transforming DNA irradiated at low temperatures. Biophys. J. *9* (1969) 510–517.

[42] J.D. Regan and J.S. Cook, Photoreactivation in an established vertebrate cell line. Proc. Nat. Acad. Sci. US *58* (1967) 2274–2279.

[43] J.D. Regan, J.S. Cook and W.H. Lee, Photoreactivation of amphibian cells in culture. J. Cell. Physiol. *71* (1968) 173–176.

[44] J.D. Regan, J.S. Cook and S. Takeda, Reptilian cells in vitro exhibit photoreactivation. Hemic Cells in Vitro, ed. P. Farnes (The Williams and Wilkins Co., Baltimore, 1969) p. 162.

[45] J.D. Regan, J.E. Trosko and W.L. Carrier, Evidence for excision of ultraviolet-induced pyrimidine dimers from the DNA of human cells in vitro. Biophys. J. *8* (1968) 319–325.

[46] M.A. Resnick, A photoreactivationless mutant of *Saccharomyces cerevisiae*. Photochem. Photobiol. *9* (1969) 307–312.

[46a] A.F. Rieck and J.S. Cook, unpublished experiment.

[47] C.S. Rupert, Photoreactivation of transforming DNA by an enzyme from baker's yeast. J. Gen. Physiol. *43* (1960) 573–595.

[48] C.S. Rupert, Repair of ultraviolet damage in cellular DNA. J. Cell Comp. Physiol., Suppl. 1 *58* (1961) 57–68.

[49] C.S. Rupert, Photoenzymatic repair of ultraviolet damage in DNA. I. Kinetics of the reaction. J. Gen. Physiol. *45* (1962) 703–724.

[50] C.S. Rupert, Photoenzymatic repair of ultraviolet damage in DNA. II. Formation of an enzyme-substrate complex. J. Gen. Physiol. *45* (1962) 725–741.

[51] C.S. Rupert, Relation of photoreactivation to photoenzymatic repair of DNA in *Escherichia coli*. Photochem. Photobiol. *4* (1965) 271–275.

[52] C.S. Rupert and S.H. Goodgal, Shape of ultraviolet inactivation curves of transforming deoxyribonucleic acid. Nature *185* (1960) 556–557.

[53] N. Saito and H. Werbin, Evidence for a DNA-photoreactivating enzyme in higher plants. Photochem. Photobiol. *9* (1969) 389–393.

[54] P. Schaeffer, Interspecific reactions in bacterial transformation. Symp. Soc. Exp. Biol. *12* (1958) 60–74.

[55] J.K. Setlow, The wavelength-dependent fraction of biological damage due to thymine dimers and to other types of lesion in ultraviolet-irradiated DNA. Photochem. Photobiol. *2* (1963) 393–399.

[56] J.K. Setlow, Effects of UV on DNA: correlations among biological changes, physical changes and repair mechanisms. Photochem. Photobiol. *3* (1964) 405–413.

[57] J.K. Setlow, Photoreactivation. Radiat. Res., Suppl. *6* (1966) 141–155.

[58] J.K. Setlow, The molecular basis of biological effects of ultraviolet radiation and photoreactivation. In: Current Topics in Radiation Research, Vol. II, eds. M. Ebert and A. Howard (North-Holland, Amsterdam, 1966) pp. 197–248.

[59] J.K. Setlow and M.E. Boling, The action spectrum of an in vitro DNA photoreactivation system. Photochem. Photobiol. *2* (1966) 471–477.

[60] J.K. Setlow and M.E. Boling, unpublished experiment.

[61] J.K. Setlow, M.E. Boling and F.J. Bollum, The chemical nature of photoreactivable lesions in DNA. Proc. Nat. Acad. Sci. US *53* (1965) 1430–1436.

[62] J.K. Setlow and F.J. Bollum, The minimum size of the substrate for yeast photoreactivating enzyme. Biochim. Biophys. Acta *157* (1968) 233–237.

[63] J.K. Setlow, M.L. Randolph, M.E. Boling, A. Mattingly, G. Price and M.P. Gordon, Repair of DNA in *Haemophilus influenza*. II. Excision, repair of single-strand breaks, defects in transformation and host cell modification in UV-sensitive mutants. Cold Spring Harbor Symp. Quant. Biol. *38* (1968) 209–218.

[64] J.K. Setlow and R.B. Setlow, Ultraviolet action spectra of ordered and disordered DNA. Proc. Nat. Acad. Sci. US *47* (1961) 1619–1627.

[65] J.K. Setlow and R.B. Setlow, Nature of the photoreactivable ultra-violet lesion in deoxyribonucleic acid. Nature *197* (1963) 560–562.

[66] J.K. Setlow and R.B. Setlow, Contribution of dimers containing cytosine to ultra-violet inactivation of transforming DNA. Nature *213* (1967) 907–909.

[67] R. Setlow, The action spectrum for the reversal of dimerization of thymine induced by ultraviolet light. Biochum. Biophys. Acta *49* (1961) 237–238.

[67a] R.B. Setlow, Cyclobutane-type pyrimidine dimers in polynucleotides. Science *153* (1966) 379–386.

[68] R.B. Setlow and W.L. Carrier, The disappearance of thymine dimers from DNA: an error-correcting mechanism. Proc. Nat. Acad. Sci. US *51* (1964) 226–231.

[69] R.B. Setlow and W.L. Carrier, Pyrimidine dimers in ultraviolet-irradiated DNA's. J. Mol. Biol. *17* (1966) 237–254.

[70] R.B. Setlow and W.L. Carrier, Formation and destruction of pyrimidine dimers in poly-nucleotides by ultra-violet irradiation in the presence of proflavine. Nature *213* (1967) 906–907.

[71] R.B. Setlow, W.L. Carrier and F.J. Bollum, Cytosine dimers in UV-irradiated poly dI:dC. Proc. Nat. Acad. Sci. US *53* (1965) 1111–1118.

[72] R.B. Setlow, J.D. Regan, J. German and W.L. Carrier, Evidence that xeroderma pigmento-sum cells do not perform the first step in the repair of ultraviolet damage to their DNA. Proc. Nat. Acad. Sci. US *64* (1969) 1035–1041.

[73] R.B. Setlow and J.K. Setlow, Evidence that ultraviolet-induced thymine dimers in DNA cause biological damage. Proc. Nat. Acad. Sci. US *48* (1962) 1250–1257.

[74] R.B. Setlow and J.K. Setlow, The proper use of short-wavelength reversal as a criterion of the importance of pyrimidine dimers in biological inactivation. Photochem. Photobiol. *4* (1965) 939–940.

[75] R.B. Setlow and J.K. Setlow, unpublished experiment.

[76] K.C. Smith, Photochemical reactions of thymine, uracil, uridine, cytosine and bromoura-cil in frozen solution and in dried films. Photochem. Photobiol. *2* (1963) 503–517.

[77] B.M. Sutherland, W.L. Carrier and R.B. Setlow, Photoreactivation in vivo of pyrimidine dimers in paramecium DNA. Science *158* (1967) 1699–1700.

[78] B.M. Sutherland and J.C. Sutherland, Mechanism of inhibition of pyrimidine dimer forma-tion in deoxyribonucleic acid by acridine dyes. Biophys. J. *9* (1969) 292–302.

[79] P.A. Swenson and R.B. Setlow, Effects of ultraviolet radiation on macromolecular syn-thesis in *Escherichia coli*. J. Mol. Biol. *15* (1966) 201–219.

[80] C.E. Terry and J.K. Setlow, Photoreactivating enzyme from *Neurospora crassa*. Photo-chem. Photobiol. *6* (1967) 799–803.

[81] J.E. Trosko and V.H, Mansour, Response of tobacco and Haplopappus cells to ultraviolet irradiation after posttreatment with photoreactivating light. Radiat. Res. *36* (1968) 333–343.

[82] J.E. Trosko and V.H. Mansour, Photoreactivation of ultraviolet light-induced pyrimidine dimers in Ginkgo cells grown in vitro. Mutat. Res. *7* (1969) 120–121.

[83] H. Werbin and C.S. Rupert, Presence of photoreactivating enzyme in blue-green algal cells. Photochem. Photobiol. *7* (1968) 225–230.

[84] D.L. Wulff, Kinetics of thymine photodimerization in DNA. Biophys. J. *3* (1963) 355–362.

[85] D.L. Wulff and C.S. Rupert, Disappearance of thymine photodimer in ultraviolet irradi-ated DNA upon treatment with a photoreactivating enzyme from Baker's yeast. Bio-chem. Biophys. Res. Commun. *7* (1962) 237–240.

CHAPTER 20

DARK REPAIR OF DNA DAMAGE

Kendric C. SMITH

Department of Radiology,
Stanford University School of Medicine,
Stanford, California 94305, USA

1. Introduction

One of the most exciting results to come out of photobiology and radiation biology in recent years is the observation, at the biochemical level, that cells can repair
damage to their DNA. The importance of this observation extends to all areas of
biological science. The genetic control of repair systems has become a subject of intensive investigation. The enzymes involved in repair are being isolated and characterized. The relationship of repair systems to normal life processes in the absence of
radiation is being assessed. The absence of one type of repair process has been correlated with the genetically controlled susceptibility to light-induced skin cancer
(xeroderma pigmentosum) and may have far-reaching implications in the field of
cancer research. A new repair system, controlled by the genes that control genetic
recombination, appears to be the major system by which cells repair X-ray induced
to their DNA. Several agents have been found that inhibit this repair system. The
use of specific inhibitors for this repair system may find application in the radiation
treatment of cancer.

The most recent development in studies on the repair of radiation damage is the
observation that there are at least two dark repair systems that differ both in their
biochemical mechanism and their genetic control, and there may well be other systems yet to be discovered. The repair systems that operate in the light (e.g., photoreactivation) have been discussed by Dr. J. Setlow [51]. This paper will be largely
restricted to those repair systems that do not require light energy to power their
biochemical reactions.

The first indication that cells might have the capacity to recover from radiation
damage was the observation that minor modifications in the handling of the cells
(e.g., growth media, temperature, etc.) had a marked effect upon the ultimate viability of irradiated cells. Thus in 1937, Hollaender and Claus [24] found that higher
survival levels of UV-irradiated fungal spores could be obtained if they were allowed
to remain in water or salt solution for a period of time before plating on nutrient
agar. Roberts and Aldous [45] extended these observations by showing that the

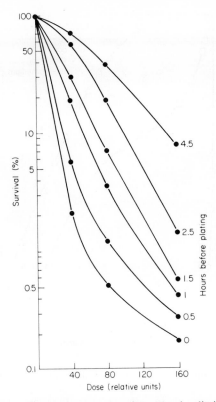

Fig. 1. Ultraviolet radiation survival curves for *E. coli* B. After irradiation the cells were suspended in a liquid medium without an energy source for the times indicated before being plated on nutrient agar. It is evident that both the slopes and the shapes of the survival curves can be altered by the post-irradiation treatment of the cells [45].

shapes of the UV survival curves for *E. coli* B could be changed quite drastically simply by holding the irradiated cells in media devoid of an energy source for various times before plating on nutrient agar (fig. 1). This phenomenon, known as liquid holding recovery, has now been shown to require the presence of intact uvr genes [12], the genes that control the first step in the excision repair system. Thus, holding *E. coli* B (and certain rec strains of *E. coli* K-12 [10]) in non-nutrient media appears to improve the efficiency of the excision repair process.

A second indication of the possible repair of radiation damage came from a study of survival curves. According to classical target theory, a shoulder on a survival curve should indicate either multiple targets or multiple hits on targets. However, very closely related mutants of *E. coli* are not expected to have a markedly different number of targets, yet their survival characteristics are markedly different (fig. 2). The shoulder on the survival curve of the more resistant strains has been

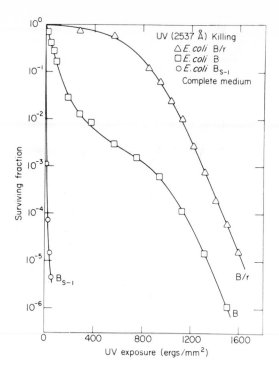

Fig. 2. Ultraviolet radiation survival curves of different mutants of *E. coli* B [21].

reinterpreted as implicating the capacity to repair [22]. The shoulder represents the dose range within which the cells can cope with the damage produced. At higher doses where the survival curve becomes steep, the repair systems have themselves either become inactivated by the radiation or the number of lesions in the DNA exceed the capacity of the repair system to cope with this damage.

Another method for studying or detecting the presence of repair systems in cells is the split-dose technique used by Elkind and co-workers [7, 8]. The rationale for this type of experiment is that if there is no repair of radiation damage then it should have little effect upon the ultimate survival of the cells whether the total radiation dose is given at one time or whether only part of the dose is given at one time and the remainder is given at some later time. However, if the survival of the cells receiving a split dose of radiation is greater than that for cells receiving the same total dose delivered at one time, then it seems reasonable to conclude that the former cells have been able to repair a portion of the first dose of radiation (fig. 3). This split-dose technique has been most widely used with mammalian cells in tissue culture although some work has been done with microorganisms [32].

Perhaps the most conclusive proof of the presence of repair systems in cells is the observation that different mutants of the same strain of bacteria show widely

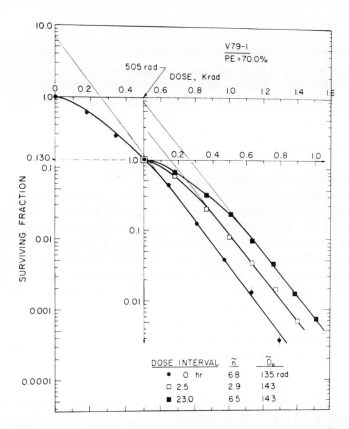

Fig. 3. X-ray survival curves for Chinese hamster cells (V79-1) using fractionated doses. When there was either 2.5 or 23.0 hr between the first dose of 505 rad and the subsequent doses of X rays, the cells were more resistant than if there was no fractionation of the dose. This shift in resistance suggests that the cells have repaired part of the damage produced by the first dose of radiation [7].

differing sensitivities to radiation (fig. 2). The location of several genes that affect the radiation sensitivity of cells have been mapped, and the biochemical deficiencies of several of these mutants have been determined (for reviews see [59, 61, 66]).

Having established that cells have the capacity to recover from radiation damage, we may speculate on the possible molecular mechanism of this recovery. To date, three modes of repair have been documented.

(1) The damaged molecule or part of the molecule may be restored to its functional state in situ. This may be accomplished by the activity of some enzymatic mechanism (e.g., photoreactivation) or it may simply result from the 'decay' of the damage to an inocuous form.

(2) The damaged section of the DNA may be removed and replaced with undamaged nucleotides to restore the normal function of the DNA.

(3) The damage, while not being directly repaired, is either ignored or bypassed and the missing information is supplied by a redundancy of information within the cell.

Before discussing examples of these three modes of recovery, we should discuss some modes of cellular recovery which have little to do with the direct repair of radiation damage. These have been combined under the heading of fortuitous recovery [59].

2. Fortuitous recovery

2.1. *Biologically undetectable damage*

It is sometimes possible for an organism to survive damage in its genetic material without any conscious recognition that the damage is present. The most obvious example would be the case in which the damage is irrelevant to the effect being measured. Grossman [14] has shown that UV damaged cytosine behaves in an RNA polymerase system in vitro as though it were uracil. If the altered base composition of the RNA synthesized on a UV-irradiated template does not result in a lethal mutation, it might even go undetected, since depending on its position in the codon triplet, such a transition might or might not result in an amino acid change in some resultant protein. Furthermore, damage to regions of the genome that are not being actively transcribed may be of little consequence to the organisms until such time as the information in those regions is required.

2.2. *Redundancy of information*

This category includes any inactivation of units of function for which a redundancy exists in the cell. Thus, for example, the inactivation of a few enzyme molecules would have no detectable effect on cell growth if there were still many undamaged enzyme molecules present in the cell.

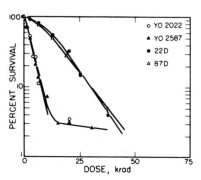

Fig. 4. X-ray survival curves of *Saccharomyces cerevisiae*; haploid cultures YO2022 and YO2587 and homozygous diploid cultures 22D and 87D [37].

2.2.1. *Polyploidy.* The effect of polyploidy on X-ray survival is quite effectively illustrated in a comparison of the survival curves for haploid and diploid strains of yeast (fig. 4). The haploid form initially exhibits the simple exponential survival curve expected for a single target inactivation process, while the diploid strain gives rise to a multi-component inactivation curve with an extrapolation number of roughly 2, consistent with two sensitive targets per cell [37]. It is important to realize that inactivation curves with shoulders may indicate either polyploidy or the presence of repair mechanisms, and it may be quite difficult to ascertain which effect is responsible. The correlation of an apparent polyploidy with cytological observations sometimes can resolve this question.

2.2.2. *Multiplicity reactivation.* This phenomenon, first observed by Luria [34], involves the cooperative effects of UV inactivated bacteriophage to produce some viable phage when the host cell is multiply infected. Multiplicity reactivation has also been demonstrated with animal viruses and it has been speculated that it may even occur between nuclei within uninfected diploid cells. Multiplicity reactivation has also been observed in phage after deleterious treatments other than UV, such as X-rays, nitrous acid, and ^{32}P decay (see Rupert and Harm [47], and references therein). The phenomenon evidently involves genetic recombination in which the random process of molecular rearrangement may result in the production of a viable genome from the undamaged components of otherwise non-viable genomes.

2.2.3. *Cross-reactivation (marker rescue).* The process known as cross-reactivation or marker rescue is essentially the same as the molecular rearrangement aspect of multiplicity reactivation. The bacteria are infected with two genetic types of phage, of which one has been UV-irradiated. Genetic markers from the UV-inactivated phage may be physcially incorporated into the genome of the unirradiated phage. The rescue of genetic markers can be demonstrated even after most of the genetic information 'donor' phage particles has been destroyed by radiation.

2.3. *Suppression of prophage induction*

The UV sensitivity of bacteria may be enhanced by the presence of UV-inducible prophage [47]. Any condition that might inhibit the induction of such a prophage would then fortuitously lead to an increased resistance of the bacteria to irradiation. It is clear that such an apparent recovery factor might bear no relation to the repair of potentially damaging photoproducts in either the prophage or the bacterial genome.

3. Repair of DNA damage in situ

3.1. *Decay of photoproducts*

The simplest mechanism of repair is the one that involves the spontaneous reversion of radiation products to the original undamaged state. Obviously the cell can have little control over this sort of restoration, but environmental conditions can have a great deal to do with it. The hydration products of the pyrimidines are known to revert spontaneously. Also several of the dimeric thymine photoproducts have been shown to be reversed by acid catalysis (for recent reviews on the photochemistry of the nucleic acids see Smith and Hanawalt [59], Setlow [53] and Smith [55]. For radiation products, with a fleeting existence, to express a biological effect it would seem that they must occur just ahead of the replication or transcription enzymes. This explanation would be consistent with the general observation that cells are more sensitive to UV if they are actively growing (e.g., replicating DNA). The *thermal* reactivation of the viability of irradiated cell (for a review see Rupert and Harm [47], may in part involve the increased decay rate of labile radiation products at higher temperatures.

Little more can be said about the relevance of radiation product decay to biological recovery until we understand more about the kinds of radiation products that can revert spontaneously. Nevertheless, one should be aware of this possible mode of recovery, particularly when considering environmental effects on cellular survival.

3.2. *Enzyme catalyzed photoreactivation*

The most thoroughly characterized cellular recovery mechanism is that of enzymatic photoreactivation, in which illumination with visible light facilitates the direct repair in situ of photoproducts produced by UV in DNA. This subject is covered in the report by J. Setlow [51].

4. Reconstruction of damaged DNA

4.1. *Evidence for excision repair*

The studies of R.B. Setlow and co-workers (for a review see [53]), provided the first experimental evidence leading to a model for the excision repair of UV-damaged DNA. Since it was found that the same number of thymine dimers were produced by a given dose of UV in the UV sensitive strain *E. coli* B_{s-1} and the resistant strain *E. coli* B/r, it seemed evident that the resistant strain must somehow be able to remove or bypass these photoproducts in order to exhibit a higher resistance to killing. The mechanism for this recovery was clarified when it was shown that the resistant strain (but not the sensitive strain) released thymine dimers from its DNA during subsequent incubation in the dark after irradiation (fig. 5). Similar results were soon reported by Boyce and Howard-Flanders [2] for resistant and sensitive strains of *E. coli* K-12. A repair mechanism was postulated in which defective regions in one of the two DNA strands could be excised and then subsequently replaced with normal nucleotides, utilizing the complementary base pairing information in the intact strand. This mechanism (fig. 6), which has come to be known colloquially as 'cut and patch', has turned out to be of widespread significance for the repair of a variety of structural defects in DNA. The existence of this mechanism also provides a logical explanation for the evolution of two-stranded DNA, which comprises a redundancy of information.

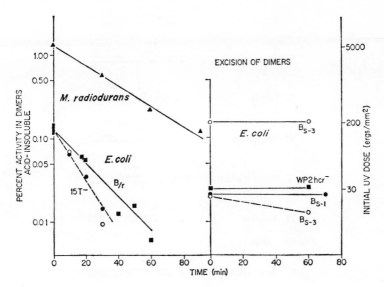

Fig. 5. *Left:* Excision of dimers from the acid-insoluble fraction of ultraviolet resistant cells at various times of incubation in growth medium after irradiation. For *Escherichia coli* 15 T⁻; ○, incubation with thymine; ●, incubation without thymine. *Right:* Lack of extensive excision, even at low doses, in hcr strains of *E. coli*. The initial doses at 265 nm are given on the right-hand ordinate [58].

Direct physical evidence for the repair replication or 'patch' step in the postulated scheme, was provided by the studies of Pettijohn and Hanawalt [40]. These studies began with attempts to isolate partially replicated fragments of bacterial chromosomes by imposing blocks to replication (i.e., UV-induced damage). Replication was followed by the use of the thymine analog, 5-bromouracil (5BU), as a density label in newly synthesized DNA, and by the subsequent analysis of the density distribution of isolated DNA fragments in a cesium chloride density gradient (fig. 7). This is essentially the method developed by Meselson and Stahl [36] and utilized by them to prove that DNA normally replicates semi-conservatively. When 5BU was used to label the DNA synthesized after UV irradiation of *E. coli* strain TAU-bar to 10^{-2} percent survival, the density pattern observed was not as expected for normal semi-conservative replication. Instead of a hybrid density band in the gradient, the initial incorporation of the 5BU label after UV resulted in no detectable shift in density from the normal parental DNA band.

Proof that the early incorporation of 5BU into DNA that results in little or no shift in density of the DNA is the postulated step of repair replication has come from a number of control experiments as follows:

(1) This mode of replication is not observed if bacteria are illuminated with visible light to allow the in situ photoreactivation of pyrimidine dimers prior to 5BU labeling [40].

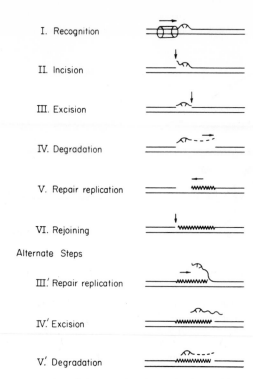

I. Recognition

II. Incision

III. Excision

IV. Degradation

V. Repair replication

VI. Rejoining

Alternate Steps

III.' Repair replication

IV.' Excision

V.' Degradation

Fig. 6. Schematic representation of the postulated steps in the excision repair of damaged DNA. Steps I through VI illustrate the 'cut and patch' sequence. An initial incision in the damaged strand is followed by local degradation before synthesis of the region has begun. In the alternative 'patch and cut' model, resynthesis step III' begins immediately after incision step II and the excision of the damaged region occurs when repair replication is complete. In either model the final step (VI) involves a rejoining of the repaired section to the contiguous DNA of the original strand [59].

(2) It is not observed following UV irradiation of the UV-sensitive strain *E. coli* B_{s-1}, which is unable to perform the excision step in the repair sequence [15].

(3) The non-conservative mode of repair replication can also be demonstrated by the use of D_2O, ^{13}C, and ^{15}N as density labels for newly synthesized DNA to rule out possible artifacts caused by the pathogenicity of 5BU [1, 19].

(4) In low dose experiments (in which viability was as high as 80%) it was demonstrated that DNA which had incorporated 5BU non-conservatively after UV irradiation could then proceed to replicate by the normal semi-conservative model [16].

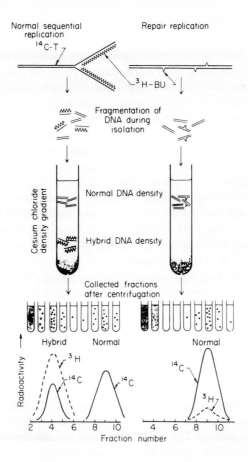

Fig. 7. Protocol for the demonstration of normal replication and repair replication of DNA in growing cells. The DNA is first radioactively labeled (e.g., by the incorporation of ^{14}C-thymine); then the cells are permitted to incorporate another radioactive label at the same time that a 'density label' is being incorporated (e.g., ^{3}H-5-bromouracil). 5-Bromouracil (5BU) is an analog of thymine that can be incorporated into DNA in place of the natural base thymine. Since 5BU is more dense than thymine it has the effect of increasing the density of the DNA fragments that contain it. This density increase is, of course, proportional to the relative amount of thymine and 5BU in the DNA. The density distribution of the isolated DNA fragments is analyzed by means of equilibrium sedimentation in the ultracentrifuge in a density gradient of cesium chloride solution. At equilibrium the DNA fragments will be found in the gradient at positions that correspond to their buoyant densities rather than to their size. This is essentially the method developed by Meselson and Stahl [36] and utilized by them to prove that DNA normally replicates semiconservatively (shown on left half of figure). Parental DNA fragments that contain short regions of repair may differ little in density from those that contain no 5BU (shown on right half of figure) (adapted from Hanawalt and Haynes [18]).

4.2. *The steps in excision repair*

4.2.1. *Recognition.* The first step in the repair process must involve the recognition of the damaged region in the DNA. The photoreactivating enzyme, of course, has been shown to be capable of recognizing pyrimidine dimers. However, unlike the photoreactivation system, the excision repair system is able to recognize a variety of structural defects in DNA which do not involve pyrimidine dimers and which do not result from UV effects. Repair replication has been observed following treatment of bacteria with the bifunctional alkylating agent, nitrogen mustard, which primarily attacks the 7-nitrogen position of guanine [17]. It has also been demonstrated following exposure of bacteria to the powerful mutagen, nitrosoguanidine [5]. Indirect evidence that still other DNA damage can be recognized and repaired comes from the finding that mitomycin C treatment leads to DNA degradation in UV resistant bacteria but not in UV sensitive strains [3]. Thus, it may not be the precise nature of the base damage that is recognized, but rather some associated secondary structural alteration in the phosphodiester backbone of the DNA. The damage recognition step may be formally equivalent to threading the DNA through a close-fitting 'sleeve' that gauges the closeness of fit to the Watson and Crick structure [18].

Enzymes have now been isolated which specifically recognize UV damaged DNA as a template for excision [29, 54, 64], while another enzyme has been isolated which specifically recognizes DNA that has been damaged by methyl methane sulfonate [9, 62].

4.2.2. *Incision.* Following the recognition of damage in DNA, a necessary prerequisite to the excision of the damaged region is the incision or production of a single strand break near the damage. The incision step may precede the excision step, although it has not been ruled out that the two might normally occur as a single enzymatic process.

The incision step has been demonstrated in cell-free extracts of *Micrococcous lysodeikticus* by Rörsch and co-workers [46] in an elegant series of experiments with the double-stranded form of bacteriophage ΦX174. The so-called replicative form of this bacteriophage, when irradiated, can be repaired in spheroplasts of wild-type *E. coli* but not in mutants defective in the recognition (and incision?) step in excision repair. However, a marked increase in biological activity was observed when the damaged DNA was first incubated in an extract of *Micrococcous lysodeikticus* before infection of the excision-defective spheroplasts. Confirmation that the extract was indeed performing the incision step in repair came from studies on the sedimentation behavior (to measure single chain breaks) of untreated and UV-irradiated phage after exposure to the extract. The actual excision of pyrimidine dimers from UV-irradiated DNA by extracts of *M. lysodeikticus* has been shown [4].

4.2.3. *Excision and repair replication.* The process of excision and replacement of damaged nucleotides may occur as separate steps or they may be carried out concurrently with a pealing back of the defective DNA strand. The fact that there is a close coordination between excision and repair replication is indicated by the fact that relative to the number of lesions produced in a cell by a given dose of UV irradiation, there are only a relatively small number of chain breaks present at any one time during the repair process [52]. This certainly rules out the concept that excision takes place throughout the whole genome before repair replication proceeds.

The known specificities of exonuclease III and the DNA polymerase from *E. coli* make these enzymes attractive candidates for excision and repolymerization, respectively. An in vitro model for the 'cut and patch' process was demonstrated in which a portion of one strand of a transforming DNA was degraded with exonuclease III with the concommitant loss in biological activity. The biological activity was subsequently restored by the action of the DNA polymerase [44].

Kelly et al. [31] have recently demonstrated exonucleolytic activity in highly purified *E. coli* DNA polymerase. A single strand break in a double-stranded DNA template is translated along the structure as nucleotides are released from the 5' phosphate end of the template, while the polymerase adds nucleotides to the 3' hydroxyl end. The 5' → 3' exonuclease activity of the polymerase also has the ability to excise mismatched sequences, including pyrimidine dimers, by hydrolyzing phosphodiester bonds in the hydrogen-bonded region on the 3' side of pyrimidine dimers or other distortions in the polynucleotide duplex. Thus, the Kornberg polymerase can perform both the cut and the patch steps in repair replication, requiring only the ligase to close the polynucleotide chain for complete repair. These results, therefore, support the 'patch and cut' model of repair (fig. 6); the simultaneous excision and replacement of nucleotides.

Two lines of evidence support the concept that different enzyme systems are involved in the normal semi-conservative mode and in the repair mode of DNA synthesis in *E. coli*. Firstly, the repair mode of synthesis is essentially unaffected at the restrictive temperature for normal DNA synthesis in temperature sensitive mutants [19]. Such mutants synthesize DNA and grow normally at 35°C but normal replication stops when the temperature is raised to 42°C, presumably because some component in the replicase complex is thermosensitive. Secondly, it has been shown that repair replication exhibits a greater selectivity for thymine over 5BU than does normal DNA synthesis, when both the natural base and its analogue are present in the culture medium [27]. Since it can be presumed that both types of synthesis utilize the same internal pool of nucleotide triphosphate precursors, the repair polymerase seems to have a more stringent requirement for thymine than does the normal polymerase.

4.2.4. *Rejoining.* The excision repair process is completed by the rejoining of the repaired segment to the continuous intact DNA strand to restore the integrity of the two-stranded molecule. Evidence for the occurrence of this step in vivo was found by an examination of the molecular weights of single stranded DNA by sedimentation in alkaline sucrose gradients following the gentle lysis of bacteria on top of the gradient (method of McGrath and Williams [35]). Thus, large single stranded DNA fragments were obtained from unirradiated cells and smaller pieces were seen shortly after irradiation. A subsequent reduction in the number of strand breaks with time during incubation after irradiation could be followed by this method [52]. It has been shown that the breaks occur only in the damaged strand [54].

An enzyme has been isolated that is specific for joining single strand breaks in a double stranded polynucleotide providing the break occurs such that there exists a 5'-phosphate in juxtaposition with a 3'-hydroxyl group [38]. This enzyme also requires a divalent cation (Mg^{++} or Ca^{++}) and DPN. This enzyme rejoins the chains through the formation of 3'—5'-phosphodiester linkages, This enzyme has been shown to repair single chain breaks in DNA produced by pancreatic DNase [67]. An ATP-dependent enzyme system with related activity has been purified from *E. coli* infected with T4 bacteriophage [13, 65]. This enzyme (polynucleotide ligase) may well be the rejoining enzyme responsible for the proposed last step in the excision repair of radiation damage. Certainly, mutants deficient in this enzyme are appreciably more radiation sensitive than wild-type strains [39].

4.3. *Generality of excision repair*

We have already discussed that this repair system is not specific only for UV-induced damage, but can also repair chemical damage resulting from the treatement of cells with alkylating agents and certain antibiotics.

Three genes are known to control the excision step in the repair of UV damage [25]. It is not known whether these three genes specify three different enzymes or whether two of the genes function only to control one enzyme. A cell that is mutant at any one of these three loci is just as UV sensitive as a cell that is mutant in all three [25].

This repair system is found in a wide variety of microorganism and in certain strains of mammalian cells in tissue culture. Excision repair has also been demonstrated in the smallest living cells, the mycoplasma [19]. The presence of a DNA repair mechanism in these cells attests to the general importance of such mechanisms for the maintenance of viability in even the simplest organisms.

The preferential removal of thymine dimers has been observed from three human cell lines (RA, RAX-10 and HeLa) in tissue culture [42] but not in mouse L-cells [33]. Rassmussen and Painter [41] found an 'unscheduled' DNA synthesis stimulated by UV in cultures of HeLa cells. Their analysis of the replicated DNA, using the 5BU density labeling method, has provided support for the interpretation that repair replication occurs in mammalian cells.

In studies on repair replication in human skin fibroblasts, Cleaver [6] made an important discovery that correlates carcinogenesis with defective repair of DNA. Fibroblasts from patients with *xeroderma pigmentosum* were found to exhibit much reduced levels of repair replication. It was suggested that the failure of DNA repair

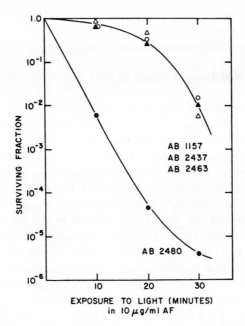

Fig. 8. Survival of mutants of *E. coli* K-12 after exposure to visible light (G.E. 'Daylight' fluorescent lamps) in the presence of acriflavine (AF). AB1157 ('wild-type'), AB2437 (uvrA6), AB2463 (recA13) and AB2480 (uvrA6, recA13) [20].

might be related to the fatal skin cancers that patients with this hereditary disease develop upon exposure to sunlight.

Since bacterial strains that are deficient in the excision process for ultraviolet-induced damage are not particularly sensitive to killing by X-irradiation, it may by hypothesized that either the excision repair system does not play a major role in the repair of X-ray induced damage in the DNA of bacteria or that the incision step is not required for the repair of radiation damage since X-rays themselves produce breaks in the DNA backbone. The repair of X-ray induced damage in DNA will be more thorroughly discussed when we turn to the dark repair system that appears to be controlled by the genes that control genetic recombination.

Numerous investigations have attempted to demonstrate the repairability of the deleterious damage produced when cells are exposed to visible light in the presence of certain dyes (photodynamic action). These attempts have largely proved unsuccessful. However, Harm [20] has recently shown that cells that are deficient in excision of UV damage and cells that are deficient in genetic recombination are just as resistant to the deleterious effects of visible light and acriflavine as is the wild-type strain, whereas the double mutant, deficient both in excision and genetic recombination, is very sensitive to killing by acriflavine and visible light (fig. 8). These results may be interpreted to suggest that the lesions produced by acriflavine and visible light are repaired to about an equal extent both by the excision repair system and the repair system that is controlled by the genes controlling genetic recombination.

5. Biochemical bypass of damaged regions of DNA

5.1. *Evidence for a new dark repair system*

Several lines of evidence suggest that the excision mode of repair is not the only mechanism by which cells repair radiation damage to their DNA in the dark. The first indication was that bacterial cells deficient both in the excision repair mode (uvr) and in genetic recombination (rec) are much more sensitive to killing by UV than are cells carrying either mutation alone (fig. 9). This suggested that certain steps in genetic recombination might be important in the repair of radiation damage [25]. The fact that uvr cells show a large recovery of viability when plated on minimal medium as compared to plating on complex medium ('minimal medium recovery') suggested that these excision deficient cells are able to repair radiation damage [11]. Thirdly, it has been demonstrated that photoproducts such as pyrimidine dimers do not permanently stop DNA synthesis in cells that are deficient in the excision mode of repair [49, 56]. Fourthly, the DNA that is synthesized immediately after UV irradiation in excision deficient cells of *E. coli* K-12 has discontinuties when assayed in alkaline sucrose density gradients. The mean length of the

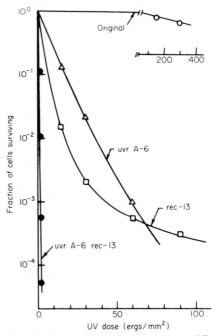

Fig. 9. Ultraviolet radiation survival curves of mutants of *E. coli* K-12 [25].

newly synthesized DNA approximates the distance between pyrimidine dimers in the parental strand. With further incubation of the cells, however, these discontinuities disappear and the DNA approximates the molecular size of that from the unirradiated control cells [49. 26]. A post-replication repair mode is thus indicated which appears to be mediated by some of the enzymes involved in genetic recombination [11, 26, 49].

5.2. Steps in post-replication repair

5.2.1. *DNA synthesis can occur on radiation damaged templates.* The effect of UV radiation upon DNA synthesis kinetics has been recently reinvestigated, and it has now been shown that pyrimidine dimers do *not* permanently inhibit DNA synthesis in cells that are deficient in the excision repair of pyrimidine dimers [49, 56]. Because of the stability of UV-induced pyrimidine dimers in excision-defective mutants, it is possible to investigate the replication of bacterial DNA containing a known number of damaged bases. Rupp and Howard-Flanders [49] have measured the molecular weight of the DNA synthesized upon damaged templates in UV-irradiated bacteria, using the technique of McGrath and Williams [35]. In this technique, bacterial protoplasts containing radioactive DNA are lysed on the top of an alkaline sucrose gradient and then sedimented in an ultracentrifuge. The distance that the DNA moves in the gradient under these conditions is proportional to the molecular weight of the single-stranded pieces of the DNA.

The DNA from excision deficient cells that have been labeled prior to UV irradiation with radioactive thymidine exhibits essentially the same sedimentation characteristics whether the cells are irradiated with UV and immediately banded in the centrifuge or whether they are allowed to incubate in growth medium for about 70 min before banding in the ultracentrifuge (fig. 10). This indicated that the parental DNA in these excision-deficient strains was not broken down nor were a significant number of single chain breaks introduced into the parental DNA during this time period. The parental DNA therefore appears to be a stable template on which DNA can be synthesized [26, 49, 60].

However, DNA synthesized by UV irradiated cells during a 10-min pulse is of lower molecular weight than DNA synthesized during 10 min in unirradiated cells [49]. Fig. 10 shows a typical alkaline sucrose density gradient sedimentation for radioactive DNA from cells that were labeled for 10 min after exposure to 0 or 63 erg/mm^2 of UV at 254 nm. The DNA synthesized during a 10 min labeling pulse in the untreated cells sediments nearly as fast as the parental DNA. In contrast, DNA synthesized after UV irradiation sediments much more slowly, indicating that it is of a lower molecular weight.

The factors that are presently known that affect the sedimentation characteristics of the DNA that is synthesized after UV irradiation are as follows:

(a) The molecular weight of the DNA synthesized after UV is a function of the dose of UV (fig. 11). The higher the dose of UV the slower the sedimentation rate

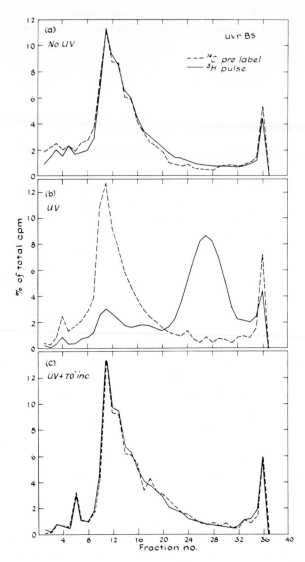

Fig. 10. Sedimentation in alkaline sucrose gradients of DNA labeled before and after UV irradiation. *E. coli* K-12 (uvrB5) were grown for several generations on ^{14}C-thymine and the ^{14}C-
thymine was removed from the medium. The cells were then pulsed with ^{3}H-thymidine for 10
min (a) before or (b) after 63 erg/mm^2 (254 nm) and (c) after 70 min of further incubation in
non-radioactive medium. The cells were protoplasted and lysed on an alkaline sucrose gradient
and spun in rotor SW 50.1 at 30,000 rpm for 105 min at 20°C in a Spinco L2-65B
centrifuge [60].

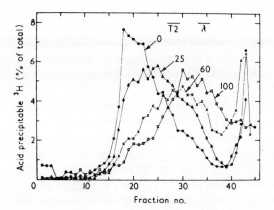

Fig. 11. Effect of several doses of UV on the sedimentation characteristics of the DNA synthesized by *E. coli* K-12 AB2500 uvrA6 after UV irradiation. The conditions are similar to those
described in fig. 10 except that the DNA was not prelabeled. Centrifugation was in a SW 50
rotor for 120 min at 30,000 rpm at 20°C in a Spinco L2 centrifuge. The positions of intact
strands of phages T2 and λ are indicated. The numbers by the arrows refer to dose of UV (254
nm) in erg/mm² [49].

and therefore smaller the molecular weight of the DNA synthesized after UV irradiation [49].

(b) Knowing the number of pyrimidine dimers produced per erg of UV radiation
and the molecular weight of the *E. coli* chromosome, one can calculate the mean
distance between pyrimidine dimers in the *E. coli* chromosome for a given dose of
UV. The average molecular weight of the DNA synthesized after UV irradiation was
found to be in close agreement with the average molecular weight of the pieces of
parental DNA that were present between the pyrimidine dimers. The obvious conclusion therefore was that the DNA was synthesized along the undamaged section
of the parental DNA but the polymerase skipped the section containing the pyrimidine dimers, thus leaving a gap in the daughter strand DNA [49].

(c) When irradiated cells were exposed to visible light under conditions that favor
photoreactivation and then pulsed with radioactive thymidine, it was found that
the size of the pieces synthesized was much larger than for the same cells prior to
the photoreactivation treatment [48, 60]. Since photoreactivation is known to repair pyrimidine dimers in situ [51], it follows that pyrimidine dimers are important
in determining the length of the DNA pieces that are synthesized after UV irradiation.

(d) If a cell that is capable of excision repair is UV irradiated and then pulsed for
10 min with radioactive thymidine, small pieces of DNA are observed to be synthesized as they are in the cells that are excision deficient [48, 60]. This indicates that
the excision repair system does not in itself interfere with the post-replication repair process.

(e) However, if an excision proficient cell is irradiated with UV and, then instead

of pulsing immediately after UV, the cells are allowed to grow for about 60 min
before being pulsed with radioactive thymidine, the newly synthesized DNA is of a
molecular weight that is comparable to that observed for the unirradiated control
cells [60]. This suggests that if excision proficient cells are given sufficient time
they can repair the damage that causes the DNA to be synthesized in short pieces
after UV irradiation.

(f) On the other hand, if excision minus cells are grown for about 90 min after
UV irradiation before being pulse labeled with radioactive thymidine, there is little
difference between the molecular weight of the newly synthesized DNA whether
pulsed immediately after UV or after the 90 min growth period after UV [49, 60].
This indicates that in the absence of the excision repair system, the lesions leading
to the synthesis of small pieces of DNA after UV irradiation are not repaired.

The implication, therefore, is that the DNA synthesized after UV irradiation contains real gaps that are presumed to be opposite the pyrimidine dimers present in
the parental template strands of the DNA [49]. Since the experiments are performed
in strong alkali, it could equally be argued that the newly synthesized DNA contains
some kind of alkaline labile bond opposite each pyrimidine dimer which then is
cleaved when it is placed in the alkaline sucrose gradients. Howard-Flanders and
co-workers [26] have performed one experiment to directly test this hypothesis in
the absence of alkali. Labeled DNA was isolated from control and irradiated cells
with phenol. It was denatured by heating for 5 min at 100°C and then centrifuged
in a neutral sucrose gradient. The pulse labeled DNA from the UV irradiated cells
sedimented more slowly in a neutral sucrose gradient than did the DNA from the
control cells. These authors concluded that it appears unlikely that the low molecular weight chains synthesized upon the damaged template are joined by alkali
labile bonds. However, since the formulation of subsequent steps in the current
working hypothesis for post-replication repair assumes the presence of real gaps in
the newly synthesized DNA, independent confirmation of the presence or absence
of real gaps is desirable.

5.2.2. Repair of defects in the daughter strands of DNA

5.2.2.1. *The chemical nature of post-replication repair.* If instead of assaying the
UV irradiated cells immediately after pulse labeling with radioactive thymidine the
cells are incubated in the presence of non-radioactive thymidine for various periods
of time before sedimentation in alkaline sucrose, the lifetime of the low molecular
weight material synthesized on the damaged template can be investigated. With a
60 min incubation in non-radioactive medium, after the post-radiation pulse, the slow
sedimenting material is converted to a fast sedimenting form that is comparable
in its sedimentation with that of the control DNA (fig. 10).

This increase in molecular weight cannot be due to degradation of the pulse
labeled DNA and a reutilization of the label for the synthesis of high molecular
weight DNA. If this were true, then the radioactivity in the small molecular weight
piece should simply disappear and radioactivity should appear directly in the high

molecular weight fraction. What is observed, however, is that there is a progressive shift in the molecular weight of the pulse labeled material with time towards the molecular weight of the unirradiated DNA [59]. This suggests that the lower molecular weight material is enzymatically joined during incubation into high molecular weight DNA. This change may reflect the action of a genetic recovery mechanism.

5.2.2.2. *Genetic control of post-replication repair*. In the introduction to this section, we listed several lines of evidence that suggest that the excision repair mode is not the only mechanism by which cells repair radiation damage to their DNA. Among these observations was the indication that certain steps in genetic recombination might be important in the repair of radiation damage. If this post-replication repair of radiation damage is mediated by some of the enzymes involved in genetic recombination [11, 26, 49] then one would expect to find recombination deficient mutants unable to repair the discontinuities in the newly synthesized DNA. This hypothesis has been investigated and it has been observed [57] that recA mutants synthesize the small pieces of DNA after UV irradiation but they are unable to repair the discontinuties in the newly synthesized DNA (fig. 12). This result lends support to the hypothesis that this post-replication repair system is mediated by the genes that control genetic recombination. However, it has been observed [58] that two other recombination deficient mutants, namely recB (fig. 13) and recC are capable of repairing the discontinuties in the DNA synthesized after UV irradiation. It is not known at this time whether these results mean that there are steps in this repair process beyond the mere repair of the discontinuties in the DNA or whether the deficiencies of the recB and recC mutants are quantitative rather than qualitative.

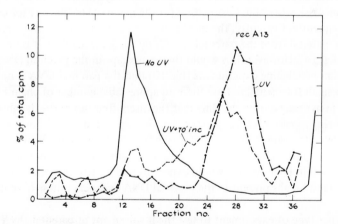

Fig. 12. The absence of post-replication repair in *E. coli* K-12 (rec A13). The conditions are similar to those described in fig. 10 except that the DNA was not prelabeled [60].

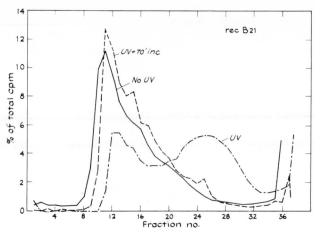

Fig. 13. The presence of post-replication repair in *E. coli* K-12 (recB21). The conditions are similar to those described in fig. 10 except that the DNA was not prelabeled [60].

Experiments in progress in this laboratory are designed to investigate this problem. Specifically, we are attempting to determine if at still higher doses of UV radiation the kinetics of repair of the discontinuities are the same in the rec[+] cells and in the recB and recC cells. If the rates prove to be different then the deficiency would appear to be quantitative rather than qualitative. We already have evidence that this is the case for the repair of single strand breaks produced in these strains by X-irradiation [30]. This point will be discussed more fully below.

5.2.2.3. *The role of parental DNA in filling the gaps in the daughter strands.* If information was not available during the time of DNA synthesis to fill in these gaps then one may ask where the information (or the material) to fill the gaps comes from subsequently. Currently, the most plausible explanation is that the gaps are filled with material from the parental strands by some mechanism of genetic recombination (fig. 14). However, this would then leave gaps in the parental DNA and it appears from prelabeling experiments (fig. 10) that the parental DNA is not broken down into small fragments, nor are there an appreciable number of chain breaks produced in uvr cells during the time that the discontinuities in the daughter strand DNA are being repaired. This suggests either that the parental DNA is not being utilized for filling the gaps in the daughter strands or that simultaneous with the transfer of material to the daughter strand the gaps in the parental strands are repaired by repair replication. Even so, one would expect to find a slight increase in the number of single strand breaks present in the parental DNA and as yet these have not been observed.

In another type of experiment to test the involvement of parental DNA in post-replication repair, excision defective cells were density labeled for several generations in a $^{13}C-^{15}N$ medium containing ^{14}C-thymine. The cells were then transferred

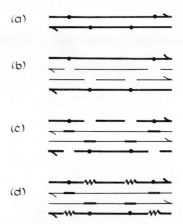

Fig. 14. A model for post-replication repair of UV-damaged DNA. (a) Dots indicate radiation lesions produced in the DNA. (b) DNA synthesis proceeds past the lesions in the parental strands leaving gaps in the daughter strands. (c) Filling of the gaps in the daughter strands with material from the parental strands by a recombinational process. (d) Repair of the gaps in the parental strands by repair replication.

to light medium without radioactive label for 30 min so that the growing point would be incorporating only light label at the time of irradiation. The cells were then exposed to 0, 20 or 50 erg/mm^2 (254 nm) and incubated in the presence of ^{3}H-thymidine in light medium for 30 min. The DNA was isolated and after heat denaturation, the single stranded DNA was centrifuged to equilibrium in a cesium chloride gradient. In the unirradiated cells the heavy ^{14}C-peak and a light ^{3}H-peak were symmetrical and well separated from one another. After irradiation the ^{14}C-peak remained heavy and quite symmetrical whereas the ^{3}H-peak became skewed (as a function of dose) towards the heavy side of the gradient (fig. 15). This indicates that in the UV irradiated cells the newly synthesized DNA has become associated in the same strand with dense label that was synthesized 30 min before irradiation. Rupp and Howard-Flanders [50] have interpreted these results as suggesting that the intermediate density material was produced by recombinational exchanges between sister duplexes, and offer this as a method by which the gaps are repaired in the DNA that is synthesized after UV irradiation.

5.3. *Generality of dark repair controlled by recombination genes*

5.3.1. *Photodynamic action.* As first indicated by Harm [20] and now confirmed by our laboratory [23], the sensitivity to killing by mutants that are deficient in either excision or in recombination are not grossly different from rec$^+$ uvr$^+$ cells in their sensitivity to killing by acriflavine and visible light. However, cells whose genotype is rec uvr are very sensitive to killing by acriflavine and visible light. This sug-

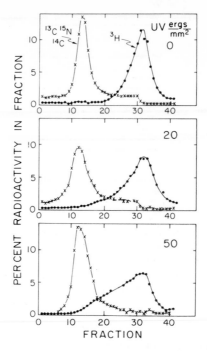

Fig. 15. Evidence for sister strand exchanges in UV irradiated but not in unirradiated *E. coli* K-12. Cells were prelabeled with ^{13}C, ^{15}N and ^{14}C-thymine; grown 30 min in ^{12}C, ^{14}N, non-radioac-tive medium; UV irradiated and then grown in ^{3}H-thymidine for 30 min; DNA isolated and de-natured and banded in CsCl gradients [43].

gests that the excision repair system and the repair system controlled by the recom-bination genes are of about equal efficiency in the repair of the damage produced in bacteria by the action of visible light in the presence of acriflavine. The photo-dynamic lesion that leads to the inactivation of bacteria cells is not known. However, our laboratory is currently investigating the possibility that these lethal lesions may be the cross-linking of DNA and protein, a process that is known to occur with a high efficiency with several different photodynamic dyes [59].

5.3.2. *X-rays.* Cells that are deficient in the excision repair of UV damage are only slightly more sensitive to killing by X-rays than are wild-type cells. However, cells that are deficient in genetic recombination are very easily killed by X-rays. This suggests that the repair of X-ray damage may be mediated by the genes that control genetic recombination. We have attempted to test this hypothesis. The major lesion produced by X-irradiation appears to be the chain break produced in DNA. Single chain breaks have been shown to be repaired in resistant but not sensitive strains of bacteria [35], whereas double chain breaks do not appear to be repaired in *E. coli*

[28]. We have therefore studied the production and the repair of X-ray induced single chain breaks in various mutants of *E. coli* K-12.

The excision deficient cells are able to repair X-ray induced single chain breaks with about the same efficiency as the wild-type cells. The recA mutants appear to be unable to repair single chain breaks. RecB and recC mutants can repair single chain breaks but not with the same efficiency with which the wild-type cells perform this function [30].

We have observed that when cells starved for amino acids for 90 min, to allow them to complete their DNA synthesis, are irradiated with X rays and incubated in the absence of amino acids, no repair of single chain breaks occurs. However, if amino acids are added back to the cultures immediately after X-irradiation, the kinetics of the repair of single chain breaks are comparable to those for cells that have never been deprived of amino acids [D.S. Kapp and K.C. Smith, unpublished observations]. This experiment does not distinguish between the possibility that either active DNA synthesis is required before single chain breaks can be repaired or that some labile protein which is lost in the absence of amino acids is required for the repair of these single chain breaks.

Although it has been reported that hydroxyurea (an inhibitor of DNA synthesis) does not inhibit the repair of X-ray induced single chain breaks in mammalian cells in culture [63] we have found it (and/or an associated impurity) to be a potent inhibitor in bacterial cells [D.S. Kapp and K.C. Smith, unpublished observations]. The addition of impure hydroxyurea not only inhibits the repair of single chain breaks produced in irradiated bacterial cells, but it also has a profound effect upon the survival of the cells after X-irradiation. No such effect of hydroxyurea on the viability of X-irradiated recA-56 cells was observed. Since these cells cannot normally repair single chain breaks, the added insult of an inhibitor of this repair process causes no additional killing of these X-irradiated cells.

If the repair of X-ray induced single chain breaks is mediated by the genes controlling genetic recombination, one may speculate as to the biochemical mechanism by which these single chain breaks are repaired. After UV irradiation, the gaps that need to be repaired are in the newly synthesized daughter strands (fig. 14). The cells therefore contain at least four strands of DNA that can be used for recombinational processes to give one viable genome. In the X-ray case, however, immediately after irradiation the breaks are in the parental strands. One may then ask where the extra DNA required for recombinational events can come from. This may come from additional nuclei within the cells since it has been shown that diploid yeast cells are more resistant to X-ray inactivation than are haploid cells [37].

6. Conclusions

Two dark repair systems are now known, differing both in their biochemical mechanism and in their genetic control.

(1) The excision repair system is controlled by the uvr genes. The steps envisioned in this system are those of recognition of the damaged region in the DNA, the cutting out of the damaged region, the patching of the hole by polymerase action, and the subsequent linking of the repaired section by the polynucleotide ligase. This repair system is not specific for UV damage, but has also been shown to repair certain kinds of chemical damage to bacterial DNA. The excision repair system appears to be of about equal importance with the repair system mediated by the genes that control genetic recombination for the elimination of UV damage in bacterial cells. Using the effects of acriflavine as an example of the photodynamic inactivation of bacterial cells, one may conclude that excision repair and recombinational repair are of about equal importance for repairing photodynamic lesions. The excision repair system appears to be of minor importance for the repair of X-ray damage to bacteria, while the recombinational repair system seems to be of extreme importance.

(2) Post-replication repair appears to be controlled by the genes that control genetic recombination. Growth conditions that favor DNA synthesis are required for this repair system to express itself. Thus, after UV irradiation, it appears that DNA synthesis proceeds past lesions in the parental DNA leaving gaps in the daughter strands opposite these lesions. Upon subsequent incubation, these gaps in the daughter strands are repaired by recombinational events whose mechanisms are yet unknown. If parental DNA is involved in the filling of the gaps in the daughter strands, then it is done by a very efficient process that also includes the filling of the gaps left in the parental strands by this process.

We should be cautioned by the fact that several years ago many people thought that the excision repair system could explain all of the repair phenomena in radiation biology. With further experimentation, however, the new dark repair system mediated by the genes that control genetic recombination has been discovered. Still other mechanisms of repair may yet be discovered.

Addendum

Since this talk was presented (September 1969), significant progress has been made in elucidating several systems for the repair of X-ray-induced DNA single-chain breaks. While rec-gene-controlled repair takes about 40 min in growth medium, a new repair system that takes only about 5 min in buffer gas been discovered which requires the action of the Kornberg DNA polymerase. There is also preliminary evidence for an ultrafast system (less than 2 min in buffer; ligase only?) for the repair of X-ray-induced single-chain breaks.

These recombination and DNA polymerase controlled repair systems have recently been reviewed (K.C. Smith in Photophysiology 6 (1971) 209). In addition, the irreversible inhibition of the rec repair system by certain drugs is described.

References

[1] D.R. Billen, R. Hewitt, T. Lapthisophon and P.M. Achey, J. Bacteriol. *94* (1967) 1538.

[2] R.P. Boyce and P. Howard-Flanders, Proc. Nat. Acad. Sci. US, *51* (1964) 293.

[3] R.P. Boyce and P. Howard-Flanders, Z. Vererbungsl. *95* (1964) 345.

[4] W.L. Carrier and R.B. Setlow, Biochim. Biophys. Acta *129* (1966) 318.

[5] E. Cerdá-Olmedo and P. Hanawalt, Mutation Res. *4* (1967) 369.

[6] J.E. Cleaver, Nature *218* (1968) 652.

[7] M.M. Elkind, in: Radiation Research, ed. G. Silini (North-Holland, Amsterdam, 1967) p. 558.

[8] M.M. Elkind and W.K. Sinclair, in: Current Topics in Radiation Research, Vol. 1, eds. M. Ebert and A. Howard (North-Holland, Amsterdam, 1965) p. 165.

[9] E.C. Friedberg and D.A. Goldthwait, Cold Spring Harbor Symp. Quant. Biol. *33* (1968) 271.

[10] A.K. Ganesan and K.C. Smith, J. Bacteriol. *96* (1968) 365.

[11] A.K. Ganesan and K.C. Smith, Cold Spring Harbor Symp. Quant. Biol. *33* (1968) 235.

[12] A.K. Ganesan and K.C. Smith, J. Bacteriol. *97* (1969) 1129.

[13] M.L. Gefter, A. Becker and J. Hurwitz, Proc. Natl Acad. Sci. US *58* (1967) 240.

[14] L. Grossman, Photochem. Photobiol. *7* (1968) 727.

[15] P.C. Hanawalt, in: Recent Progress in Photobiology, ed. E.J. Bowen (Blackwell, Oxford, 1965) p. 82.

[16] P.C. Hanawalt, Nature *214* (1967) 269.

[17] P. Hanawalt and R.H. Haynes, Biochem. Biophys. Res. Commun. *19* (1965) 462.

[18] P.C. Hanawalt and R.H. Haynes, Scientific American *216* (1967) 36.

[19] P.C. Hanawalt, D.E. Pettijohn, E.C. Pauling, C.F. Brunk, D.W. Smith, L.C. Kanner and J.L. Couch, Cold Spring Harbor Symp. Quant. Biol. *33* (1968) 187.

[20] W. Harm, Biochem. Biophys. Res. Commun. *32* (1968) 350.

[21] R.H. Haynes, Photochem. Photobiol. *3* (1964) 429.

[22] R.H. Haynes, in: Physical Processes in Radiation Biology, eds. L. Augenstein, R. Mason and B. Rosenberg (Academic Press, New York, 1964) p. 51.

[23] O. Hidalgo-Salvatierra and K.C. Smith, unpublished observations (1969).

[24] A. Hollaender and W. Claus, Bull. Nat. Res. Council, Nat. Acad. Sci. US *100* (1937) 75.

[25] P. Howard-Flanders and R.P. Boyce, Radiation Res. Suppl. *6* (1966) 156.

[26] P. Howard-Flanders, W.D. Rupp, B.M. Wilkins and R.S. Cole, Cold Spring Harbor Symp. Quant. Biol. *33* (1968) 195.

[27] L. Kanner and P.C. Hanawalt, Biochim. Biophys. Acta *157* (1968) 532.

[28] H.S. Kaplan, Proc. Nat. Acad. Sci. US *55* (1966) 1442.

[29] J.C. Kaplan, S.R. Kushner and L. Grossman, Proc. Nat. Acad. Sci. US 63 (1969) 144.

[30] D.S. Kapp and K.C. Smith, Radiation Res. *42* (1970) 34.

[31] R.B. Kelly, M.R. Atkinson, J.A. Huberman and A. Kornberg, Nature *224* (1969) 495.

[32] J. Kiefer, Photochem. Photobiol. *11* (1970) 37.

[33] M. Klimek, Neoplasma *12* (1965) 599.

[34] S.E. Luria, Proc. Nat. Acad. Sci. US *33* (1947) 253.

[35] R.A. McGrath and R.W. Williams, Nature *212* (1966) 534.

[36] M. Meselson and F.W. Stahl, Proc. Nat. Acad. Sci. US *44* (1958) 671.

[37] R.K. Mortimer, Radiation Res. *9* (1958) 312.

[38] B. Olivera and I.R. Lehman, Proc. Nat. Acad. Sci. US *57* (1967) 1426.

[39] C. Pauling and L. Hamm, Proc. Nat. Acad. Sci. US *60* (1968) 1495.

[40] D. Pettijohn and P. Hanawalt, J. Mol. Biol. *9* (1964) 395.

[41] R.E. Rasmussen and R.B. Painter, J. Cell Biol. *29* (1966) 11.

[42] J.D. Regan, J.E. Trosko and W.L. Carrier, Biophys. J. *8* (1968) 319.

[43] D. Reno and P. Howard-Flanders, (1969) unpublished observations kindly made available to this author.
[44] C. Richardson, R. Inman and A. Kornberg, J. Mol. Biol. 9 (1964) 46.
[45] R.B. Roberts and E. Aldous, J. Bacteriol. 57 (1949) 363.
[46] A. Rörsch, P. van de Putte, I.E. Mattern and H. Zwenk, in: Radiation Research, ed. G. Silini (North-Holland, Amsterdam, 1967) p. 771.
[47] C.S. Rupert and W. Harm, Advan. Radiat. Biol. 2 (1966) 1.
[48] W.D. Rupp, Abstracts of the Fifth International Congress on Photobiology (1968) p. 143.
[49] W.D. Rupp and P. Howard-Flanders, J. Mol. Biol. 31 (1968) 291.
[50] W.D. Rupp and P. Howard-Flanders, Abstracts of the Biophysical Society Meeting (1969) A-20.
[51] J.K. Setlow (1972) chapter 19 of these proceedings.
[52] R.B. Setlow, in: Radiation Research, ed. G. Silini (North-Holland, Amsterdam, 1967) p. 525.
[53] R.B. Setlow, Prog. Nucleic Acid Res. Mol. Biol. 8 (1968) 257.
[54] R.B. Setlow, W.L. Carrier and J.K. Setlow, Biophys. J. 9 (1969) A-57.
[55] K.C. Smith, Radiation Res. Suppl. 6 (1966) 54.
[56] K.C. Smith, Mutation Res. 8 (1969) 481.
[57] K.C. Smith and A.K. Ganesan, Abstracts of the Biophysical Society Meeting (1969) A-20.
[58] K.C. Smith and A.K. Ganesan, Abstracts of the Pacific Slope Biochemical Conference (1969) p. 13.
[59] K.C. Smith and P.C. Hanawalt, Molecular Photobiology: Inactivation and Recovery (Academic Press, New York, 1969).
[60] K.C. Smith and D.H.C. Meun, J. Mol. Biol. 51 (1970) 459.
[61] B.S. Strauss, Current Topics in Microbiology and Immunology 44 (1968) 1.
[62] B. Strauss, M. Coyle and M. Robbins, Cold Spring Harbor Symp. Quant. Biol. 33 (1968) 277.
[63] S. Swada and S. Okada, Abstracts of the Biophysical Society Meeting (1969) A-201.
[64] Y. Takagi, M. Sekiguchi, S. Okubo, N. Nakayama, K. Shimada, S. Yasuda, T. Nishimoto and H. Yoshihara, Cold Spring Harbor Symp. Quant. Biol. 33 (1968) 219.
[65] B. Weiss and C.C. Richardson, Proc. Nat. Acad. Sci. US 57 (1967) 1021.
[66] E.M. Witkin, Brookhaven Symp. Biol. 20 (1968) 17.
[67] S.B. Zimmerman, J.W. Little, C.K. Oshinsky and M. Gellert, Proc. Nat. Acad. Sci. US 57 (1967) 1841.

CHAPTER 21

GROWTH DELAY AND PHOTOPROTECTION INDUCED BY NEAR-ULTRAVIOLET LIGHT*

John JAGGER

Division of Biology, University of Texas at Dallas,
Dallas, Texas 75230, USA

1. Introduction

Ultraviolet radiation lies just beyond the violet end of the visible spectrum (fig. 1). It is convenient to divide the region into those wavelengths that lie above 300 nm (near-ultraviolet radiation; near-UV) and those that lie below 300 nm (far-ultraviolet radiation; far-UV). Choice of the dividing line is not arbitrary, but arises from several important facts: (1) the solar spectrum at the surface of the earth cuts off at about 300 nm, (2) most solids that are transparent to visible light are also transparent down to about 300 nm, but are opaque at lower wavelengths, and (3) radiation below 300 nm is strongly absorbed by proteins and nucleic acids, resulting in major lethal and mutagenic effects in biological systems, while radiation above 300 nm is little absorbed by these compounds and is much less deleterious (fig. 1). (For further discussion of both physical and biological aspects, see Jagger [19].)

The near-UV region shares some characteristics with its neighboring spectral regions: like far-UV, it can produce major chemical changes (illustrated by the bleaching action of sunlight) but, like visible light, it is not damaging to living systems at low doses. Nevertheless, it is an important region of the spectrum, since it represents the most energetic radiation that living organisms normally encounter in large amounts. About 3% of the total sunlight reaching the earth lies in the near-UV region. Natural induction of sunburn and skin cancer by these wavelengths shows that this amount of radiation represents a biologically significant dose. At high altitudes the fraction of near-UV in sunlight is much greater than at sea level; it has been suggested that this flux may be partly responsible for the stunted growth of alpine plants [26].

All organisms that near-UV can penetrate can be strongly affected by it. At sufficiently high doses, a wide variety of biological effects are observed (see [26] for references). Some of these, such as killing (fig. 1) and mutation, are usually associ-

*This work was carried out under U.S. Public Health Service Research Grants AI 06971, AM 14893, and GM 13234.

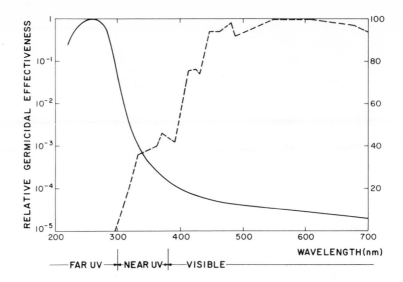

Fig. 1. The near-ultraviolet and related regions. Solid line, left ordinate: action spectrum for killing of *Escherichia coli* (Luckiesh [31]). Broken line, right ordinate: relative photon flux of solar radiation at zenith in temperate zone (Cadle and Allen [9]). Note that left ordinate is logarithmic and right is linear.

ated with far-UV action; the mechanisms of their induction by near-UV are unknown. Other effects, such as inhibition of certain developmental processes in plants and animals, are more specific for near-UV.

Probably the most characteristic effect of near-UV is inhibition of growth. This is an easily recognized phenomenon and, since it is widespread and usually induced by relatively low and non-lethal doses, it may be considered to be the primary effect of near-UV on biological systems. Furthermore, growth inhibition appears to be the mechanism by which near-UV produces a 'photoprotection' from far-UV killing. In this paper, we discuss growth delay and its relationship to photoprotection. The primary object is to understand the mechanisms by which near-UV affects the growth of cells.

2. Growth delay

Inhibition of growth by near-UV has been observed in bacteria [17, 22, 23], fungi [6, 12, 27], protozoa [12], algae [12, 28], higher plants (most recently by [26, 27, 32]), and animal (Hela) cells [26].

Growth inhibition is usually observed in a population of cells or in a multicellular organism, and may be caused by (1) reduction of ability of individual cells to grow, (2) inhibition of cell division, resulting in decreased population growth, or (3) killing of part of a population, which induces an apparent delay in growth of the population. The last factor (killing) is generally negligible at the doses used in growth-delay studies. But it is difficult to distinguish between the first two factors. In most systems, there appears to be a simultaneous, but temporary, effect on both growth and division. Also, where a *single dose* of near-UV is given (usually experiments with single-celled organisms), one typically observes a *growth delay* from which the population eventually recovers. In other studies, however (especially with higher organisms), there is a *continuous exposure* to near-UV radiation; under such conditions, growth delay is constantly being imposed, with the net result that a long-term *lowering* (or complete cessation) *of the growth rate* is observed. These considerations make it clear that one must be cautious in comparing studies on inhibition of growth in different systems.

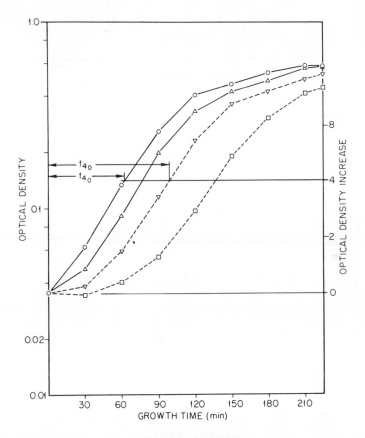

Fig. 2. Growth delay of *E. coli* B in nutrient broth induced by radiation of 334 nm. The optical density of the growing suspension on a logarithmic scale (left) and the optical density increase on a logarithmic scale (right) are plotted against time. Symbols correspond to the following doses of radiation (erg/mm^2): circles, 0; triangles, 139,000; inverted triangles, 278,000; squares, 416,000. This is a typical experiment. (Jagger et al. [22]).

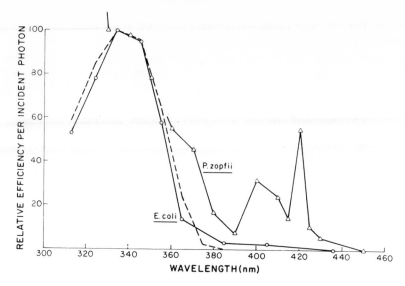

Fig. 3. Action spectrum for growth delay in the bacterium *E. coli* B (solid line, circles), and action spectrum for photoprotection in *E. coli* B (broken line) (Jagger et al [22]). The solid line with triangles is the action spectrum for inhibition of growth in the alga *Prototheca zopfii* (Epel and Krauss [12]). To avoid confusion, the line is not drawn through the triangles at 340 and 350 nm. The triangle at 320 nm would be off scale at a relative efficiency of 184.

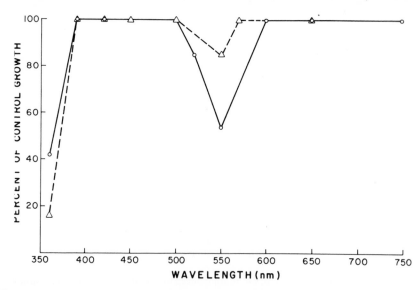

Fig. 4. Action spectra for suppression of growth in *Parthenocissus tricuspidatus* crown gall cultures (solid line, circles; Klein [25]) and HeLa monolayer cultures (broken line, triangles; Klein and Edsall [26]). Running the *Parthenocissus* line to 100% at 500 nm is justified by knowledge of the band-pass of the filters used.

If a population of the bacterium *Escherichia coli* B is starved in the logarithmic growth phase, illuminated with monochromatic near-UV light, and then placed in nutrient broth, it will experience a growth delay, as shown in fig. 2. At 334 nm, the delay is about 1–2 hr for a dose around 300,000 erg/mm^2, which induces less than 10% killing. The population appears to recover completely from the growth inhibition, since the growth curves regain the control slope after about two divisions.

A measure of growth delay in such a system is the ratio of the time for control cells to undergo about two divisions (four-fold optical density increase) to the time for illuminated cells to do the same (arrows on fig. 2). A plot of the logarithm of this ratio as a function of dose gives a straight line [22]. The slopes of such lines will differ for different wavelengths, and one can draw an action spectrum based upon the relative slopes. Fig. 3 shows such an action spectrum (solid line). It is very narrow, with a peak around 340 nm. Fig. 3 also shows the action for inhibition of growth in the colorless alga *Prototheca zopfii*. It is similar to the bacterial spectrum below 380 nm, but shows major peaks at 400 and 420 nm that do not occur in the bacterial spectrum.

There is evidence that growth delay in some plant and animal systems shows an action spectrum similar to the one we have obtained in bacteria. Thus, inhibition of growth occurs at 360 nm, but not at 400–500 nm, in *Parthenocissus* crown-gall tissue cultures and in HeLa tissue cultures (fig. 4). (The effect around 550 nm appears to involve a different mechanism and is anyway beyond the range of our discussion.)

3. Photoprotection

Most cells can repair a large fraction of the potentially lethal damage they suffer upon irradiation by far-UV. This damage consists principally of dimerization of adjacent pyrimidines in either strand of DNA. There are at least two systems known in bacteria that can enzymatically remove or bypass most of such damage in the dark. This is called 'dark repair'. Most organisms can achieve in addition another kind of enzymatic repair, in which pyrimidine dimers complex with an enzyme (photoreactivating enzyme) and are then monomerized when the complex absorbs light in the range 300–500 nm. This phenomenon is called 'light repair' or 'photoreactivation' (for review, see [36]). It is not unusual for bacteria to be able to repair 99% of their far-UV damage by dark repair and 75% by photoreactivation.

It is also possible to increase the survival of some cells by irradiation with near-UV before the far-UV treatment. This is called 'photoprotection' (for review, see [18]). ('Protection' is caused by a pre-treatment, 'reactivation' by a post-treatment.) Photoprotection (PP) is sometimes as large an effect as photoreactivation. The action spectrum for PP is shown in fig. 3 (broken line); the range of effective wavelengths is narrower than for photoreactivation and is restricted to the near-UV. The action spectrum for PP in *E. coli* is similar to that for growth delay in *E. coli*, which of course suggests some connection between two phenomena. It had early been observed that bacteria often survive irradiation better under conditions where they grow more slowly [2, 3], and it is now believed that this occurs because the slower growth permits more time for enzymatic dark repair of the damaged DNA. We therefore proposed some years ago that the mechanism of PP is to induce a growth delay which permits more time for dark repair [22].

If this idea were correct, then it should not matter whether the growth delay is induced immediately before or immediately after far-UV irradiation. In the latter case, however, it would be confused with the usual enzymatic photoreactivation (which we call 'Type I PR'). Clear separation of the two phenomena became possible with the isolation of a bacterium lacking active photoreactivating enzyme. In this strain (*E. coli* B phr⁻), we were able to show a photoreactivation that occurred only at the wavelengths normally effective in PP [20]. In addition, the doses required were the same as for PP and, as in PP, there was little dependence upon temperature and dose rate of the light. These findings showed that the two effects operated by the same mechanism, and we called the new effect 'indirect photoreactivation' or 'Type II PR'.

Proof that the mechanism of Type II PR is different from that of Type I was provided by experiments on splitting of thymine dimers. In this work, cells of *E. coli* were labelled with tritiated thymidine. They were then exposed to far-UV, given a photoreactivation treatment, and immediately hydrolyzed and assayed for thymine dimers. Type I PR should remove dimers (by photoenzymatic splitting) in such an experiment. Type II PR, however, should have no effect on dimers, since it does not operate by splitting them, and not enough time was provided in the experiments for dark-repair processes to operate. As shown in fig. 5, in *E. coli* B phr⁻, which has no photoreactivating enzyme, there was no dimer loss after photoreactivation treatment at 405 nm (dimer concentration = 99% of that with no photoreactivation), a wavelength effective in Type I, but not Type II, PR. The same result was obtained with PR at 334 nm (dimer concentration = 104%), where Type II PR is effective (fig. 5). Since, under the conditions of the experiment, good biological photoreactivation was observed, it is clear that Type II PR does not operate by splitting thymine dimers. Its mechanism is therefore different from that of Type I PR. These experiments also showed (fig. 5) that, in *E. coli* B, which has photoreactivating enzyme, there was extensive dimer splitting at 405 nm (dimer concentration = 22%), where only Type I PR would be expected to operate, but considerably less dimer splitting at 334 nm (dimer concentration = 43%), where both types can operate (biological PR was the same at both wavelengths). This showed that photoreactivation in *E. coli* B is a mixture of the two types at shorter wavelengths.

Further work supported the idea that PP operates through the induction of a growth delay. We showed that (1) both division and growth are delayed in *E. coli* B by near-UV, and the action spectrum is the same for the two effects [34]. We have also found (2) that Type II PR occurs in *E. coli* B/r only if the cells are in the logarithmic growth phase [21]. The apparent reason for the latter finding is that B/r has a very efficient dark-repair system that can function optimally if a short growth

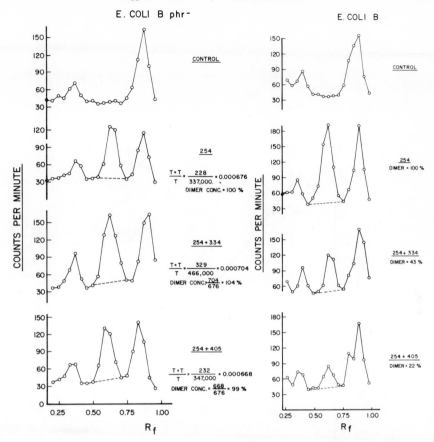

Fig. 5. Radioactive counts (ordinate) obtained in a single experiment (left) with *E. coli* B phr⁻ and a single experiment (right) with *E. coli* B. Points represent activity in 5-mm strips from paper chromatograph obtained with ammonium sulfate, molar sodium acetate, and isopropyl alcohol (80 : 18 : 2). Profiles second from top show thymine dimers induced by UV (254 nm) irradiation (400 erg/mm²). Profiles third from top show these photoproducts after PR treatment (approx. 4 × 10⁵ erg/mm²) at 334 nm, and bottom profiles after PR treatment (similar doses) at 405 nm. Broken lines indicate assumed background count, based on observation of control profiles (top). Sample calculations at left show how dimer concentration was determined from ratios of radioactive counts. Dimer concentration is normalized to 100 per cent for no PR treatment (254 nm only). (Jagger, Stafford, and Snow [21]).

delay is provided. Such a growth delay occurs normally upon plating of stationary-phase bacteria, so there is no need for the additional growth delay imposed by a near-UV treatment. In the logarithmic phase, however, the cells begin growth almost immediately after plating, so that the imposed growth delay is useful. Finally, (3) we may note the good correlation between the extent of PP and the extent of near-UV-induced growth delay in a variety of bacterial genera (table 1).

One final experiment would seem to complete the argument. If PP operates by

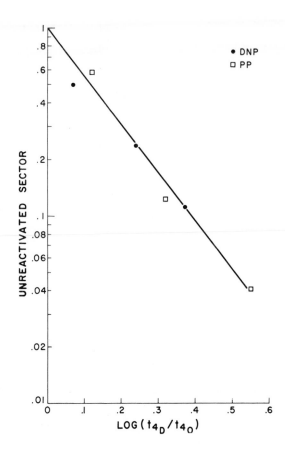

Fig. 6. Correlation of radiation recovery efficiency with growth delay in *E. coli* B. 'Unreactivated sector' (the fraction of the reactivable damage that remains unreactivated) is plotted on a logarithmic ordinate versus the logarithm of the 'growth-delay factor', t_{4D}/t_{40} (see fig. 2) on the abscissa. A growth-delay factor of 0.5 was obtained with a photoprotection (PP) dose of 1.7×10^5 erg/mm² at 334 nm and a 2,4-dinitrophenol (DNP) concentration of 2.2×10^{-4} M in the nutrient-broth plating medium. Average of three experiments (Lakchaura, unpublished).

Table 1
Quinone content and ultraviolet response of bacteria

Organism	Coenzyme Q*	Vitamin K*	Photo-protection	Near-UV-induced growth delay	Liquid-holding recovery
Escherichia coli B	++	++	+++	+++	+++
Pseudomonas aeruginosa	+++	0	+++	+++	+
Enterobacter aerogenes	++	0	+++	+++	++
Staphylococcus epidermidis	0	+++	+	+	0
Streptomyces griseus	0	++	0	+	0
Lactobacillus casei	0	0	0	0	

*From Bishop et al[5], except data for *S. griseus,* which is from Lester and Crane [30].

inducing a growth delay, then similar effects should be observed regardless of how the growth delay is induced, provided the inducer has no deleterious side effects. For reasons outlined below, we chose 2,4-dinitrophenol (DNP), a compound that uncouples oxidative respiration and phosphorylation (fig. 8). This was placed in the plating medium of *E. coli* B that had been irradiated with far-UV. Excellent recovery from far-UV killing was observed. Growth delay also was induced by DNP. Fig. 6 shows that the degree of recovery from far-UV killing (ordinate) is correlated (linear relationship) with the extent of growth delay (abscissa), for both PP and DNP, and the correlation factor is the same for the two agents (same slope). That two such completely different agents should give rise to this correlation we take as good evidence that PP operates primarily by inducing a growth delay.

The question of the mechanism of PP thus becomes a part of the much more general question of the mechanism of induction of growth delay by near-UV.

4. Mechanism of induction of growth delay

Insight into the mechanism of near-UV-induced growth delay in bacteria was provided by A.F. Brodie and co-workers. Kashket and Brodie [23] showed that *continuous* illumination of *E. coli* W with black light (broad-band near-UV) induced a lower growth rate when the cells were incubated in glucose, and an almost complete cessation of growth when they were incubated in succinate, the latter inhibition being largely removed upon addition of glucose (fig. 7). A similar cessation of growth was found with *Pseudomonas aeruginosa*, an obligate aerobe, regardless of substrate. These findings pointed to the oxidative respiratory system as the probable site of near-UV damage leading to growth delay. (We have since found that *single* near-UV doses in these systems cause growth delays but not the indefinite cessation of growth found with continuous illumination.)

The oxidative respiratory system in mitochondria is outlined in fig. 8. It can be

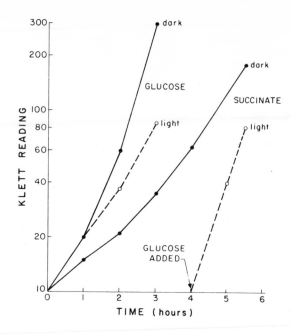

Fig. 7. Growth of *E. coli* W in dark (solid lines) or under continuous near-UV illumination (broken lines) in glucose and succinate minimal media. The plate count of the illuminared succinate culture increased only 50% in 4 hr. (Kashket and Brodie [23]).

divided *functionally* into two more specialized systems, which are *physically* intermingled (see Racker [35] for general discussion). One of these is an *electron-transport chain* running from NADH (or sometimes succinate or malate) through flavoproteins, quinones, and cytochromes to oxygen. In terms of reducing power, this represents a downhill flow. The other system conducts *oxidative phosphorylation.* It taps off the energy lost during electron transport at three sites (one associated with NAD-linked flavoprotein, one with cytochrome c, and one with cytochrome a), and utilizes the energy to convert adenosine diphosphate (ADP) to adenosine triphosphate (ATP). The mechanism by which this is accomplished is still unknown, although reasonable suggestions can now be made [15, 38] . Oxidative phosphorylation provides the primary source of chemical energy for the aerobic organism, and interference with it would certainly decrease the growth rate. (Details of the electron-transport chain are somewhat different in bacteria, but the general scheme is the same.)

Oxidative respiration (electron transport to oxygen) often is tightly coupled to phosphorylation, such that a high concentration of ATP inhibits respiration (and may even reverse electron transport), while a high concentration of ADP stimulates it. The coupled systems are normally operating at a suboptimal rate. Moderate damage *solely* to the electron-transport system or *solely* to the phosphorylation

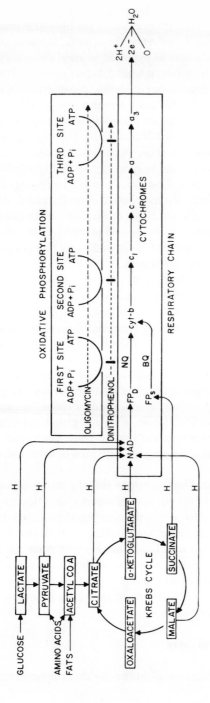

Fig. 8. Schematic diagram of oxidative respiratory system in mitochondria (adapted from Racker [35]). Sugars, proteins, and fats are partially metabolized to produce some ATP in the absence of oxygen (fermentation) and then, in mitochondria, enter the Krebs cycle, in which they are broken down to carbon dioxide, while transferring hydrogen to nicotinamide-adenine dinucleotide (NAD) to produce the highly reduced compound NADH. Electrons are then transported along the chain from NADH through flavoproteins (FP), naphthoquinone (NQ), and cytochromes b, c and a to oxygen to form water. An alternate pathway is from succinate through a flavoprotein and a benzoquinone (BQ) to the cytochromes. The respiratory chain is coupled to oxidative phosphorylation, to produce large amounts of ATP, at three points, all of which are uncoupled by 2,4-dinitrophenol (DNP).

system would not be expected to lower phosphorylation very much, since the coupled systems would speed up, under the pressure of the high ADP concentration, and this would largely overcome the effect of the damage. Moderate damage to the *coupling mechanism*, however, would be expected to lower phosphorylation with little effect on respiration. This, in fact, is what 2,4-dinitrophenol (DNP) does.

Brodie and Ballantine [8] showed that oxidative phosphorylation in crude extracts of *Mycobacterium phlei* is considerably more sensitive to near-UV than respiration (fig. 11b). Thus, a dose that greatly lowers phosphorylation has only a small effect on respiration. This indicates that near-UV is affecting the coupling of the two systems, rather than a single component of one or the other system.

We have attempted to relate growth delay to damage to the coupling mechanism. In preliminary experiments we have shown that a large growth delay can be induced by near-UV in *E. coli* B/r at doses that have no effect on the rate of oxygen uptake. (At higher doses, respiration is affected.) This shows that growth delay can be induced without lowered respiration. Since lowered phosphorylation would certainly induce growth delay, these experiments indicate indirectly that growth delay is caused by damage to the coupling mechanism.

In other experiments, (see Lakchaura [29]), we have attempted to mimic photoprotection with various inhibitors of the respiratory system. We first chose sodium cyanide, which is known to complex with cytochromes. Placed in the plating medium of far-UV-irradiated cells of *E. coli* B, this compound induced a large growth delay but only a small recovery from UV killing. This apparently is due to the fact that cyanide also inhibits dark repair [16] at the concentrations we had to use for good growth delay. Since cyanide stops both respiration and phosphorylation, this result is consistent with the idea that growth delay is produced by lowered phosphorylation, but the experiment does not prove very much. Similar experiments with DNP did not, however, suffer from this complication, and good growth delay, as well as good recovery from far-UV killing, was observed (fig. 6). These results indicate that damage to the coupling mechanism is the cause of both growth delay and photoprotection.

These findings should not lead one to believe that the chromophores for near-UV-induced growth delay are not in the electron-transport chain. Most of the electron-transport components are also involved in the coupling to phosphorylation (fig. 8).

In summary, mechanisms involved in the coupling of respiration to phosphorylation appear to be the targets for near-UV-induced growth delay and photoprotection. Specific chromophores, however, may involve almost any of the electron-transport components. Most of these components absorb near-UV quite effectively (a property related to their ability to transport electrons). We now move to a consideration of the most likely chromophores.

5. The chromophore for growth delay

Considerable information is now available on the chromophores for near-UV

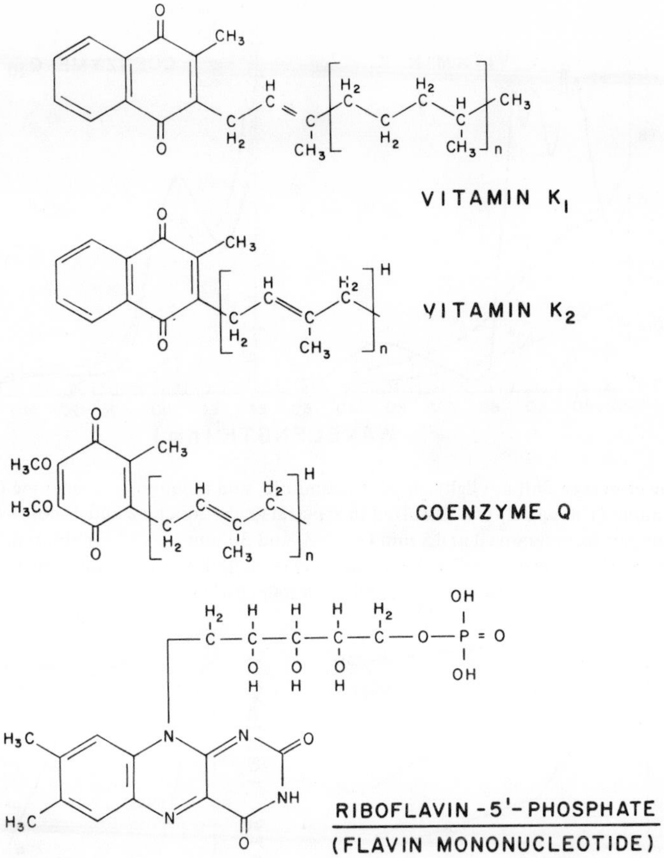

Fig. 9. Formulae of some near-ultraviolet chromophores in the respiratory system. Vitamin K compounds are naphthoquinones. Coenzyme Q is a benzoquinone.

action. Shortly after the isolation of the naphthoquinone vitamin K_1 (see fig. 9), its sensitivity to sunlight was demonstrated by Almquist [1], and Ewing et al. [13] showed that 366 nm light was highly effective in the inactivation. Brodie and Ballantine [8] explored this effect in detail and showed that vitamin K_1 is drastically affected (figs. 10a and 11a) by irradiation in vitro with black light (mostly around 360 nm; see Jagger [19]), while flavin mononucleotide (FMN) is about half as sensitive (fig. 11a). They also showed that, in crude extracts of *Mycobacterium phlei*, irradiation at 360 nm rapidly inactivates oxidative phosphorylation and more slowly lowers respiration (fig. 11b). Addition of unirradiated FMN or flavin adenine dinucleotide (FAD) restores respiration but not phosphorylation; addition of vitamin K_1 restores both activities. Working with whole cells of *E. coli* W, Kashket and Brodie [23] showed that near-UV irradiation causes a drastic loss of the benzoquinone coenzyme Q (fig. 9), and probably also of vitamin K_2, as well as a smaller loss of FMN:

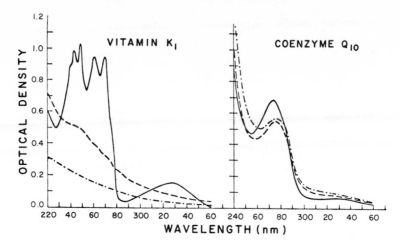

Fig. 10. The effects of 360 nm light on pure vitamin K_1 and ubiquinone-coenzyme Q. The naphthoquinone (100 μg/ml) was dissolved in spectral-grade isooctane and exposed to light at 360 nm. Samples were removed at 15 min (– – – –) and 30 min (– · – · – · –), diluted, and assayed spectrophotometrically. The benzoquinone (100 μg/ml) was dissolved in ethanol and treated in a similar manner (Brodie [7]).

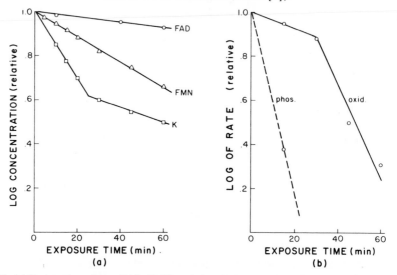

Fig. 11. (a) Destruction of pure FAD, FMN, and vitamin K_1 by light at 360 nm. Each coenzyme (10^{-5} M) was exposed to light for the intervals indicated. The remaining concentration of FAD (circles) and FMN (triangles) was measured by fluorescence and that of vitamin K_1 (squares) by absorption at 249 nm. The bimodal inactivation of vitamin K_1 is typical. (b) Destruction of oxidation (solid line) and phosphorylation (broken line) after exposure of extracts of *Myco-bacterium phlei* to light at 360 nm. The crude extract (20 mg of protein per ml) was exposed to light for the intervals indicated, and the reaction carried out at 30°C for 7 minutes (Brodie and Ballantine [8]).

the cytochromes and FAD are undamaged. Later work [24] on cell-free extracts of *E. coli* W showed that electron transport from NADH to cytochrome *c* can be inactivated by near-UV and restored only by addition of unirradiated vitamin K_2, while electron transport from succinate to cytochrome *c* (see fig. 8) is inactivated by near-UV and restored only by addition of coenzyme Q. Finally, Fujita et al. [14] have shown that irradiation of membrane fragments of *Micrococcus lysodeikticus* with near-UV inhibits oxidation and phosphorylation owing to destruction of a membrane-bound naphthoquinone.

This body of work shows that near-UV easily destroys naphthoquinones, either in vitro or in bacteria, and that their replacement by unirradiated naphthoquinones in particulate systems can restore both electron transport and oxidative phosphorylation. Similar findings have been made in mammalian mitochondria [4]. There is thus little doubt that naphthoquinones, which operate in the NADH pathway to oxygen, are chromophores for near-UV action, in particular for effects on respiration and on growth (which requires phosphorylation).

Benzoquinones, which operate in the succinate pathway to oxygen, appear to be less important targets. They are not as sensitive to near-UV in vitro (fig. 10). As noted above, however, they are readily destroyed in *E. coli* cells, and, in *E. coli* extracts (see above) and in mammalian mitochondria [4], electron transport from succinate to cytochrome *c* can be restored after near-UV irradiation by adding back coenzyme Q. In *M. phlei* extracts, however, electron transport from succinate cannot be restored by adding back coenzyme Q or any other known cofactor. Recent work of Murti and Brodie [33] shows that a water-soluble factor can be obtained from *M. phlei* or from mammalian mitochondria that is resistant to heat and to near-UV. When still combined with protein at an earlier stage of purification, however, it is very sensitive to near-UV. This factor restores electron transport from succinate in irradiated particles from *M. phlei*. The chemical nature of the factor is unknown; it shows an absorption maximum at 405 nm (oxidized) or 415 nm (reduced). In summary, it appears that benzoquinone in some systems, and the unknown factor in other systems, can be a near-UV target in the electron-transport pathway from succinate. There is no evidence, however, that adding back benzoquinone or the unknown factor can restore oxidative phosphorylation. It may be that naphthoquinones play a direct role in both respiration and phosphorylation, while benzoquinones may be involved only in respiration.

Flavins are also less important targets. The work noted above shows that FMN is less sensitive to near-UV than naphthoquinone, both in vivo and in vitro, and its addition back to irradiated systems restores only respiration. FAD is quite resistant to near-UV in vitro and in vivo, although it does sometimes restore respiration in irradiated systems.

The cytochromes apparently can be targets for near-UV action, but they are probably much less important than the quinones. Beyer [4] found that oxidative phosphorylation in rat liver mitochondria is inactivated by near-UV and can be restored by adding back cytochrome *c* alone, if succinate is the substrate, but vitamin K_1 is also required if glutamate is the substrate. In bacterial systems, vitamin K functions in that part of the electron-transport chain that is bypassed by the succinate system, so this finding might be explained by assuming that both vitamin K and cytochrome c are targets. One difficulty with this explanation is that it is not yet clear [10] that naphthoquinone occurs in mammalian systems (benzoquinone is found). Furthermore, Kashket and Brodie [23] found no effect of near-UV on the cytochromes of *E. coli*. Epel and Butler [11] have found that 'blue light' from a mercury arc lamp inhibits respiration in the colorless alga *Prototheca zopfii*, and spectrophotometric measurements on whole cells show destruction of the cytochromes. Essentially similar findings were made with yeast and with beef-heart mitochondria. Furthermore, cytochrome-like structure is seen in the action spectrum for growth inhibition of *P. zopfii* (fig. 3). In summary, it is clear that cytochromes can be affected by near-UV, but they appear not to be very sensitive, and may be important chromophores only in some systems.

In evaluating this work on near-UV chromophores, it should be recognized that most workers have not used a variety of radiation doses, nor have they generally recorded their doses in terms of an absolute amount of monochromatic light. The failure to find effects on some target molecules may result simply from having given too low a dose. Considering all the data, we conclude that, in general, the sensitivity of near-UV targets falls in the sequence naphthoquinone > benzoquinone > flavin > cytochrome. Since oxidative phosphorylation and growth delay are both quite sensitive to near-UV, it seems likely that the quinones are normally involved. Of course, there may be considerable differences among different biological systems, and our suggested sensitivity sequence can serve only as a guideline.

In our laboratory, we have made some preliminary measurements of the action spectrum for destruction of vitamin K_2 suspended in isooctane. Fig. 12 shows the action spectrum (points) for destruction of vitamin K_2, the absorption spectrum (solid line) of vitamin K_2, and the action spectrum for growth delay in *E. coli* B (from fig. 3). The absorption and action spectra for vitamin K_2 are seen to be reasonably similar, as one would expect if the quantum yield is independent of wavelength. There is also a similarity to the action spectrum for growth delay. Furthermore, the doses required for growth delay and for vitamin K_2 destruction are similar. These data show that destruction of vitamin K_2 could account for the observed growth delay in bacteria, and perhaps also in animal and plant tissues (fig. 4). (Our data for the *E. coli* benzoquinone are incomplete, but suggest a poorer correlation.) The only cofactor other than the quinones that shows an absorption spectrum anything like the action spectrum for growth delay in *E. coli* is NADH, but this compound is very resistant to near-UV in vitro (unpublished data). (Other action-spectrum studies [37] have shown that quinones are *not* the primary targets for far-UV-induced growth delay, which occurs at very much lower doses.)

In a further attempt to pinpoint quinones as major targets for near-UV-induced growth delay, we compared growth delay in a variety of bacteria with differing quinone contents (table 1). The results show that species lacking both benzo- and

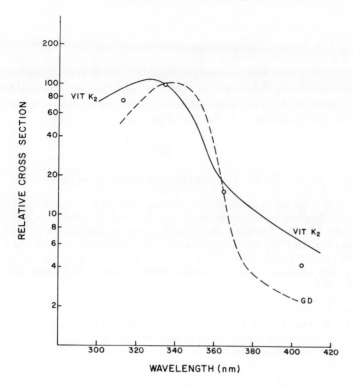

Fig. 12. Action spectrum for destruction of vitamin K_2 upon irradiation in isooctane (circles) compared with the absorption spectrum of the vitamin (solid line). The action spectrum for growth delay in *E. coli* B is shown for comparison (broken line).

naphthoquinones exhibit no growth delay, a finding consistent with the hypothesis of a quinone target. However, the data do not show a good correlation between degree of growth delay and quinone content, and they show a better correlation with high benzoquinone content than with high naphthoquinone content. These findings are inconclusive. There are many difficulties in obtaining significant data of this type, the most important one being that the normal growth rates of these bacteria vary greatly (*Lactobacillus*, with no quinones, grows very slowly) and therefore only fast-growing bacteria will require a copious supply of ATP and thus show a marked growth delay if phosphorylation is lowered.

In general, however, these experiments (action spectrum and comparative studies) do support the hypothesis that quinones are the major chromophores for near-UV-induced growth delay and for photoprotection. Proof of this hypothesis would be the ability to annul growth delay or photoprotection in vivo by adding back intact quinones. We have not yet been able to do this.

Acknowledgements

Many of the experiments here reported from our laboratory were conducted by Dr. B. Lakchaura or Messrs. P. Liou and L. Stanberry. The excellent technical assistance of Mrs. Jeanne Snow is gratefully acknowledged, as well as stimulating conversations with my colleagues, Drs. D. Creed, B. Lakchaura, and H. Werbin.

References

[1] H.J. Almquist, Chemical and physical studies on the antihemorrhagic vitamin. J. Biol. Chem. *117* (1937) 517–523.

[2] T. Alper and N.E. Gillies, 'Restoration' of *Escherichia coli* strain B after irradiation: its dependence on suboptimal growth conditions. J. Gen. Microbiol. *18* (1958) 461–472.

[3] T. Alper and N.E. Gillies, The relationship between growth and survival after irradiation of *Escherichia coli* strain B and two resistant mutants. J. Gen. Microbiol. *22* (1960) 113–128.

[4] R.E. Beyer, The effect of ultraviolet light on mitochondria. II. Restoration of oxidative phosphorylation with vitamin K_1 after near-ultraviolet treatment. J. Biol. Chem. *234* (1959) 688–692.

[5] D.H.L. Bishop, K.P. Pandya and H.K. King, Ubiquinone and vitamin K in bacteria. Biochem. J. *83* (1962) 606–614.

[6] W.H. Brandt, Morphogenesis in *Verticillium*: effects of light and ultraviolet radiation on microsclerotia and melanin. Can. J. Bot. *42* (1964) 1017–1023.

[7] A.F. Brodie, Isolation and photoinactivation of quinone coenzymes. In: Methods in Enzymology, eds. S.P. Colowick and N.O. Kaplan, Vol. 6 (Academic Press, New York, 1963) pp. 295–308.

[8] A.F. Brodie and J. Ballantine, Oxidative phosphorylation in fractionated bacterial systems. II. The role of vitamin K. J. Biol. Chem. *235* (1960) 226–231.

[9] R.D. Cadle and E.R. Allen, The photochemistry of the lower atmosphere is dominated by atoms, molecules, and free radicals. Science *167* (1970) 243–249.

[10] F.L. Crane and H. Low, Quinones in energy-coupling systems. Physiol. Rev. *46* (1966) 662–695.

[11] B. Epel and W.L. Butler, Cytochrome a_3: destruction by light. Science *166* (1969) 621–622.

[12] B. Epel and R.W. Krauss, The inhibitory effect of light on growth of *Prototheca zopfii* Kruger. Biochim. Biophys. Acta *120* (1966) 73–83.

[13] D.T. Ewing, F.S. Tomkins and O. Kamm, The ultraviolet absorption of vitamin K_1 and the effect of light on the vitamin. J. Biol. Chem. *147* (1943) 233–241.

[14] M. Fujita, S. Ishikawa and N. Shimazono, Respiratory chain and phosphorylation site of the sonicated membrane fragments of *Micrococcus lysodeikticus*. J. Biochem. (Japan) *59* (1966) 104–114.

[15] D.O. Hall and J.M. Palmer, Mitochondrial research today. Nature *221* (1969) 717–723.

[16] W. Harm and K. Haefner, Decreased survival resulting from liquid-holding of U.V.-irradiated *Escherichia coli* C and *Schizosaccharomyces pombe*. Photochem. Photobiol. *8* (1968) 179–192.

[17] A. Hollaender, Effect of long ultraviolet and short visible radiation (3500–4900 Å) on *Escherichia coli*. J. Bacteriol. *46* (1943) 531–541.

[18] J. Jagger, Photoprotection from far ultraviolet effects in cells. In: Advances in Chemical Physics, Vol. VII (Interscience, New York, 1964) pp. 584–601.

[19] J. Jagger, Introduction to Research in Ultraviolet Photobiology (Prentice-Hall, Englewood Cliffs, N.J., 1967).

[20] J. Jagger and R.S. Stafford, Evidence for two mechanisms of photoreactivation in *Escherichia coli* B. Biophys. J. *5,* (1965) 75–88.

[21] J. Jagger, R.S. Stafford and J.M. Snow, Thymine-dimer and action-spectrum evidence for indirect photoreactivation in *Escherichia coli*. Photochem. Photobiol. *10* (1969) 383–395.

[22] J. Jagger, W.C. Wise and R.S. Stafford, Delay in growth and division induced by near ultraviolet radiation in *Escherichia coli* B and its role in photoprotection and liquid holding recovery. Photochem. Photobiol. *3* (1964) 11–24.

[23] E.R. Kashket and A.F. Brodie, Effects of near-ultraviolet irradiation on growth and oxidative metabolism of bacteria. J. Bacteriol. *83* (1962) 1094–1100.

[24] E.R. Kashket and A.F. Brodie, Oxidative phosphorylation in fractionated bacterial systems. X. Different roles for the natural quinones of *Escherichia coli* W in oxidative metabolism. J. Biol. Chem. *238* (1963) 2564–2570.

[25] R.M. Klein, Repression of tissue culture growth by visible and near visible radiation. Plant Physiol. *39* (1964) 536–539.

[26] R.M. Klein and P.C. Edsall, Interference by near ultraviolet and green light with growth of animal and plant cell cultures. Photochem. Photobiol. *6* (1967) 841–850.

[27] R.M. Klein, P.C. Edsall and A.C. Gentile, Effects of near ultraviolet and green radiation on plant growth. Plant Physiol. *40* (1965) 903–906.

[28] W. Kowallik, Wachstumstemmung von Prototheca in Licht. Flora Allg. Bot. Zg *156* (1965) 231–235.

[29] B. Lakchaura, Chemical reactivation of *Escherichia coli* B from radiation damage. Ph. D. Thesis, Univ. of Oklahoma (1969).

[30] R.L. Lester and F.L. Crane, The natural occurrence of coenzyme Q and related compounds. J. Biol. Chem. *234* (1959) 2169–2175.

[31] M. Luckiesh, Applications of Germicidal, Erythemal and Infrared Energy (Van Nostrand, Princeton, N.J., 1946).

[32] H. Mohr, Wirkungen kurzwelligen Lichtes. Handbuch Pflanzenphysiol. *16* (1961) 439–531.

[33] C.R.K. Murti and A.F. Brodie, New light-sensitive cofactor required for oxidation of succinate by *Mycobacterium phlei*. Science *164* (1969) 302–304.

[34] S.L. Phillips, S. Person and J. Jagger, Division delay induced in *Escherichia coli* by near-ultraviolet radiation. J. Bacteriol. *94* (1967) 165–170.

[35] E. Racker, The membrane of the mitochondrion. Scientific American, February (1968), 32–39.

[36] C.S. Rupert and W. Harm, Reactivation of photobiological damage. In: Advances in Radiation Biology, Vol. 2., eds. L. Augenstein and R. Mason (Academic Press, New York, 1966) pp. 1–81.

[37] H. Takebe and J. Jagger, Action spectrum for growth delay induced in *Escherichia coli* B/r by far-ultraviolet radiation. J. Bacteriol. *98* (1969) 677–682.

[38] J.H. Wang, Oxidative and photosynthetic phosphorylation mechanisms. Science *167* (1970) 25–30.

CHAPTER 22

ANALYSIS OF PHOTOENZYMATIC REPAIR OF
UV LESIONS IN *E. COLI* BY LIGHT FLASHES

Walter HARM*

Division of Biology,
University of Texas at Dallas,
Dallas, Texas 75230, USA

1. Introduction

Kelner [9] and Dulbecco [1] discovered that irradiation of cells or viruses with ultraviolet light < 300 nm ('UV') causes lethal damage which can often be repaired by post-treatment with wavelengths in the range 310–480 nm ('light'). This phenomenon was called 'photoreactivation' (PR). Although at least two mechanisms are now known to cause increased UV survival as a result of post-treatment with light [8], the most common and usually the most effective one is *photoenzymatic repair,* for which in this paper the abbreviation PR will be used exclusively. The involvement of an enzyme ('photoreactivating enzyme' or PRE) was strongly suggested from work by Goodgal et al. [2], who demonstrated that UV-irradiated transforming DNA of *Haemophilus influenza* can be photoreactivated in vitro in the presence of an extract from *E. coli* cells. Similar extracts obtained from *H. influenzae* cells, being themselves not photoreactivable, were ineffective in the in vitro assay. The requirement for a cell factor explained the observation [1] that UV-irradiated bacteriophages can be photoreactivated only after having infected their host cells.

In the following years, Rupert [11, 12, 13] characterized in more detail the enzymatic nature of the in vitro PR, working with *H. influenzae* transforming DNA and a partly purified PRE-containing extract from yeast cells. He showed that the repair process can be described by the following reaction scheme:

$$E + S \underset{k_2}{\overset{k_1}{\rightleftharpoons}} ES \xrightarrow[\text{light}]{k_\text{p}} E + P; \tag{1}$$

*The experimental work was supported by U.S. Public Health Service Research Grants GM-12813 and GM-13234 from the National Institute of General Medical Sciences and by a Research Career Development Award, GM-34963.

where E is a molecule of the photoreactivating enzyme, S is the substrate (a photo-repairable UV lesion in irradiated DNA), ES is the enzyme—substrate complex, and P is the repaired lesion. Only the right half of the reaction ('photolytic step') requires photoreactivating light; the left half ('complex formation') can occur in the dark. There is good evidence now that the photorepairable lesions are cyclobutyl pyrimidine dimers in DNA and that the repair consists of splitting the dimers into the original bases.

Experimental work in the past on PR of phage or *E. coli* suggested that the above reaction scheme also holds for PR in vivo, although relatively few experimental criteria were formerly applicable (for reference, see [14]). It is the aim of this article to summarize recent results which provide further evidence for the validity of the above scheme for PR in vivo and which give us detailed information about it. These results were obtained in experiments in which PR was achieved by a single intense light flash, or a sequence of flashes. This technique was developed by Helga Harm and C.S. Rupert, using the in vitro PR system of *Haemophilus* DNA and yeast PRE [4]; under certain conditions it was also found applicable to *E. coli* cells [7]. The advantage of this technique is that the number of PRE-substrate complexes present in the cellular DNA can be determined at any time, thereby permitting independent studies of complex-formation and of the photolytic step.

2. General procedures

Experimental. Most of the results summarized here were obtained with stationary phase cells of *E. coli* strain B_{s-1} or of mutants of this strain, selected for their increased content of PRE. They were UV-irradiated with a low-pressure mercury-vapor lamp, emitting chiefly the wavelength 254 nm. In a few cases UV-irradiated T1 phages were used. The very high UV sensitivity characteristic of these biological systems is important for the following reason. The number of UV lesions repaired by a single flash cannot possibly exceed the number of PRE molecules present in the cell, which — as we will see — is very small. Therefore, appreciable PR by a single flash is expected to occur only in systems sufficiently UV-sensitive for a slight difference in the number of UV lesions to affect greatly the survival.

A single light flash was usually applied by simultaneous discharge of 4 electronic flash units (Yashica PRO 50) designed for photographic use. They were mounted at right angles to each other so that the sample, placed in the center, received the light from four directions. A colorless plastic filter or a blue glass filter was used to eliminate wavelengths that might cause inactivation. Only where explicitly stated, fewer than 4 units were discharged simultaneously. To obtain maximal PR, samples were illuminated with white fluorescent light in a 37°C waterbath. Monochromatic light was supplied by a Bausch & Lomb grating monochromator, operated with a Philips SP 500 high-pressure mercury-vapor lamp.

Unless otherwise stated, all manipulations were carried out at room temperature (approximately 23°C) and under yellow fluorescent light to exclude any uncontrolled PR.

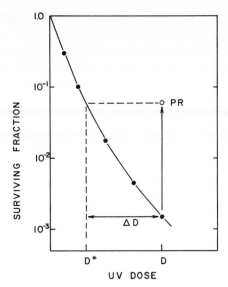

Fig. 1. Quantitative determination of the PR effect by the dose decrement ΔD. The PR survival after a UV dose D (open circle) is the same as the dark survival (solid line connecting the filled circles) after a dose D^*. Therefore, the difference $D - D^*$ (= ΔD) expresses the amount of dose whose effect is annulled by the PR treatment. The PR effect can also be expressed by the photoreactivated section (= $\Delta D/D$), or by the photoreactivable section (= $\Delta D_{max}/D$).

Quantitative evaluation. The PR obtained by either flash photolysis or continuous illumination is expressed in terms of the dose decrement ΔD. This is the portion of the UV dose which is in effect annulled by the photoreactivating treatment. Thus $\Delta D = D - D^*$, where D is the UV dose actually applied and D^* is the UV dose which would give the same survival without PR treatment as the UV dose D with PR treatment, as illustrated in fig. 1. Since the number of UV lesions (i.e. pyrimidine dimers) produced is for all practical purposes proportional to the UV dose, the number of UV lesions repaired should be proportional to ΔD.

A UV dose of 1 erg/mm^2 of 254 nm UV radiation produces about 6.5 pyrimidine dimers in an *E. coli* chromosome [15]. Since about 85% of the primary UV lesions in *E. coli* B_{s-1} are photorepairable, we suppose that the number of substrate molecules S in a stationary phase cell is $0.85 \times 6.5 \times D$ (where D is measured in erg/mm^2), and a dose decrement of ΔD erg/mm^2 corresponds to repair of $6.5 \times \Delta D$ substrate molecules. The substrate remaining unrepaired should then be $6.5 (0.85 D - \Delta D)$. Evaluation of the experimental results described in this article is based on these simple relations. We should be aware that the figure 6.5 pyrimidine dimers for a dose of 1 erg/mm^2 is an approximation. Also, it is possible that all (not only 85%) of the dimers serve as substrate for the PRE. Even if future work would necessitate slight corrections of the values above, this would have little effect on the essential results and conclusions described in this paper.

3. Results and discussion

3.1. *Efficiency of single light flashes*

We have been able to show that a single light flash is sufficient to repair essentially *all* enzyme—substrate complexes ('ES complexes') present when the flash is given. The evidence is twofold: (1) the PR obtained in UV-irradiated T1 phage is equally extensive after a normal 4-unit flash and after a flash of only 0.75, 0.50, or 0.25 the normal light intensity, as shown in fig. 2. Thus the PR effect is not limited by the amount of light delivered, but only by the number of ES-complexes present. (2) At sufficiently low UV doses, where the number of PRE molecules exceeds the number of substrate molecules a single flash leads to the same maximal PR effect which would otherwise be obtained with continuous illumination. This is best demonstrated in mutant cells with increased PRE content (cf. fig. 4b). It is likewise found for flash PR of *Haemophilus* transforming DNA in vitro [4], where an excess of PRE over UV lesions can be obtained without difficulty.

3.2. *Kinetics of complex formation*

The high UV sensitivity of strain B_{s-1} permits creating all of the substrate for the PRE by irradiation of the cells for a very short period of time (of the order of 1 sec). This enables us to measure with high accuracy the kinetics of formation of ES-complexes, which is reflected by the extent of PR obtained when a single flash

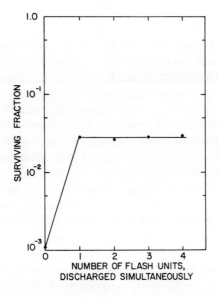

Fig. 2. Efficiency of light flashes of different intensities. Phage T1 irradiated with 280 erg/mm² received, after infecting B_{s-1} cells, a single light flash from either 1, 2, 3 or 4 flash units discharged simultaneously; the resulting survival is shown on the ordinate.

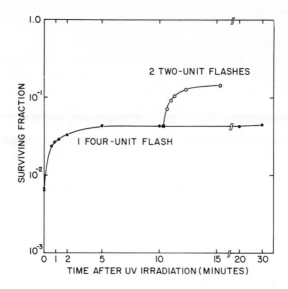

Fig. 3. Kinetics of ES-complex formation in B_{s-1} cells. Cells were irradiated with 4.8 erg/mm²
and subject to flash PR at the times indicated on the abscissa. Filled symbols: samples were
kept dark until a single 4-unit flash was given. Open symbols: samples received a 2-unit flash
at 10 min 20 sec and a second 2-unit flash at the indicated times.

is applied at various times t after irradiation. Results of such experiments are shown
in fig. 3. The PR effect increases first rapidly with the time elapsed between UV
irradiation and flash, but it takes about 5 min to reach the equilibrium for complex
formation, which is determined by k_1 and k_2.

Fig. 3 shows further that, in accordance with the statements in section 3.1, a
2-unit flash applied 10 min after irradiation leads to the same PR effect as does the
normal 4-unit flash at that time. However, if a second 2-unit flash is given after the
first one, the total PR effect increases, although the total light dose equals that of
a normal 4-unit flash. The explanation is fairly obvious: Photolysis of ES-complexes
by the first flash liberates PRE, which thus can form new complexes to be repaired
by the second flash. As the data indicate, the kinetics of this second 'round of complex-formation' resemble the first one. Therefore, in order to obtain maximal PR
with a single flash or with several flashes, a period of at least 5 to 10 min must
elapse between UV irradiation and the first flash, or between consecutive flashings.

3.3. *Estimation of the number of PRE molecules in the cell*

As pointed out earlier, the PR effect expressed by the dose decrement ΔD permits calculating the average number of repaired UV lesions per cell. Because of the
very high repair efficiency of a single flash, this number is virtually identical with
the number of ES-complexes present at the time of the flash. To obtain an estimate

of the number of PRE molecules per cell, we have studied ΔD in B_{s-1} as a function of the UV dose, allowing sufficient time for maximal complex-formation after irradiation with varying UV doses. The following results were obtained [7]

UV dose (erg/mm^2):	1.6	3.2	4.8	all doses from 6.4 to 24.0
ΔD (erg/mm^2):	1.3	2.0	2.6	3.0 ± 0.2

It is evident that ΔD reaches a maximum of 3 erg/mm^2, corresponding to the presence of about 20 ES-complexes per cell. This value must be limited by the amount of PRE, since it cannot be exceeded even at a several-fold increase in the amount of substrate (see fig. 5). Therefore, we conclude that the number of PRE molecules in these cells is approximately 20.

A comparison of the flash PR effect in phage T4v^-x^-, infecting either B_{s-1} or B/r cells, indicated that B/r has about the same number of PRE molecules as has B_{s-1}. However, by the same criteria various derivatives of the *E. coli* strain K12 seem to have even less PRE, i.e. roughly 10 to 15 molecules per cell [7].

3.4. *Mutants with increased PRE content*

Since 20 is a rather low molecule number for an enzyme species in the cell, we expected mutations to occur which affect in some way the regulatory mechanism controlling the amount of PRE produced. We were able to isolate three mutants with increased PRE content [6]; fig. 4b shows the UV survival, flash PR, and maximum PR for one of them. The corresponding curves for the parental strain B_{s-1} are presented in fig. 4a.

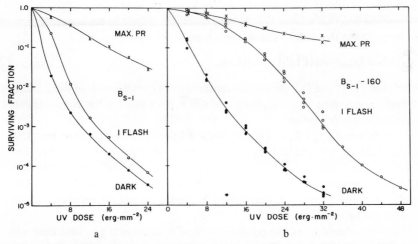

Fig. 4. PR effects in stationary-phase cells of strain B_{s-1} (left) and of strain B_{s-1} −160 as a function of UV dose. The ordinate shows the survival of cells which were either kept dark (●), or were subject to PR by a single flash (○), or by continuous illumination for 50 min (×). The B_{s-1} values are the geometric means of 2 to 5 experiments.

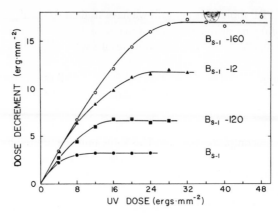

Fig. 5. Dose decrement obtained with 1 flash in cells of B_{s-1} and three mutant strains, as a function of UV dose.

To determine for the three mutants the number of PRE molecules per cell, their dose decrement after a single flash was measured as a function of the UV dose. The data are plotted together with the ΔD values for the parental B_{s-1} in fig. 5. Each strain reaches a characteristic maximal dose decrement which remains constant over a considerable range of dose, from which the number of PRE molecules per cell can be calculated to be 43, 75, and 110 for the mutant strains $B_{s-1}-120$, $B_{s-1}-12$, and $B_{s-1}-160$, respectively. Strain $B_{s-1}-160$ was used in some of the following experiments, where the greater number of PRE molecules was advantageous.

3.5. Determination of the reaction constant k_1

From the reaction scheme (eq. 1) we expect any change in the concentration of ES-complexes in the dark to follow the equation

$$(d[ES/dt])_{dark} = k_1[E][S] - k_2[ES], \tag{2}$$

where [ES], [E], and [S] are the concentrations of complexes, free PRE, and free substrate, respectively, For simplicity, we will express these as the average number of molecules per cell volume.

According to eq. (2), k_1 can be determined from the following experiment. A cell suspension is UV-irradiated and immediately afterwards given a rapid sequence of flashes, one 1-unit flash every half or every one second. This provides that virtually every complex formed is repaired within half a second or one second. The last flash is given at varying times t after UV irradiation and the PR effect is measured by the survival.

Under these conditions, the concentration of complexes at any time during the interval from 0 to t is small. Therefore, since k_2 is not large (as will be seen later), we can neglect the product $k_2[ES]$ in eq. (2), and [E] approximates $[E]_0$, i.e. the

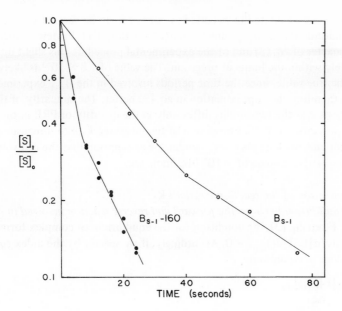

Fig. 6. Determination of k_1 for *E. coli* B_{s-1} and $B_{s-1}-160$. The concentration $[S]_t$ of photo-repairable lesions remaining unrepaired after time t, relative to the concentration $[S]_0$ of photo-repairable lesions initially present, is plotted as a function of time. Light flashes in rapid sequence during the time interval $0 \rightarrow t$ provide that repair and liberation of PRE molecules occurs very soon after formation of ES-complexes. The UV dose was 8 erg/mm^2.

concentration of free enzyme when no substrate is present. Furthermore, since each ES-complex formed and repaired corresponds to the disappearance of 1 UV lesion S, the equation simplifies to

$$-d[S]/dt \approx k_1[E]_0[S],\tag{3}$$

which corresponds to a pseudo-first order reaction. Upon integration we obtain

$$[S]_t/[S]_0 \approx \exp(-k_1[E]_0t),\tag{3a}$$

where $[S]_0$ is the substrate concentration before PR and $[S]_t$ is the substrate concentration remaining after time t.

Fig. 6. shows the results of such experiments. The decrease of $[S]_t/[S]_0$ versus t is not strictly exponential; the shapes of the curves suggest that roughly one half of the lesions form complexes considerably faster than the remainder. The initial slopes correspond to a weighted average of k_1 for the faster and slower forming complexes. As expected from eq. (3a), these slopes are approximately proportional to $[E]_0$, which is 20 molecules/cell volume in the case of B_{s-1} and 110 molecules/cell volume in the case of $B_{s-1}-160$.

From fig. 6 we calculate the weighted average of k_1 to be about 1.8×10^{-3} cell volumes/molecule·sec for B_{s-1}, and 1.5×10^{-3} for B_{s-1}-160. In view of the approximative character of eq. (3) and of the experimental procedure this slight difference is presumably within the limits of precision. The value of 1.8×10^{-3} is likely to be closer to the true value, since the time periods involved in the B_{s-1} experiments are longer and therefore the approximation in eq. (3) better. The similarity of the two values suggests that the two strains differ only in the quantity of PRE in the cell, but not in the properties of PRE. Conversion of the measured k_1 into dimensions conventional for enzyme kinetics gives — under the assumption that the cell volume is $\approx 10^{-15}$ liter [10] — a value of $\approx 10^6$ liter/mole·sec.

3.6. Determination of the reaction constant k_2

3.6.1. Equilibrium between the forward and reverse reaction involved in complex formation.
From eq. (2), the condition for the equilibrium of complex formation in the dark is $(d[ES]/dt)_{dark} = 0$. Accordingly, if we specify by the index *eq* the concentrations in equilibrium,

$$k_2 = \frac{k_1 [E]_{eq} [s]_{eq}}{[ES]_{eq}}, \tag{4}$$

or

$$k_2 = \frac{k_1 ([E]_0 - [ES]_{eq})([S]_0 - [ES]_{eq})}{[ES]_{eq}}. \tag{4a}$$

Since k_1 is known, $[E]_0$ and $[S]_0$ is given by the experimental conditions, and $[ES]_{eq}$ is determined from the flash PR effect obtained, we are able to calculate k_2, provided we use conditions where $[ES]_{eq}$ is reasonably large but neither approximating $[E]_0$ nor $[S]_0$. Such calculations would be fairly inaccurate for B_{s-1}, since the differences involved in eq. (4a) are small, whereas one obtains reasonably consistent k_2 values around 5×10^{-3} sec^{-1} for B_{s-1}-160 in the dose range 12 to 20 erg/mm^2. For example, at a dose of 16 erg/mm^2, $[S]_0 = 88$ and $[ES]_{eq} = 80$. With $k_1 = 1.8 \times 10^{-3}$ and $[E]_0 = 110$ we obtain a k_2 value of 5.4×10^{-3} sec^{-1}.

3.6.2. Presence of competing substrate.
In order to determine k_2 independently of k_1, the following experiment was designed. Unirradiated host cells were singly infected with UV-irradiated phage T1. After maximum formation of ES-complexes in the phage DNA, the infected cells are irradiated in order to create UV lesions in the bacterial DNA at a 20-fold to 30-fold excess over those in the phage DNA. This does not require a high dose, because the amount of bacterial DNA is very large compared to the amount of phage DNA. Because of the high competition by the irradiated bacterial DNA any PRE molecule dissociating in the dark from an ES-complex in phage DNA has presumably a very small chance to form again a com-

plex in the phage DNA. The resulting decrease in the number of ES-complexes in phage DNA can be calculated from the PR effect obtained by a single flash at various times after the host cell irradiation.

Under the conditions of this experiment, its is expected that the product $k_1[E][S]$ in eq. (2) can essentially be neglected, so that

$$(d[ES]/dt)_{dark} \approx -k_2[ES],$$ (5)

or

$$\left(\frac{[ES]_t}{[ES]_0}\right)_{dark} \approx \exp(-k_2 t).$$ (5a)

This approximation is invalid for large t, since even for $t \to \infty$ the fraction $[ES]_t/[ES]_0$ is expected to have still a finite value, because a small number of ES-complexes should be present in phage DNA after the new equilibrium for complex formation is reached. Therefore, a better approximation should be

$$\left(\frac{[ES]_t - [ES]_{t\,max}}{[ES]_0 - [ES]_{t\,max}}\right)_{dark} \approx \exp(-k_2 t).$$ (5b)

In fig. 7, the expression on the left side of eq. (5b), using 120 min as t_{max}, is plotted versus the time. As found for the complex formation, the curve shows upward concavity, reflecting an appreciable heterogeneity among the complexes with respect to their stability in the dark. Considering the initial slope only, we obtain a weighted average of k_2 of approximately 1.7×10^{-3} sec^{-1}.

This value is considerably lower (although of the same order of magnitude) than the k_2 value calculated from the dark equilibrium as described in section 3.6.1. A possible reason for the difference would be that the results plotted in fig. 7 refer to complexes formed in T1 DNA. Although this DNA resembles *E. coli* DNA in base composition and nearest-neighbor frequencies, there might nevertheless be dissimilarities in the stability of ES-complexes. However, it seems more likely that the difference reflects some principal limitation in accuracy related to the experimental approach, as discussed in the following paragraph.

3.6.3. *Difficulties concerning the determination of* k_2. The following facts seem at present to make an accurate determination of k_2 problematic.

(1) k_1 is not constant for all lesions and k_2 is not constant for all complexed lesions. In order to calculate k_2 from the equilibrium of complex formation in the dark one would have to know — besides the total variability of k_1 and of k_2 — the interrelation between the variations: e.g. whether the fastest forming complexes are the most labile ones, or the most stable ones.

(2) Because of the very low molarity of the reactants in the cell, their distribution within the cell is not likely to be random at all times: therefore the applicabil-

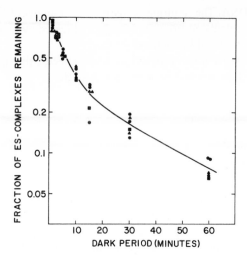

Fig. 7. Determination of k_2. T1 phages irradiated with 280 erg/mm² were infecting $B_{s-1}-160$ cells and allowed to form a maximum number of ES-complexes. Excessive competing substrate was then produced by irradiation of the infected cells with dose of 32 (•), 64 (▲) or 96 (■) erg/mm², and the decrease in the number of ES-complexes in phage DNA was studied as a function of time (see text for further details).

ity of eq. (5) becomes questionable. For example, shortly after dark dissociation of an ES-complex, the likelihood of the liberated enzyme molecule re-attaching to the original lesion should be higher than for attachment to any other lesion, if the distance to the original lesion is much smaller. As a consequence, our k_2 value obtained in the presence of excessive competing substrate would be too low. Similarly, eq. (4a) might not correctly describe the equilibrium of complex formation, since it requires the distribution of uncomplexed PRE molecules and UV lesions within the cell being random. From the equation we would again calculate a k_2 value too low, if shortly after dissociation of an ES-complex the local concentration $[E]'$ in the neighborhood of a lesion is higher than it appears from the average concentration $[E]$ within the cell. One should obtain a more correct estimate of k_2 if one could prevent re-attachment of a dissociated PRE molecule to the same UV lesion, e.g. by adding caffeine at a high concentration. As mentioned in section 3.10, one obtains a k_2 value of about 10^{-2} sec^{-1} under these conditions.

3.7. Determination of the photolytic constant k_p

In the Michaelis-Menten scheme for an enzymatic dark reaction, the constant k_3 characterizing the step ES → E + P has the dimension of reciprocal time. Since in the PR reaction such a time constant would depend on the light intensity I present at the time of the reaction, the constant characterizing the photolytic step in the PR scheme (eq. 1) is actually k_3/I, which has been called k_p. It has thus the

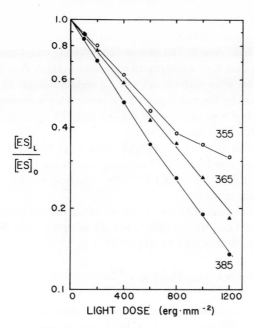

Fig. 8. Determination of the photolytic constant k_p in $B_{s-1}-160$ cells at the wavelengths 355, 365 and 385 nm. The concentration ES-complexes remaining after illumination, relative to the concentration of ES-complexes present before illumination, is plotted as a function of the light dose L. The cells had received a UV dose of 8 erg/mm².

dimension of a reciprocal dose, and varies with the wavelength of the photoreactivating light.

In order to determine k_p, we have carried out experiments on the repair of ES-complexes as a function of the dose of light at the wavelengths 355, 365, 385 nm. By using the mutant strain $B_{s-1}-160$, irradiated with 8 erg/mm² UV, we were able to provide conditions where, at the beginning of the PR treatment, more than 95% of the substrate originally present is in complexed form. It is important to measure k_p under these conditions, because less extensive complex formation would lead to an underestimation of k_p, since a photon effective in repairing a complex would have been wasted if the substrate was not complexed at that time.

Fig. 8 shows the results of such experiments. Since the reaction ES → E + P is first order, the fraction of complexes $[ES]_L$ remaining after illumination with the dose L, relative to the total number of complexes originally present, $[ES]_0$, should be expressed by the function

$$[ES]_L/[ES]_0 = \exp(-k_pL), \tag{6}$$

provided that complexes behave alike with respect to k_p. $[ES]_0$ is determined from

the dose decrement that can be obtained with maximum PR, and $[ES]_L =$
$[ES]_0 - [ES]_{repaired}$ where $[ES]_{repaired}$ is determined from the dose decrement
observed after the light dose L. The plot of $[ES]_L/[ES]_0$ on a logarithmic scale
versus the incident dose L of photoreactivating light in fig. 8 shows that not all of
the ES-complexes are alike with respect to k_p; a certain fraction of them photolyses
slower than reflected by the initial part of the curve. This is obvious for illumination with 355 nm, but it was also found for 365 and 385 nm in experiments where
ordinate values below 0.10 were obtained.

From the light dose at which $[ES]_L/[ES]_0$ is e^{-1} (or 36.8%), k_p is calculated
according to eq. (6) to be 1.75×10^{-3} mm^2/erg for the most effective wavelength
385 nm, and to be 1.37×10^{-3} and 1.13×10^{-3} mm^2/erg for the wavelengths
365 nm and 355 nm, respectively.

Knowing k_p, we can calculate for a given wavelength λ the product of the molar
extinction coefficient ϵ of the complexes and the quantum yield Φ of the reaction.
A conversion similar to that used by Rupert [13] is

$$\epsilon\Phi\,[\text{liter/mole·cm}] = k_p\,[\text{mm}^2/\text{erg}] \times \frac{5.2 \times 10^9}{\lambda\,[\text{nm}]}.$$

From our k_p we obtain an $\epsilon\Phi$ value of 2.4×10^4 liter/mole·cm for the most
effective wavelength 385 nm, and 1.95×10^4 for the wavelengths 365 nm and
355 nm respectively. Although it is not possible with these methods to determine
ϵ and Φ separately, the figures obtained for the product tell us that both values
must be high. ϵ must be at least as high as the product $\epsilon\Phi$, because Φ cannot exceed
1. On the other hand, Φ must be fairly high (probably between 1 and 10^{-1}) because
it would be very unlikely that ϵ is higher than 10^5 liter/mole·cm.

3.8. Temperature dependence of the reaction constants

Both the k_1 and k_2 experiments were carried out in the cold room (+5°C) and
the incubation room (+37°C), besides at room temperature. The results showed
that both reaction constants have a positive temperature dependence. The Arrhenius
plot (fig. 9) shows that the logarithms of the reaction constants decrease linearly
with the reciprocal of the absolute temperature. The greater slope for k_1 corresponds to an activation energy of approximately 11 kcal/mole, the lesser slope for
k_2 corresponds to approximately 4.5 kcal/mole. However, for reasons discussed in
section 3.6, all k_2 values might actually be higher than shown here; therefore the
activation energy calculated for the k_2 reaction might not be correct.

The temperature dependence of k_p was not tested for monochromatic light.
Earlier experiments with light flashes of limiting intensity in vivo [7] and in vitro
[4] had shown that the photolytic effect at +2°C and at +37°C is virtually the
same for complexes formed at 37°C. Thus k_p is temperature-independent in this
range. At temperatures below 0°C, k_p decreases drastically, but even at −196°C a
small response to a single light flash can be observed [3].

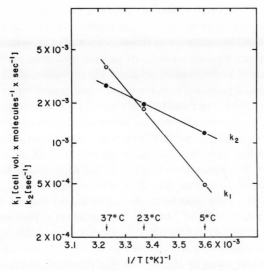

Fig. 9. Temperature dependence of k_1 and k_2. Experimental values of k_1 and k_2 are plotted on a logarithmic ordinate against the reciprocal of the absolute temperature T. The activation energy A can be calculated from the slopes, according to the Arrhenius equation.

$$A \, [\text{cal/mole}] = R \, \frac{\ell n(k_x/k_y)}{(1/T_y) - (1/T_x)},$$

where the indices x and y represent two different temperatures and R is the gas content, or 1.99 cal/°K·mole.

3.9. Heterogeneity of UV lesions and complexes

Several years ago J.K. Setlow [16] had found that under photoreactivating treatment the different types of pyrimidine dimers ($\widehat{TT}$, $\widehat{CT}$, and $\widehat{CC}$) disappear at different rates. Our curves determining k_1, k_2, and k_p are consistent with this observation. It is likely, though not proven, that the heterogeneity observed in all three reaction steps is due to the presence of different types of dimers (and possibly different neighboring bases for a given type of dimer). Yet nothing can be said about the correlation between k_1, k_2 and k_p within a given class of dimers. For genetic reasons it seems very unlikely that the PRE molecules themselves are heterogeneous.

3.10. Use of flashes for studying inhibition of PR by caffeine

If UV-irradiated B_{s-1} cells are suspended in buffer containing 1.6% w/v caffeine and illuminated with continuous white light, their rate of PR is only about $\frac{1}{5}$ to $\frac{1}{10}$ that of the same cells suspended in buffer. In contrast, the rate of PR for B/r cells is essentially the same in the presence and absence of caffeine. The rate of PR for UV-irradiated T1 phage, after infecting either B_{s-1} or B/r cells, is strongly affected by caffeine. This indicates that the caffeine is taken up by both bacterial strains and that the unirradiated cells do not differ in anything that is relevant for the inhibition of PR by caffeine. Investigation of this effect is still in progress, but some results shall be mentioned here because (a) study of caffeine inhibition by the flash technique will presumably help in understanding details of the formation of ES-complexes and (b) such a study should provide a basis for understanding dark repair inhibition by caffeine and its greatly varying extent in different biological systems (for ref., see [5]).

Of several possible hypotheses, only the following one can explain the difference in PR inhibition between B_{s-1} and B/r. We assume that the ratio [formation time] : [life time] of the ES-complexes is critical for the extent of PR inhibition, expressing by 'formation time' the average time required for a PRE molecule to form a complex, and by 'life time' the average time elapsed between formation of a complex and its photolysis. This ratio must be much smaller for B/r than for B_{s-1}, since after UV irradiation to a survival of 10^{-3} to 10^{-4} the number of pyrimidine dimers is higher by a factor 50 to 100 in B/r cells. It is obvious from results reported further below that caffeine affects only the complex formation, but not the photolysis.

This hypothesis was tested by altering the light intensity, which amounts to an alteration of the ratio mentioned above. As expected, one finds at very high light intensity (a sequence of 120 1-unit flashes, 0.5 sec apart) PR inhibition in B/r cells:

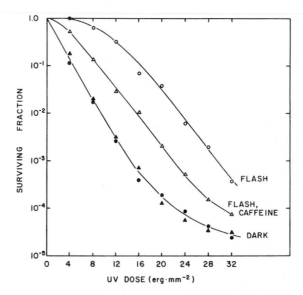

Fig. 10. Effect of caffeine (1.6% w/v) on the equilibrium of ES-complex formation in strain $B_{s-1}-160$. The UV survival curves in the dark are about the same in the presence (▲) and in the absence (●) of caffeine. The equilibrium of complex formation (characterized by the extent of PR obtained with a flash) is considerably lower in the presence (△) than in the absence (○) of caffeine.

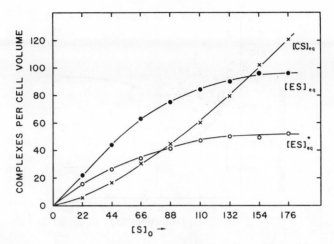

Fig. 11. Equilibrium concentrations of complexes in the absence and in the presence of 1.6% (w/v) caffeine. $[ES]_{eq}$, $[ES]_{eq}^*$ and $[CS]$ are calculated according to eq. (7a) from the results shown in fig. 10, and plotted as a function of $[S]_0$, the concentration of substrate produced by UV irradiation.

the rate in the presence of caffeine is reduced to about 0.5 of that in the absence of caffeine. On the other hand, inhibition in B_{s-1} cells is far less pronounced when the light intensity is reduced to about 1/100 of the usual intensity.

The flash photolysis technique permits one to determine at any UV dose the PR corresponding to the equilibrium of complex formation in the dark. Such results are presented in fig. 10. They show that the number of ES-complexes in the equilibrium, which according to eq. (4) is expressed by $[ES]_{eq} = k_1[E]_{eq}[S]_{eq}/k_2$, is lowered in the presence of caffeine. We may presume that k_1 and k_2 remain essentially unchanged when caffeine is added, but $[S]_{eq}$ should become $[S]_0 - [ES]_{eq} - [CS]_{eq}$, if we expressed by $[CS]_{eq}$ the concentration of substrate which in the equilibrium is prevented by caffeine from reacting with the PRE ('caffeine-substrate complexes'). Thus, denoting the various concentration in the presence of caffeine by [ ]*, we can write

$$[ES]_{eq}^* = \frac{k_1}{k_2}([E]_0 - [ES]_{eq}^*)([S]_0 - [ES]_{eq}^* - [CS]_{eq}), \qquad (7)$$

or

$$[CS]_{eq} = [S]_0 - [ES]_{eq}^* - \frac{k_2}{k_1}\frac{[ES]_{eq}^*}{[E]_0 - [ES]_{eq}^*}. \qquad (7a)$$

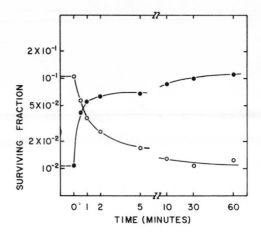

Fig. 12. Loss of ES-complexes in $B_{s-1}-160$ cells after addition of caffeine at 1.6% w/v (○), and formation of ES-complexes in $B_{s-1}-160$ cells after removal of caffeine (●), both as a function of time. The values are geometric means of 3 experiments, in which cells were irradiated with 16 erg/mm².

From the results in fig. 10 the values for $[CS]_{eq}$ as well as for $[ES]_{eq}$ and $[ES]_{eq}^*$ have been calculated and plotted as a function of $[S]_0$, as shown in fig. 11. As expected, $[ES]_{eq}$ and $[ES]_{eq}^*$ level off at quite different values, whereas $[CS]_{eq}$ continues to increase. We can see that any dose $[ES]_{eq}^* + [CS]_{eq}$ is close to $[S]_0$, i.e. essentially all of the substrate is either complexed with PRE or blocked by caffeine. We can see that in the lower dose range, where the PRE is in excess of substrate, the probability of a substrate molecule being present in the equilibrium as an ES-complex is considerably higher than its probability of being present as an CS-complex, in spite of the presumably much higher number of caffeine molecules in the cell*. This indicates a much greater affinity between PRE and UV lesions compared to caffeine and UV lesions.

This notion is supported by the following simple experiment. UV-irradiated $B_{s-1}-160$ cells were allowed to reach the dark equilibrium of ES-complex formation either in the presence or in the absence of caffeine. In one case, the caffeine was diluted at least 100-fold and the formation of additional ES-complexes was studied as a function of time by the flash technique. In the other case, caffeine was added at high concentration and the decrease in the concentration of ES-complexes in the cells was studied in the same manner.

The results of such experiments are shown in fig. 12. Evidently the formation of CS-complexes is a fast process, as indicated by the rapid decrease in the number

---

*If the caffeine concentration inside the cell is similar to the caffeine concentration in the medium (1.6% w/v), a cell should contain about 10^7 to 10^8 caffeine molecules.

of ES-complexes after addition of the caffeine. However, most of the CS-complexes are fairly unstable, since after diluting out the caffeine most of the additional ES-complexes are formed within 1 to 2 min.

The decrease in the number of ES-complexes after adding caffeine can also be used as a basis for calculating k_2, provided that dark dissociation of an ES-complex is prerequisite for the caffeine inhibition. The k_2 value thus calculated is roughly 10^{-2} sec^{-1}, which is still a factor of 2 higher than the k_2 value calculated in section 3.6.1 from the dark equilibrium.

Further investigations on the nature and kinetics of formation of 'CS-complexes' by the flash photolysis technique might also provide further insight into the mechanism of dark-repair inhibition by caffeine in *E. coli*.

4. Summary

The mechanism of photoenzymatic repair of UV lesions in *E. coli* DNA has been studied, using short intense light flashes. It is shown that one such flash photolyses virtually all of the enzyme—substrate complexes (ES-complexes) present at that moment; therefore, their number is reflected by the photoreactivation effect obtained. This permits studying separately the formation of ES-complexes in the dark and the photolytic step.

The formation of ES-complexes in B_{s-1} cells at room temperature reaches its dark equilibrium about 5 min after creation of the substrate. The maximum number of ES-complexes formed is about 20, even in the presence of excess substrate. It is therefore concluded that the average number of molecules of the photoreactivating enzyme (PRE) in a B_{s-1} cell is about 20. Similar or lower figures are obtained for other *E. coli* strains. However, three mutants have been isolated from strain B_{s-1} which contain approximately 43, 75 and 110 PRE molecules per cell. The mutant containing 110 PRE molecules (B_{s-1}–160) is useful in experiments where a high number of ES-complexes is desirable.

The dark reaction constants k_1 and the photolytic constant k_p, characterizing the various steps of the photoenzymatic repair process, have been determined independently of each other. Presumably the values obtained for k_1 and k_p are reasonably accurate. However, different experimental approaches have led to somewhat varying values for k_2; possible reasons for this are discussed. k_1 and k_2 show a positive temperature dependence in the range +5 to +37°C, whereas k_p does not. A heterogeneity of the lesions (and complexes) is observed with respect to all three reaction constant.

Flash photolysis shows that the formation of ES-complexes is inhibited in the presence of caffeine. This is due to interaction of caffeine with the UV lesions (i.e. formation of 'caffeine-substrate complexes'), rather than by interaction with PRE. Further investigation of this phenomenon might also be helpful for interpreting caffeine inhibition of dark-repair processes.

References

[1] R. Dulbecco, Experiments on photoreactivation of bacteriophages inactivated with ultra-violet radiation. J. Bacteriol. *59* (1960) 329–347.

[2] S.H. Goodgal, C.S. Rupert and R.M. Herriott, Photoreactivation of *Hemophilus influenzae* transforming factor for streptomycin resistance by an extract of *Escherichia coli* B. In: The Chemical Basis of Heredity, eds. W.D. McElroy and B. Glass (John Hopkins Press, Baltimore, 1957) pp. 341–343.

[3] H. Harm, Analysis of photoenzymatic repair of UV lesions in DNA by single light flashes. III. Comparison of the repair effects at various temperatures between +37 and −196°C. Mutation Res. *7* (1969) 261–271.

[4] H. Harm and C.S. Rupert, Analysis of photoenzymatic repair of UV lesions in DNA by single light flashes, I. In vitro studies with *Haemophilus influenzae* transforming DNA and yeast photoreactivating enzyme, Mutation Res. *6* (1968) 355–370.

[5] W. Harm, Differential effects of acriflavine and caffeine on various ultraviolet-irradiated *Escherichia coli* strains and T1 phage. Mutation Res. *4* (1967) 93–110.

[6] W. Harm, Analysis of photoenzymatic repair of UV lesions in DNA by single flashes. IV. Mutations affecting the number of photoreactivating enzyme molecules in *E. coli* cells. Mutation Res. *8* (1969) 411–415.

[7] W. Harm, H. Harm and C.S. Rupert, Analysis of photoenzymatic repair of UV lesions in DNA by single light flashes, II. In vivo studies with *Escherichia coli* cells and bacteriophage, Mutation Res. *6* (1968) 371–385.

[8] J. Jagger and R.S. Stafford, Evidence for two mechanisms of photoreactivation in *Escherichia coli* B. Biophys. J. *5* (1965) 75–88.

[9] A. Kelner, Effect of visible light on the recovery of *Streptomyces griseus* conidia from ultraviolet irradiation injury. Proc. Nat. Acad. Sci. US *35* (1949) 73–79.

[10] S.E. Luria, The bacterial protoplasm: Composition and organization. In: The Bacteria, Vol. I., eds. I.C. Gunsalus and R.Y. Stanier (Academic Press, New York, N,Y., 1960) pp. 1–34.

[11] C.S. Rupert, Photoreactivation of transforming DNA by an enzyme from baker's yeast. J. Gen. Physiol. *43* (1960) 573–595.

[12] C.S. Rupert, Photoenzymatic repair of ultraviolet damage in DNA. I. Kinetics of the reaction. J. Gen. Physiol. *45* (1962) 703–724.

[13] C.S. Rupert, Photoenzymatic repair of ultraviolet damage in DNA. II. Formation of an enzyme–substrate complex. J. Gen. Physiol. *45* (1962) 725–741.

[14] C.S. Rupert and W. Harm, Reactivation after photochemical damage. In: Advances in Radiation Biology, Vol. 2, eds. L.G. Augenstein, R. Mason and M.R. Zelle (Academic Press, New York, N.Y., 1966) pp. 1–81.

[15] W.D. Rupp and P. Howard-Flanders, Discontinuities in the DNA synthesized in an excision-deficient strain of *Escherichia coli* following ultraviolet irradiation. J. Mol. Biol. *31* (1968) 291–304.

[16] J.K. Setlow, Photoreactivation. Radiation Res. Suppl. *6* (1966) 141–155.

CHAPTER 23

REMARKS ON SOME BIOLOGICAL AND EVOLUTIONARY PROBLEMS OF PHOTOREACTIVATION AND DARK REPAIR

Albert KELNER

Department of Biology, Brandeis University,
Waltham, Massachusetts, USA

1. Introduction

This lecture represents an attempt to understand the biological meaning of DNA repair. From the beginning of the work on photoreactivation I wondered about the meaning of this phenomenon, which seemed to deal only with an artificial laboratory radiation, UV of wavelength 2537 Å, not found in the earth's biological environment for at least 600 million years.

Many experimenters in repair must have had thoughts similar to those I expressed in 1960:
'....We ask whether such a widespread phenomenon, found in most of the living world, would exist unless it either had survival value in the past, or the same reactions (causing repair to genetic elements) had some counterpart in non-irradiated cells. Photoreactivation may be one of several mechanisms whose functions is to maintain the genetic structure of the cell — a stability essential to life as we know it.... The discovery of something homologous to photoreactivation in the normal cell, and understanding its role in evolution would be a fitting part of research in photoreactivation' [12].

I have also been puzzled by the biological meaning of dark repair, and tried to understand its place in the biological scheme of things.

The general field of DNA repair (which includes photoreactivation and dark repair) has unusual biological significance. Adopting the spirit of evolutionists such as Dobzhansky and Simpson, one would ask what made repair evolve; what natural selection forces first brought it into the living world; what forces kept it there over the aeons, and are keeping it there even now? Are there new biological concepts suggested by DNA repair? And so on.

2. The setting

A preliminary version of part of the material in this paper is found in Kelner [13]. We will not attempt to review the general field of DNA repair, since others at this meeting are doing so, and excellent reviews are already in the literature, for example the recent book by Smith and Hanawalt [23]. The following two concepts seem, however, most important for our biological theme.

(1) DNA is an extraordinarily sensitive molecule, vulnerable to damage by a variety of agents. The fact that 0.2 erg/mm^2 of UV can kill a bacterial cell highly deficient in repair [9] is sufficient evidence that DNA in vivo is very sensitive.

(2) Just as DNA is sensitive, so repair is very efficient [9, 13, 20]. Ninety percent of the damage produced by low doses of UV are repaired. The curves shown by Wacker at this meeting for acetone-sensitized dimerization, and its complete photo-reversal illustrates how efficient is repair.

One might consider that the sensitivity of DNA *required* the evolution of repair; or conversely, that repair *allowed* the evolution of a DNA whose sensitivity is a reflection of its function. We would think both evolved in parallel.

3. The selective damaging agent

Since we are considering natural selection, we ask what *causes* the damage which is repaired. The damaging agent must exist in the natural environment in which the organism has evolved.

The theoretical possibilities are that the damaging agents can be:

(a) *Intracellular,* perhaps nucleases, or endogenously produced DNA-active chemicals, or just 'spontaneous' accidents. There is as yet little evidence for these.

(b) *Extracellular,* perhaps

(1) chemicals found in the natural environment of the cell, or

(2) radiation. These may be either ionizing or UV radiation, or from evidence presented at this meeting, even visible light.

We should also consider whether UV or other damaging radiations may have nothing directly to do with the evolution of repair. Protection against UV may be just a pleiotropic effect of enzymes such as DNA ligase necessary for the *normal* replication of DNA (e.g., see Kornberg [14]; Pauling and Hamm [18]. Under this view, DNA ligase, for example, evolved not because of its role in repair but because it was part of a complex genetic system (DNA and RNA, and mechanisms of replication, transcription, translation) which was superior to possible alternative genetic systems, and thus was selected for during evolution. This may at least partly explain dark repair, but probably not photo-repair. All of the foregoing possibilities may be operable in evolution.

4. Radiations

Let us now consider radiations as selective forces. Dark and photo-repair are considered together, for they often have the same biological end-result: repair of UV damage. And so cells having either type of repair share a common advantage. Of course the mechanisms, and probably the evolutionary history, are different.

Ionizing radiation. Repair of damage by ionizing radiation is relatively inefficient [9] and the intensity of ionizing radiation to which organisms are exposed in nature is very low. Yet it may be more important for evolution of repair than is realized, as will be discussed later. Cleaver [2] also considered ionizing radiation as a possible selective agent in repair.

UV. In early days of repair research the UV usually studied had a wavelength of about 2600 Å, at the nucleic acid absorption peak, a wavelength not now reaching the earth, and therefor not evolutionary active for at least 600 million years [1].

The question of how long a time UV at wavelengths shorter than 2900 Å has been absent from the earth, is worth examining in detail. From the calculations of Berkner and Marshall [1] 2600 Å no longer reached the earth when the oxygen content of the atmosphere reached about 1% of the present atmospheric level. Just how long ago this was is uncertain but according to Berkner and Marshall the 1% level was reached about the start of the Cambrian, about 600 million years ago. 2900 Å, which is certainly also a damaging wavelength, no longer reached the earth's surface when the oxygen concentration had reached about 10% present atmospheric level, about the Silurian–Devonian, or about 400–425 million years ago [1]. So, according to these calculations UV of 2600–2900 Å has not been evolutionary important for the last 400–600 million years.

Rupert [20], however, showed that contemporary, natural sunlight contains wavelengths capable of damaging DNA in vitro, and this damage could be repaired by photoreactivation.

Harm [7] has shown that contemporary sunlight inflicts damage in bacteria repairable by either dark or photo-repair. The active damaging wavelength he considered to be 3000 Å, of which a trace is found in sunlight. 3000 Å is at the long wavelength tail end of the DNA absorption curve. In very repair-deficient bacteria such as *E. coli* B_{S-1}, or especially AB-2480, about 25 sec of Texas sunlight kills 70% of AB-2480 cells; and even at 9 o'clock on a December morning, ½ hr of sunlight kills 30%. Noontime sunlight in October in Texas is equivalent in its cell-killing action to 1 erg/mm^2 min of 2537 Å.

So, wavelengths important for natural selection for repair range from about 2600 Å to those in the lowermost limit of *contemporary* natural sunlight.

The fact that laboratory cultures of coli are killed by sunlight does not however mean necessarily that it is a natural selection force for colie. An agent of natural selection should be in the natural environment of a species long and continuously enough to affect its reproductive rate. If organisms live in the dark, then sunlight cannot select for repair in them. Or, if cells are never exposed to a sufficient dose of radiation to damage them, or have masking or other agents to protect the nuclei from damage, sunlight cannot be an important evolutionary factor.

Yet sunlight must be at least *one* factor, even today, in natural selection for repair. Cleaver [2] has shown that cells of human beings with the genetic disease Xeroderma pigmentosum are defective in repair of damage to DNA caused by UV, while normal cells can repair such damage, as evidence by repair replication.

5. Evolution of repair

Repair is present in most cells today. Our first question in examing its evolution is, *when was the earliest it is likely to have evolved? How close to the origin of life?*

As in all questions relating to the origin of life we can only make reasonable judgments, based on what we know about repair, and present theories on the origin of life.

Strong UV from the sun dominated the environment when life began, say about 4 billion years ago. The absence of oxygen made the atmosphere transparent to wavelengths ranging from about 2400 Å, (or shorter) to the visible and portions of the infrared. The UV intensity (2400–2900 Å) on the earth's surface was about 5000 erg/cm^2 sec, a dose rate much like that used in UV experiments today (foregoing figures adapted from Sagan [21]). The transparency of sea water to short wave UV varies; using the calculations of Hulbert [11] one meter transmits 10^{-6} of 2600 Å, or 2 or 3% of 3000 Å. Sagan and others estimate far more transparency for the primeval sea. But certainly the air, the surface, and upper layers of the sea, had DNA-damaging short-wave UV, near-UV, and visible radiation. Ionizing radiation Sagan [21] estimates to be only an order of magnitude more than today.

UV is important in the origin of life in at least two ways. First, as one of the chief sources of energy for chemosynthesis of carbon compounds, which were the precursors of proteins and nucleic acids. The general picture is that the synthesis of organic molecules from CH_4, NH_3, and H_2O occurred in the atmosphere. These organic molecules concentrated in the oceans, where they underwent further reactions. They escaped destruction by UV because they diffused down under the surface, and accumulated so much that the ocean became a 1% 'soup' of organic molecules. (This scheme adapted from Miller and Horowitz [16].)

The second importance of UV is that it endangered early pre-cells and cells by damaging their nucleic acids. (See Sagan [21] and De Ley [4] for discussions of this factor.)

At some stage during the evolution of life, replicating bodies containing nucleic acids were formed. In fact it was the nucleic acids which replicated and coded for proteins, including enzymes. At this stage these early bodies were certainly vulnerable to damage by UV, and selection for repair mechanisms might have begun. If we accept the concept of the extreme vulnerability of the DNA molecule, then we have to consider how life as we know it could have even started. (The same problem arises for life on almost any planet, unless shielding makes UV not a danger.) Any schema would have to separate the prebiotic synthesis probably requiring some sort of radiation, from the early cells which are *damaged* by this radiation.

People studying the origin of life have usually handled the UV-cell-damage problem by postulating that life arose 10 to 13 m under the surface of the sea. This thickness of water absorbs the short-wave UV. Sagan [21] says the UV intensity (2400–2900 Å) had to be attenuated to less than 10^{-3} erg/cm^2 to prevent complete destruction of cell clones. He calculates that at least *40 m* of water were needed to protect early replicating units. And since these deep-lying 'cells' were dependent on abiotically synthesised nutrients drifting down from above, they could not be too deep. So life perhaps began from 40 to 100 m beneath the ocean's surface. The majority of workers believe less protection was needed, the figures 10–13 m being cited (e.g., Berkner and Marshall [1] ; Ponnamperuma [17]).

We think the field of DNA repair suggests some revisions of these ideas.

First, the sensitivity of DNA without repair is far greater than was realized when these ideas were formulated. If repair-deficient coli is killed by 0.2 erg/mm^2 [9] it is more than 1200 times more sensitive than *E. coli* B/r. Sagan [21] uses 10^3 to 10^4 erg/mm^2 as the mean lethal dose in his calculations. 0.2 erg would be more accurate. Thus the shielding by water would have to be 1000 times greater than originally thought.

Berkner and Marshall [1] use the figure, 1 erg/cm^2 sec (50 Å band width) of UV in estimating the maximum intensity of UV at which life could flourish. But at an intensity of 1 erg/cm^2 sec (2537 Å) a 37% survival dose for extremely repair-deficient bacteria would be accumulated in 20 sec. Even allowing for masking pigments, or other biological shielding of cell nuclei (e.g., large cells), cells without repair would find difficulty in surviving.

It must be noted, however, that the additional protection would be achieved by one or two meters more of seawater, and therefore the problem of shielding early life from UV may not be serious. Nevertheless UV is more dangerous to early cells than has perhaps been realized, and this fact must be considered in constructing theories on the origin of life.

Of course, early DNA may not have been so vulnerable as it is today; it might have had a different structure, more polyploidy, more resistance to damage. But we do not known that, although it is almost certain it did not have the same structure as now.

The question is whether enough UV reached early cells, even with the postulated masking conditions, to make for powerful natural selection for repair very early in evolution. The answer is I think, yes, especially if we consider the amount of UV reaching the cells need have been very little, the damage slight, to have strong selection for repair. Those who have worked with mutation know that an environment — say containing an antibiotic — that is only *partially* inhibiting, results in more certain appearance of resistant mutants than an environment inhibiting totally the parent cells. Also, in evolution, where huge time scales are involved, a low selection rate operating over a long time is the more usual situation, especially for evolution of complex characters.

Sagan [21] in his calculations asked what UV dose would destroy the clone — not let newly-arisen life survive at all. We ask further whether even if life survived, and with the radiation-screening postulated, would there still be selection for repair.

We have so far almost ignored ionizing radiation. Given the vulnerability of DNA, and the fact that even slight damage is important for evolution—ionizing radiation may well have selected for repair.

To summarize this section of the paper, selection for both masking of UV, and for repair, might have started very early. How important repair would have been depends on the efficiency of masking, the distance under the surface of the sea early cells lived, etc. Life may have started at even greater depths than has been postulated. More experiments are needed before one can judge intelligently how early repair evolved. And I think that both repair and the extreme vulnerability of DNA must be considered very seriously in any hypothesis on the origin of life and its early evolution.

The latest in evolution repair could have started.

If we cannot specify how early repair started, the *latest* it could have arisen is perhaps either the time when photosynthesis started, or, alternatively, when DNA evolved its present structure and physiology.

We can imagine early life in the pre-Cambrian, living in the depths, illuminated relatively feebly by the sun. With the emergence of photosynthesis a new selection factor arose, one which pulled the cells up toward more light, while at the same time, the UV from the sun tended to keep the cells deep under the surface. One factor up; the other down. A dilemma. The selection for repair was accentuated. Thus repair may have made possible more efficient photosynthesis.

Then, as oxygen accumulated slowly in the atmosphere, the UV intensity gradually diminished, and cells rose toward the surface.

There is ample evidence that contemporary photosynthetic organisms are photo-reactivable. The chloroplast itself photoreactivates in *Euglena* [22] and this organism contains the PR enzyme [5].

The period in history when DNA evolved to its present state is much less certain than when photosynthesis arose. DNA as it is *now* suffers breaks to achieve recombination. Normal DNA with its associated enzymes can repair damage by UV. DNA and genetic systems have evolved just as have enzymes. Whenever genetic systems reached their present state they certainly could repair UV. Perhaps photosynthesis *hastened* or at least influenced the evolution of DNA into its present form.

The very important relationship of the evolution of repair and evolution of DNA is discussed again later.

As photosynthesis flourished, oxygen accumulated in the atmosphere, the ozone layer screened out wavelengths under about 2900–3000 Å. This ozone layer was once supposed to settle the problem of UV, but we know now that wavelengths present even in contemporary sunlight damage DNA – so evolution of repair and masking have continued *even after* the formation of an oxygen-containing atmosphere. Moreover, as Berkner and Marshall [1] showed, oxygen accumulated slowly, reaching its present atmospheric level only in the Carboniferous, some 300 million years ago.

The picture we have is the gradual raising of life from the depths, and the evolution of repair not only in the algae, but in protozoa and metazoa. Fish as well as invertebrates have not only dark, but even photo-repair [3].

The important point is I think that the very gradual reduction in UV intensity as the oxygen built up over hundreds of millions of years resulted in a strong selection for light for photosynthesis — coupled perhaps with continued evolution of DNA and its enzymes toward its present form — all resulting in ideal conditions for evolution toward increased repair.

6. Repair and the invasion of land by life

A remarkable thesis was advanced by Berkner and Marshall [1] : the invasion of land by life was dependent on the ozone layer. Until the damaging UV was screened out sufficiently (by oxygen to 10% present atmospheric level) organisms could not leave the water for the unprotected land. This stage in evolution was attained about when the oxygen reached 10% atmospheric level [1]. Animals were preceded by plants, which according to our own hypothesis, had repair which helped protect them in the increased UV light on land. Berkner's thesis is we think supported by the fact that contemporary amphibia have photo-repair (and almost certainly dark repair). An evolution toward repair of UV damage has thus evolved in them. And we can *add* to Berkner and Marshall's thesis, that if indeed invasion of the land was dependent even *in part* on screening out of short-wave UV by ozone, it was *also* dependent on evolution of PR and dark repair. Not only the invasion of land, but at least early stages of life on land depended in part on repair, *despite* the ozone layer for even contemporary sunlight contains DNA-damaging UV.

7. Repair and contemporary life

We look around us today, and find almost every ecological niche inhabited, from the ocean depths to the air. Dark repair would presumably be present universally, because DNA as it evolved required repair enzymes, which would *incidentally* repair radiation and other damage. But with efficient outer layers of cells in metazoa screening inner cells from sunlight, we wonder if photorepair, at least, having played its role in all of life, is no longer needed except in plants, and in those bacteria and protozoa inhabiting light-rich environments.

Cook, Regan, and co-workers in a series of experiments in which they examined different animal phyla for the PR enzyme, found it to be universally distributed in all lower and higher forms (see Cook [3] for a summary of this work). In the metazoa, PR enzyme was present in amphibians, reptiles, birds, and on up to the marsupial mammals. Only the placental mammals lacked the enzyme. This wide distribution in the advanced animals, many of which have most efficient light-screening outer integuments, such as the kangaroo, is remarkable. The PR enzyme is found even in internal organs — the liver, kidney, and even the brain. It is also found in cell cultures, as well as in natural tissues. While some assays for the enzyme are difficult, and some questionable evidence remains, we will assume that PR is widely distributed throughout the living world (except the placental mammals) and are found in internal differentiated tissues, whether these are light-struck or not.

8. Why is repair still present in life today?

This raises important problems not only for repair, but for evolution and general biology. For our discussion we will specifically ask why the *PR enzyme* in cells which never in their normal life are illuminated by any light, let alone UV. To make the problem more general we include in these organisms bacteria such as *E. coli*, normally inhabiting the intestine, and never exposed to light except perhaps briefly during transfer from one animal to another, and even then presumably only to dim light.

The explanations that may be considered are:

(1) The PR enzyme is a *vestigial* enzyme. The enzyme does not harm the cell, even if it does no good. It was needed back in the amphibian stage, and had just remained since.

Evolutionary theory says a truly neutral gene cannot remain very long [15; see also 24], certainly not hundreds of millions of years. Eventually it is replaced by an advantageous gene. And even if it could survive natural selection, a gene not undergoing natural selection would mutate away, and little by little lose its specific structure, and hence its ability to code for an active enzyme. As we see it, all genes are at any moment in time, either on their way into the gene pool, or on their way out.

More reasonable would be *selection at certain life stages*. Gametes of marine animals for example are often subject to sunlight and skyshine, even if the adults are not, and would be selected then. But this can hardly be the explanation for reptiles, or kangaroos, or the nocturnal opposums.

We also ask why internal tissues should have the PR enzyme. For a biological principle assumed by many to hold is that differentiated tissues are differentiated not only morphologically, but also enzymatically. That is, all enzymes in the brain have a function in the brain, and other enzymes coded in the DNA of the animal are suppressed in the brain. And so selection for the PR enzyme in the skin, or in the larval stage, should not explain the presence of active enzyme in the liver.

(2) We therefore conclude that all the foregoing explanations are wrong or partly wrong, or else some important concepts of evolution, genetics, or general biology need revision. We are left with the most reasonable explanation being that the PR enzyme in cells living in the dark *does have a vital function*; that the PR enzyme (if I may borrow a concept from genetics) is pleiotropic and has at least two activities: one splitting of pyrimidine dimers, and the other one vital for the marsupials down. What this other function is remains to be seen, but it almost certainly has to do with DNA.

There is no obvious reason connected with habitat why marsupials should have the PR enzyme, and not placental mammals.

The placental mammals do have dark repair, as was mentioned before in connection with sunlight and Xeroderma pigmentosum.

Xeroderma pigmentosum illustrates well the extreme vulnerability of DNA to UV, the sensitivity of the cell nucleus, and the inadequacy of the usual masking by melanin.

The connection of the lack of repair to cancer in this disease opens up a new area in which repair is important.

And lastly — since we are all human beings and interested in our evolutionary origin — the characteristic hairlessness of the human body, a trait some biologists think important for human characteristics, would have been impossible without repair. Whether that is so or not, we human beings ourselves do represent the most clearcut example yet of natural selection for repair.

9. General genetic and biological aspects of repair

Having brought the discussion of the evolution of repair to the present time in evolution, we turn to the relation of repair to (1) evolution of DNA, (2) control of mutation rate, (3) evolution of recombination, and (4) repair as a perhaps unique example of intracellular, as contrasted to extracellular, repair.

Evolution of DNA. The physiology and structure of DNA, with its strand breakage and rejoining, would serve to select for repair enzymes. As one example, polynucleotide ligase, implicated in normal DNA replication, and selected evolutionary for this function in normal DNA, is therefore present in cells, and can serve for repair of UV damage [see 14, 10, 18]. On the other hand evolution of repair enzymes perhaps due in part to natural selection for repair of radiation damage, would *allow* the evolution of a DNA as we know it. Both processes are seen as evolving together and interacting with each other. As a corollary to this, whatever environment helped select for repair, such as perhaps the marine environment of early photosynthetic organisms, hastened the evolution of DNA.

We would ask whether *any* molecule containing genetic information would of necessity have such stringent requirements for maintaining its structure exact and unchanging, that repair would *have* to be evolved. Of course to a relatively lesser degree, proteins must retain specific structure unchanged for enzymatic activity — but the requirement is less strict, and moreover many molecules of an enzyme are always made, so loss of one is less harmful. Evolutionary reasons dictate that one or only few copies of genetic molecules per cell are formed.

The long DNA molecule with its vulnerability must be maintained intact so that gene complexes can be transmitted to daughter cells as a group, with no part of the information lost. It is the evolution of gene complexes that is important in evolution, rather than single genes. For many such reasons any information-containing genetic molecule would be likely to require a repair system.

Operating against this idea is that repair-deficient bacteria apparently grow normally so long as they are not exposed to DNA-damaging radiation or chemicals. Such bacteria may not be completely repairless, and there are some indications that they may grow poorly after all, even in the abscence of UV (e.g., [6]).

In conclusion, while we do not know how serious repair deficiency would be in evolution, nevertheless the evolution of DNA would seem a most important factor in the evolution of repair.

Control of mutation rate. The mutation rate of organisms cannot be too high or low. Too low eventually slows evolution; too high, destroys organisms. Repair may be the only *general* mechanism known which influences mutation under 'natural' conditions. 'Natural' here means: in nature, exposed to UV of the sun even in modern times. (For other mechanisms less general affecting spontaneous mutation rate, see Herskowitz [8]. For a discussion of solar UV as a possible natural mutagenic agent during the Precambrian, see De Ley [4].)

UV-induced mutation may be caused by an error occurring during the process of repair itself as in one of the types of recombinational repair. UV-induced mutation is prevented by photoreactivation, excision repair, and other types of recombination repair [25].

A role for repair in spontaneous mutation remains to be shown. In organisms whose genetic material or gametes or gonads are exposed to UV, repair keeps the mutation down. Of course it does not do so in higher forms with internal fertilization, and light-protected gonads. Whether absence of dark repair would raise the mutation rate in such organs, is for future study.

And so we can envision that repair has served to lower mutation rate, which may have been raised in part in earlier times by UV. What causes the 'spontaneous' mutation in all cells, other than the unsatisfactory term, 'errors' we do not know.

Recombination. Since recombination involves repair, we can say as with DNA in general, recombination causes selection for repair; repair allows recombination to evolve. They go hand-in-hand. By recombination is meant recombination of DNA strands in lower forms. Presumably crossover in chromosomes of higher forms would also involve repair. There is no reason yet to involve repair in sexual fusion of nuclei. And in bacteria the evolutionary role of recombination, e.g., conjugation, lysogenization and transduction, and probably transformation, as a source of genetic variability is uncertain.

Both mutation and recombination increase variability among organisms, and reduction in such variation is not immediately harmful to organisms. But too great reduction in variability increases the probability of extinction of the species. Certainly any factor influencing variability will itself be subject to selection.

Intracellular repair. To turn to a more physiological aspect, repair perhaps is the only general instance of intracellular repair in cells. Damage to some of the cells in a bacterial population is usually overcome by death of the cells and renewal of the population by survivors. The same occurs substantially in metazoa. Repair may be considered only one aspect of intracellular regulation — as, for example, feedback control of enzyme synthesis. But this and similar cases involve regulation of the quantity of individual molecules, not resynthesis specifically or reconstruction of a specific part of a molecule, or even organelle.

In DNA repair there is a specific *intra-molecular repair,* surely a unique biological phenomenon, and one which has arisen because DNA is itself a unique entity.

10. Comments on future evolution of repair

Perhaps as technology develops, the environment will be flooded with DNA-active chemical pollutants. These will endanger all life, from man in his cities, to fish in the ocean depths.

If so, that aspect of repair discussed little, repair of damage by chemicals, will be selected for. But alas, instead of the hundreds of thousands of years required for great changes in environment in the past, allowing the slow pace of evolution to change the animal, man's culture acts with explosive rapidity.....

11. Conclusion

In conclusion, we think repair has probably been intimately involved with the evolution of life, from early times to the present, and can be considered part of the evolution of information-containing molecules, or more basically, genetic systems. It helps to stabilize and regulate the genetic information against damage by external, and perhaps internal, agents. It makes possible recombination, in the bacteria, at least. The natural selection factors include radiation, and perhaps other damaging agents. Its evolution has gone hand-in-hand with the evolution of DNA. Selection factors yet unknown may also be present. Its place in the scheme of things is becoming clear.

A more conservative view, especially for the PR enzyme, is that maybe after all 3000 Å is the damaging wavelength in evolution, and may have been from the beginning. And the presence in light-protected tissues of the PR enzyme is due to the fact that differentiated tissues do not have complete specialization of enzymes according to the function of the tissues; or the PR enzyme is a vestigial enzyme — a chemical fossil.

But we do not think the concepts of evolution and biology are wrong, and we do think the many paradoxes and inconsistencies are important, suggesting new discoveries to come.

References

[1] L.V. Berkner and L.C. Marshall, History of the major atmospheric components. Proc. Nat. Acad. Sci. US *53* (1965) 1215–1226.

[2] J.E. Cleaver, Defective repair replication of DNA in Xeroderma pigmentosum. Nature *218* (1968) 652–656.

[3] J.S. Cook, Photoreactivation in animal cells. In: Photophysiology, Vol. 5, ed. A.C. Giese (Academic Press, New York).

[4] J. De Ley, Molecular biology and bacterial phylogeny. In: Evolutionary Biology, eds. T. Dobzhansky et al., Vol. 2 (Appleton-Century-Crofts, New York, 1968) pp. 103–156.

[5] J. Daimond, J.H, Schiff and A. Kelner, 1969. Photoreactivating enzyme from *Euglena gracilis* var. *Bacillaris* and a mutant lacking chloroplast DNA. Plant Physiol. *44* (1969) pp. 9–10 (abstract)

[6] K. Haefner, Spontaneous lethal sectoring, a further feature of *Escherichia coli* strains deficient in the function of *rec* and *uvr* genes. J. Bacteriol. *96* (1968) 652–659.

[7] W. Harm, Biological determination of the germicidal activity of sunlight. Radiat. Res. *40* (1969) 63–69.

[8] I.J. Herskowitz, Genetics, 2nd ed. (Little Brown, Boston, 1965) pp. 383–390.

[9] P. Howard-Flanders, DNA repair and genetic recombination: studies on mutants of *Escherichia coli* defective in these processes. Radiat. Res., Suppl. *6* (1966) 156–184.

[10] P. Howard-Flanders, DNA repair. Ann. Rev. Biochem. *37* (1968) 175–200.

[11] E.O. Hulbert, The penetration of ultraviolet light into pure water sea water. J. Opt. Soc. Amer. *17* (1928) 15–22.

[12] A. Kelner. Historical background to the study of photoreactivation. In: Progress in Photobiology, eds. B.C. Christensen and B. Buchman (Elsevier, Amsterdam, 1960) pp. 276–278.

[13] A. Kelner, Biological aspects of ultraviolet damage, photoreactivation, and other repair systems in microorganisms. In: The biologic effects of ultraviolet radiation, ed. F. Urbach (Pergamon Press, New York, 1969) pp. 77–82.

[14] A. Kornberg, Active center of DNA polymerase. Science *163* (1969) 1410–1418.

[15] E. Mayr, Animal Species and Evolution (Harvard Univ. Press, Cambridge, 1963) p. 207.

[16] S.L. Miller and N.H. Horowitz, The origin of life. In: Biology and the Exploration of Mars, eds. C.S. Pittendrigh et al., Publ. 1296, Nat. Ac. Sci. (National Res. Council, Washington, U.S., 1966) pp. 41–69.

[17] C. Ponnamperuma, Ultraviolet radiation and the origin of life. In: Photophysiology, Vol. 3, ed. A.C. Giese (Academic Press, New York, 1968) pp. 253–267.

[18] C. Pauling and L. Hamm, Properties of a temperature-sensitive, radiation-sensitive mutant of *Escherichia coli*. Proc. Nat. Acad. Sci. US *64* (1969) 1195–1202.

[19] A. Rook, D.S. Wilkinson and F.J.G. Ebling, Textbook of dermatology (Blackwell-Davis, 1968) pp. 62–64.

[20] C.S. Rupert, Photoreactivation of ultraviolet damage. In: Photophysiology, Vol. 2, ed. A.C. Giese (Academic Press, New York, 1964) pp. 283–327.

[21] C. Sagan, On the origin and planetary distribution of life. Radiat. Res. *15* (1961) 174–192.

[22] J.A. Schiff and H.T. Epstein, The continuity of the chloroplast in *Euglena*. In: Reproduction: Molecular, Subcellular, and Cellular, ed. M. Locke (Academic Press, New York, 1965) pp. 131–189.

[23] D.C. Smith and P.C. Hanawalt, Molecular photobiology - Inactivation and recovery (Academic Press, New York, 1969).

[24] G.L. Stebbins, Processes of Organic Evolution (Prentice-Hall, 1966).

[25] E.M. Witkin, Ultraviolet-induced mutation and DNA repair. Ann. Rev. Microbiol. *23* (1969) 487–514.